Hartmut Zabel

Medical Physics

Also of Interest

Medical Physics.
Volume 1: Physical Aspects of the Human Body
Hartmut Zabel, 2023
ISBN 978-3-11-075691-3, e-ISBN (PDF) 978-3-11-075695-1,
e-ISBN (EPUB) 978-3-11-075698-2
Volume 1, Volume 2 and Volume 3 also available as a set:
Set-ISBN 978-3-11-076102-3

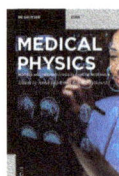

Medical Physics.
Volume 2: Physical Aspects of Diagnostics
Hartmut Zabel, 2023
ISBN 978-3-11-075702-6, e-ISBN (PDF) 978-3-11-075709-5,
e-ISBN (EPUB) 978-3-11-075712-5
Volume 1, Volume 2 and Volume 3 also available as a set:
Set-ISBN 978-3-11-076102-3

Medical Physics.
Models and Technologies in Cancer Research
Anna Bajek, Bartosz Tylkowski (Eds.), 2021
ISBN 978-3-11-066229-0, e-ISBN (PDF) 978-3-11-066230-6,
e-ISBN (EPUB) 978-3-11-066234-4

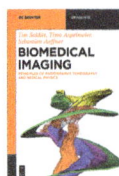

Biomedical Imaging.
Principles of Radiography, Tomography and Medical Physics
Tim Salditt, Timo Aspelmeier, Sebastian Aeffner, 2017
ISBN 978-3-11-042668-7, e-ISBN (PDF) 978-3-11-042669-4,
e-ISBN (EPUB) 978-3-11-042351-8

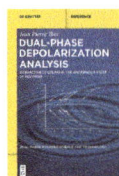

Dual-Phase Depolarization Analysis.
Interactive Coupling in the Amorphous State of Polymers
Jean Pierre Ibar, 2022
ISBN 978-3-11-075669-2, e-ISBN (PDF) 978-3-11-075674-6,
e-ISBN (EPUB) 978-3-11-075684-5

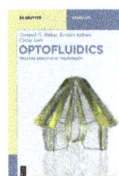

Optofluidics.
Process Analytical Technology
Dominik G. Rabus, Cinzia Sada, Karsten Rebner, 2018
ISBN 978-3-11-054614-9, e-ISBN (PDF) 978-3-11-054615-6,
e-ISBN (EPUB) 978-3-11-054622-4

Hartmut Zabel

Medical Physics

Volume 3: Physical Aspects of Therapeutics

2nd Edition

DE GRUYTER

Author
Prof. Dr. Dr. h. c. Hartmut Zabel
Ruhr-Universität Bochum
Fak. für Physik und Astronomie
44780 Bochum
hartmut.zabel@ruhr-uni-bochum.de

ISBN 978-3-11-116867-8
e-ISBN (PDF) 978-3-11-116873-9
e-ISBN (EPUB) 978-3-11-116906-4

Library of Congress Control Number: 2022951753

Bibliographic information published by the Deutsche Nationalbibliothek
The Deutsche Nationalbibliothek lists this publication in the Deutsche Nationalbibliografie;
detailed bibliographic data are available on the internet at http://dnb.dnb.de.

Preface

The second edition of the textbook on Medical Physics is now published in three instead of two volumes. This is because the new edition is not only corrected and updated but also considerably expanded. The enhancement is primarily due to many new features designed to increase the usefulness of this textbook as a learning companion to regular courses. Each chapter concludes with a detailed summary, questions, exercises, and a self-assessment of the knowledge gained. All chapters contain info boxes for in-depth information on specific topics and math boxes for deriving central mathematical concepts. References to current literature and further reading recommendations lead to reviews in the subject area. Answers to questions and solutions to tasks can be found in the appendix.

The textbook is aimed at bachelor's or master's students in the first year of medical physics. It is recommended that students first acquire some basic knowledge of modern physics before proceeding with reading this Medical Physics textbook.

The range of topics in Medical Physics is extensive. For this reason, most textbooks in this field have been written by several authors. Covering all topics on your own is both a challenge and an opportunity to design the various chapters as coherently as possible and to relate them to one another. I would not have been able to meet this challenge without giving respective lectures.

Compared to the first edition, the chapters have been rearranged to allow a clear distinction between the physical aspects of the human body in the first volume, the physical aspects of diagnostics in the second volume, and the physical aspects of therapy in the third volume.

The first volume begins with classical mechanics, including forces, moments, and energy, concepts applied to the human body. The overarching importance of the action potential and signal transmission in the body's functioning is presented in chapters five and six, followed by a discussion of those organs, systems, and senses that have a clear connection to physics, such as the cardiovascular and the respiratory system. The last two chapters deal with the primary sensory organs of the body, the visual sense, and the auditory sense.

The second volume deals exclusively with imaging methods, distinguishing between those without ionizing radiation (ultrasound, endoscopy, magnetic resonance imaging) and those with ionizing radiation (X-rays, SPECT, PET). As an introduction to these radiological chapters, essential facts about X-ray and nuclear physics are presented, the interactions of radiation with matter are explained, and measures for radiation protection are discussed.

The third volume is devoted to the physical fundamentals of radiotherapeutic procedures using X-rays, protons, neutrons, and γ-rays. These chapters are introduced by first comparing the radiation response of benign cells and malignant tumor cells. The last two chapters deal with laser processes and highlight the

https://doi.org/10.1515/9783111168739-202

physics of nanoparticles in diagnostics and therapy. An additional chapter on medical statistics rounds off the third volume.

All these additions, guidance, and exercises will hopefully make studying this medical physics textbook a valuable and informative companion to your regular coursework. Questions can be directed to the editor or author and will be answered promptly. Corrections are very welcome and will be posted on the book's website.

Bochum, December 14, 2022

Acknowledgments

My first thanks go to all those who pointed out errors and made suggestions for improvements to the first volume. Constructive criticism is always very helpful in improving the text, correcting mistakes, and clearing up misunderstandings. I am very grateful to Dr. Alexey Saphoznik, who took the time to read the entire first volume and suggested many modifications. My special thanks go to Professor Birge Kollmeier (University of Oldenburg), who made valuable recommendations for improving the content, particularly concerning sound and sound perception. I would also like to acknowledge my ophthalmologist Dr. Elbracht-Hülseweh, from whom I learned a lot during many years of treatment. Thanks also go to my colleagues who helped and guided me during the preparation of the first edition.

My special thanks go to the editorial staff at de Gruyter Verlag and, in particular, to Kristin Berber-Nerlinger, who encouraged me in the first place to prepare a second edition and gave me valuable advice on the implementation of this project. Nadja Schedensack helped in all stages, and Kathleen Prüfer did an excellent editorial job. I am very grateful to the entire publishing team of de Gruyter.

Last but not least, I would like to thank my family, and in particular, Rosemarie, who accompanied this project with much patience, understanding, and encouragement.

https://doi.org/10.1515/9783111168739-203

Contents

Part E: Diagnostics and therapeutics beyond radiology

Part F: **Measuring and statistics**

Appendix

Part D: **Radiotherapeutical methods**

1 Radiobiology basics

Important parameters

Number of chromosome pairs	23
Number of DNA bases	4
Telomere length	~50
Most effective LET	100 keV/μm (100 eV/nm)
LET of 10 MeV protons	4.7 keV/μm
Dose of single fraction	8–10 Gy
Total dose for cancer treatment	~60 Gy
RBE of x-rays	1
RBE of protons	2–3

1.1 Introduction

The aim of this chapter is to provide a basic understanding of the cell's life cycle and the difference between normal and cancerous cells. This information is instrumental for preparing a radiation treatment plan to fight the proliferation of cancer cells with photons, charged particles, or neutrons, as discussed in Chapters 2–5. In addition, we discuss here the relationship between radiation dose and cell survival rate, which is pivotal for the following chapters on radiation therapy. More details on the cell cycle can be found in standard biology [1] or physiology textbooks listed under "Further reading."

Genes and a number of messenger proteins determine when and how quickly cells grow, when they divide (mitosis) and when they die (apoptosis). Some cells only live very briefly, such as the cells in the epithelial layer of the intestine (1.5 days). Other cells have a fairly long life span, such as the skin epidermis cells (20 days) or the red blood cells (up to 60 days). Some cells live long like osteocytes (25–30 years), and some cells never die like nerve cells. The cells that die over time need to be replaced. Those who never die cannot be recovered even if injured, which is the problem with spinal cord injuries. For unknown and uncontrolled reasons, a normal cell can suddenly turn into a dysfunctional cell that overrides all life cycle rules and death of a "normal" cell. Then the cell becomes cancerous. Radiation therapy is currently one of three methods of fighting cancer cells; the other two methods are chemotherapy and surgery. These procedures can also be referred to as "burning," "poisoning," and "cutting." Although radiation treatment of tumor cells has seen tremendous development and refinement in recent years, it may not be the ultimate solution to fighting this disease. At the same time, enormous progress has been made in understanding the cell "machinery" so that biochemical cancer therapy through targeted drugs or immunizations is foreseeable in the near future. The Nobel Prize in

https://doi.org/10.1515/9783111168739-001

Medicine 2018 was awarded to James P. Allison and Tasuku Honjo "for their discovery of cancer therapy by inhibiting negative immune regulation" [2]. Immune therapy of cancer is a novel and promising direction. But for the meantime, radiation treatment of cancer remains the second-best solution.

1.2 Life cycle of cells

The normal life cycle of cells can be subdivided into two main phases: interphase and mitotic phase (M); see Fig. 1.1.[1] The interphase and mitotic phase constitute one cell cycle. During the interphase, normal cell activity occurs, including growth and DNA replication. The interphase occupies 90% of a cell cycle. The remaining 10% go to the mitotic phase, where the cell divides itself into two identical copies. After cell division is completed, both cells have the exact same genetic information, and the interphase starts over again, simultaneously in both new cells.

The life cycle of cells begins with the growth phase G1, in which the cell enlarges and synthesizes new proteins. Toward the end of this phase, a checkpoint (CP1) controls stop or go to the next phase. In case of stop, the cell goes into a rest phase G0. It remains there until called upon, for instance, for repairing injuries. Most body cells are in the rest phase G0 and fulfill their specified duties. The cell will complete the cycle and divide if the checkpoint gives a green light. For cell regulation, the checkpoint is

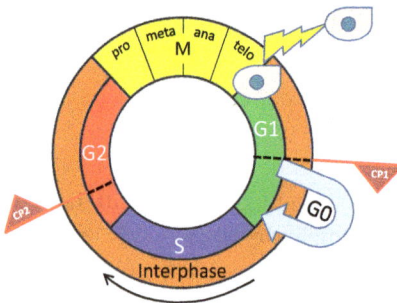

Fig. 1.1: Life cycle of a cell. G1, growth phase; G0, rest phase; CP1, first check point in G1 phase; S, replication phase; G2, second growth phase; CP2, second check point in G2 phase; M, mitosis. The mitosis or cell division phase is subdivided into four subphases: prophase, metaphase, anaphase, and telophase. Red flags indicate checkpoints. At the end of the telophase, two identical new daughter cells are formed.

---

1 If familiar with the cell biology, this section can be skipped. However, Section 1.4 is essential for the following chapters on radiotherapy.

essential. If the stop signal is overwritten in any way, uncontrolled replication of the cell can result, as we will see later.

The total genetic information is contained in any cell and is encoded in double-stranded DNA molecules. The genetic information of humans, the genome, is written in 46 *chromosomes*, each one containing one long DNA molecule. These 46 chromosomes come in 23 homologous pairs; one of each pair is inherited from the mother and father. Both chromosomes in a pair contain essentially the same genetic information governing the basic function of organs. This is true for the first 22 chromosomes. However, the 23rd pair is different for female (XX) and male (XY), determining the sex. Three pairs of chromosomes in their G1-phase are sketched in Fig. 1.2(a), representative of all others.

During the S-phase, cells replicate their genetic material, ensuring that each daughter cell will receive an exact copy during the mitotic phase. The two strands are copied in opposite directions (leading and lagging strands) during DNA replication using different transcription methods (for more information see Infobox I). After replication is completed, the DNA condenses into chromosomes. The chromosomes are a heavily coiled and folded-up version of long DNA strands. Only in this condensed phase can chromosomes be recognized by an optical microscope. Otherwise, the thin DNA chains do not offer sufficient contrast to be recognized. Each duplicated chromosome has two *sister chromatids* (joined copies of the original chromosome), which separate during cell division. The sister chromatids are attached to each other by a *centromere*; see Fig. 1.2(b). The *centromere* is a narrow "waist" of the duplicated chromosome, where two chromatids are most closely attached. Centromeres are important for organizing the associated chromatids in the cell and for their mechanical separation in the mitotic phase. Once separated by cell division, the chromatids are called *chromosomes*.

Most cells replicate with different rates during the human life span. Each cell contains the complete genetic code stored in 46 chromosomes. During replication, exact copies of all chromosomes are generated. Subsequently, the original cells are divided into two, each one containing one identical copy.

Infobox I: DNA replication

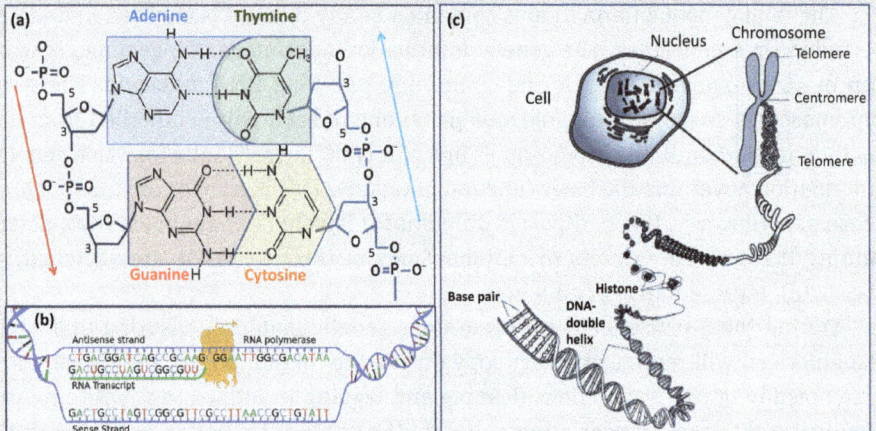

Figure source: https://en.wikipedia.org/wiki/Base_pair; https://de.wikipedia.org/wiki/Chromo
som; https://en.wikipedia.org/wiki/DNA_replication

Deoxyribonucleic acid (DNA) is the main molecule in cells responsible for storing and reading genetic information that supports all life, both bacterial and human. DNA is a linear polymer consisting of two strands arranged in a double helix (panel c). These strands are made up of subunits called nucleotides (panel (a)). Each nucleotide contains a phosphate, a five-carbon sugar molecule, and a nitrogenous base. The bases in DNA are adenine ("A"), thymine ("T"), guanine ("G"), and cytosine ("C"). They are always arranged in pairs: adenine pairs with thymine (A–T), cytosine pairs with guanine (C–G). The A–T pair has two hydrogen bonds, the C–G pair has three hydrogen bonds. Therefore, the pairs cannot be mixed, they fit like a key and lock. A strand uniquely defines the base sequence, which can be any of the four letters ACGT. Furthermore, the strands have opposite directions on their sugar/phosphate backbone running from 5′ to 3′ on one strand, and from 3′ to 5′ on the opposite strand, where the numbers count the carbon atoms in the sugar ring; 5′ is attached to a PO_4^- group, 3′ is attached to a OH^- group. This asymmetry is important for the replication. The double strand is twisted and forms a left-handed double helix[2]. The helix is further coiled and folded to form chromosomes, which are pulled away in the cell's nucleus, awaiting replication in the S-phase.

Replication requires the DNA double strand to open up like a zipper (panel (b)). The enzymatic opening of the H-bonds (helicase) is done locally in a scissor-like fashion supported by ATP. In the opened part, single-strand-binding proteins hinder nucleotides from rebinding via their strong H-bonds until the replication is finished. The RNA primase gives direction where to start the replication. Then the RNA polymerase generates a transcription of the base sequence. The semiconservative replication runs from 5′ to 3′ on the leading strand. As the opposite lagging strand is 3′ to 5′ oriented, replication is limited to short sequences (Okazaki fragments) that are fused together in a second revision step.

---

2 The structure of the double helix was analyzed and described by Rosalind Franklin (1920–1958), James D. Watson (1928), and Francis Crick (1916–2004). The latter two received the Nobel Prize in Medicine for their discovery in 1962.

While the zipped part extends, supercoils are formed in front and on the backside. Another enzyme, called topoisomerase, solves this topological problem and straightens out the DNA for continued replication.

In summary, the following six steps result in DNA replication:
1. Unwinding of the double helix; 2. splitting of the double strand and breaking hydrogen bonds; 3. primers provide starting points for the replication; 4. DNA polymerase synthesizes complementary bases, resulting in a new double-stranded DNA; 5. on the opposite strand, replication works piecewise; and 6. DNA polymerase closes the gaps with complementary bases.

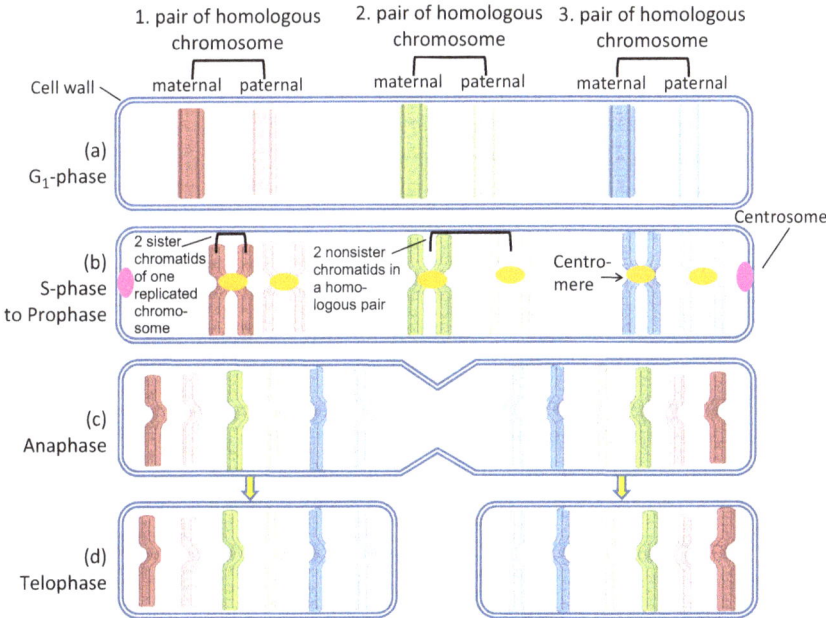

Fig. 1.2: Cell division starting from the G1-phase to the end of the M-phase. (a) In the G1-phase, all 46 chromosomes appear in homologous pairs; only three representative ones are sketched. (b) In the S-phase DNA replication and synthesis takes place, which condenses to sister chromatids in the prophase at the beginning of the M-phase. (c) In the anaphase, the sister chromatids are pulled apart to opposite halves of the cell. (d) In the final telophase, complete separation and generation of two new identical cells occur.

The complex topological copying process of DNA offers many opportunities for miscoding. Therefore, another checkpoint confirms the proper replication at the end of the S-phase and in the G2-phase. Mistakes are repaired, but if the DNA damage is too extensive, the cell development is hindered from proceeding to the next stage, the mitotic phase. Instead, the cell prepares itself for apoptosis, i.e., a programmed cell death. All cell fragments are disassembled and disposed of to avoid any further harm.

If the checkpoint in phase G2 is passed, the cell will further grow before entering the mitotic phase (M). The M-phase is subdivided into *prophase, metaphase, anaphase*, and *telophase*. In the prophase, spindle fibers attach to the centromeres on one end and to two anchor points on opposite ends of the cell, called centrosomes. In the metaphase, the chromatids align in the center of the cell. In the anaphase, the spindle fibers pull the chromatids apart (Fig. 1.2(c)), each sister moving to the opposite end of the cell. And finally, in the telophase, the cell divides by cleavage, forming two new daughter cells (Fig. 1.2(d)). The original cell content is completely taken over by the two daughter cells. One cell life cycle is completed, and two new life cycles can now begin.

Once the G1/S checkpoint has been passed, the remaining typical cell cycle time is about 10–40 h. Completion of the G1-phase takes about 30% of the cell cycle time, the S-phase 50%, G2-phase 15% and the M-phase only 5%, i.e., between 0.5 and 2 h.

At each end of a chromosome, there is a special DNA segment called telomere [3]. Telomeres protect the integrity of chromosomes during cell division and act simultaneously as a countdown clock. With each cell division, the telomere becomes shorter. After about 50 divisions, the remaining telomere length loosens its protective capability. Once this critical point is reached, the cell is commanded to commit apoptosis (suicide). The cell division limit is known as the Hayflick[3] limit, according to the anatomist Hayflick, who described it first [4]. Later, it was found that the length of telomeres and, therefore, the lifetime of cells depend on genetic factors and human factors such as lifestyle and stress management [5].

Normal cells contain an extensive feedback system inside and outside for keeping a dynamic balance between mitosis and apoptosis. Cells are also able to sense their surroundings and respond to changes. For instance, if a cell senses that it is surrounded by other cells, it will stop dividing. In this way, cells will grow when needed but stop when in contact with other cells. In case of a gap due to injury, cells fill in the gap left by a lesion but stop dividing after sealing the gap. Cancer cells do not exhibit contact inhibition. They grow even when they are surrounded by other cells, causing a mass of cells to form.

---

3 Leonard Hayflick (*1928), American microbiologist and gerontologist.

Infobox II: Apoptosis or necrosis: what is the difference?

Cell death can occur in two different ways: first, via programmed cell death (apoptosis); or second, as sudden and accidental cell death, called necrosis. Necrosis may follow as the result of external impact (injury), ischemia due to insufficient blood circulation, extreme temperatures, or any other type of physical trauma causing cell lesions. Injuries usually cause loss of cell membrane integrity. This leads to cell swelling due to the influx of extracellular fluid. Then lysosomes will burst and release disintegrating enzymes that initiate cell digestion. In contrast, in case of apoptosis, an intracellular program is activated that leads to a disassembly of cells and recycling of its parts [6].

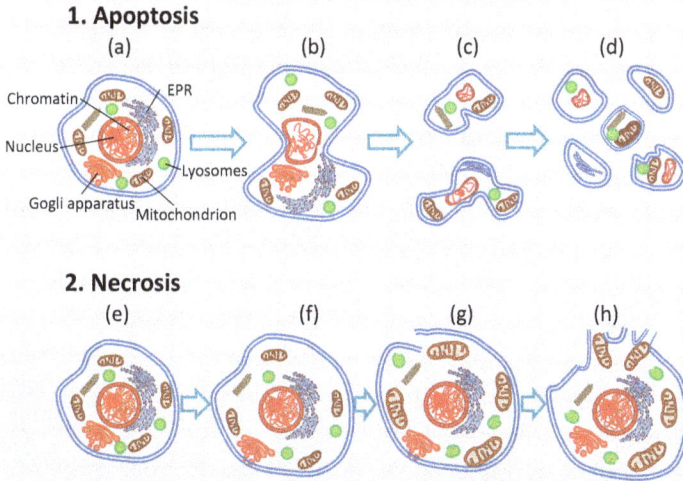

1. Apoptosis

(a)   (b)   (c)   (d)

Chromatin   EPR

Nucleus

Gogli apparatus   Lyosomes   Mitochondrion

2. Necrosis

(e)   (f)   (g)   (h)

This figure compares schematically cell death via apoptosis and necrosis. The upper panels show the different steps of apoptosis: (a) normal cell, some main components are indicated; (b) change of the nucleus while mitochondria remain intact; (c) fragmentation of cell; apoptotic debris are deposited and recycled. The lower panels show the different stages that lead to necrosis: (e) normal cell; (f) reversible swelling of the cell; (g) irreversible swelling of the cell after injury of the cell membrane and change of mitochondrial structure, while nucleus and chromatin structure remain intact; (h) destruction of the cell membrane and break up of the cell content (adapted from [6]).

1.3 Tumor cells

1.3.1 General remarks

The World Health Organization has classified abnormal masses of tissue called *neo-* (new) *plasms* (formation or creation) into four main groups: *benign neoplasms*, in situ neoplasms, *malignant neoplasms*, and neoplasms of uncertain or unknown development. *Malignant neoplasms*, simply called *cancers*, exhibit cell growth and division that exceeds normal tissue and is uncoordinated. Growth and division continue excessively even after eliminating the stimuli which evoked the original change. The newly grown cells have virulent or adverse properties in the body [7]. The paradigm in cancer research is to identify a close connection between *carcinogenesis*, the process that cancer initiates and develops, and alterations or mutations of the genome.

Whatever the cause is for damaged or mutated DNA, if such alterations are not stopped in time, the damaged DNA may replicate and act as an *oncogene*, i.e., a gene that causes cancer. These oncogenes can avoid apoptosis, evade surveillance by the immune system, and override telomerase, which otherwise would stop cell replication after 50 divisions.

In normal cells, at least two important known checkpoints are located at CP1 in G1/S and CP2 in G2/M involving tumor suppressors. The CP1 checkpoint calls for a stop when the DNA is damaged or when the surrounding signals that cell growth and cell division are not required. After replication and before division, checkpoint CP2 again controls any irreparable DNA damage that may have occurred during the S-phase. During carcinogenesis, these checkpoints fail to prevent abnormal cells from going through the cell cycle. Abnormal cells that are not eliminated by the checkpoints pile up within otherwise normal tissue. If those remain only at the original site, the lump is called a *benign tumor* or *primary tumor*. *Malignant* or *secondary* tumors occur if the abnormal cells invade surrounding tissues and metastasize by exporting cancer cells to other body parts.

The checkpoint CP1 is controlled by a protein named "p53" acting as a tumor suppressor. Because of its control function, p53 has been described as the "guardian of the genome," providing stability and preventing mutations of the genome. Accordingly, the gene TP53 responsible for coding the protein p53 has been named the tumor suppressor gene. By binding to the DNA, the protein p53 has many additional control functions. It can activate DNA repair in case of damage or alternatively command apoptosis if the DNA damage turns out to be irreparable. In cancerous cells, it was found that the gene TP53 contains numerous mutations. If the TP53 gene is damaged, tumor suppression via p53 control is severely inhibited. Cancer research pays much attention to the functionality and the frequency of mutations in the tumor suppressor gene and how this affects p53 concerning various types of cancer. An overview is given in [8], and a recent update on the state of the field can be found in [9].

Cancerous cells do not need to divide faster than normal cells. However, as soon as the balance between cell division (mitosis) and cell death (apoptosis) is disturbed, such that cells can further divide and grow, cancer is initiated. Hence, cancer is defined as uncontrolled cell division and unlimited cell life.

Cancer may be the interplay of genetic and environmental effects. Cancer researchers have found that 80–90% of malignant tumors are caused by external environmental factors [10], such as asbestosis, air pollution, ionizing radiation, inadequate nutrition, virus infection (i.e., cervical cancer), or psychic stress weakening the immune system. Heredity associated with cancer might include mutations directly linked to cells' neoplastic transformation and general health conditions such as the immune system.

There have been attempts to understand cancer as one of multiple intrinsic robust states of the endogenous cellular network, i.e., communication network between cells [11]. The dynamical interaction among endogenous agents[4] like hormones can generate many locally stable states for long periods of time with normal biological functionality, such as cell growth, cell division, and apoptosis. The statistical nature may cause a sudden transition from one stable state to another. If the switched local state is not optimized to the interest of the entire organism, the organism becomes "sick." It is hypothesized that tumor cells belong to one of those robust states that have suddenly switched, like in a phase transition of first order [11].

1.3.2 Cancer cell development

Typically tumors develop in the following six stages illustrated in Fig. 1.3:
(a) Mutation
(b) Extension
(c) Attack
(d) Penetration
(e) Angiogenesis
(f) Invasion and distribution

The starting point is a cell containing DNA material, which has transformed a normal gene into an oncogene, overriding all checkpoints. The indicated red cell in panel (a) undergoes cell division, although at this location and at this time, there is no call for replacement of an existing cell. The second stage is characterized by replicating cancerous cells without control (b). After about a million cell divisions, the tumor is the size of a pinhead, too small to be detected by any imaging modalities discussed in Chapters 1–3 and 8–10 of Volume 2. In the course of further mutations and replications, the tumor cells start attacking the connective tissue (c). Up to this point, the tumor is still locally confined. This stage is sometimes also called a *primary tumor*. A primary tumor may be present in the body for months or years before clinical symptoms develop. Mortality during primary tumor is minimal, with the exception of brain tumors.

Further attack in step (d) leads to penetration of the connective tissue. From this point on, the cancerous cell mass is no longer confined but still too small to be detected. As the uncontrolled cell growth needs oxygen and nutrients for further growth, the cancer cells send out messages that initiate the advancement and distribution of blood vessels into the cancerous region (e). This step is called *angiogenesis* and marks the transition from a *benign* to a *malign tumor*. With oxygen supply, the tumor can now grow rapidly. With a billion of cells, the *malignant tissue* or *malign tumor* has reached the size of a grape and can be detected by imaging methods. In the last step (f), cancer cells invade the circulation, and from there, other organs of the body are

---

4 Those are substances or compounds that origin from within the body.

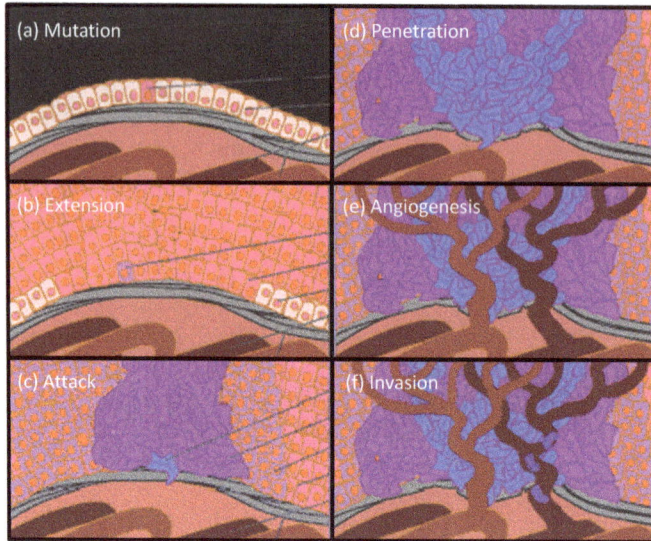

Fig. 1.3: Different stages of cancer development. For explanations see text (adapted from www.worl dofteaching.com).

affected. This process is called *metastasis* and is responsible for the formation of *secondary tumor* cells at other locations. Metastases are more commonly causing death. Alternatively, tumor cells can also spread via the lymph system.

1.3.3 Tumor growth

Since cell division is a binary process, the number of cancer cells is, simply speaking, expected to grow exponentially. Thus, the tumor volume $V(t)$ should double at constant time intervals Δt_d. For small tumors, a linear relationship

$$\log(V(t)) \sim t \tag{1.1}$$

has been observed, with an average doubling time Δt_d of about 2–3 months. The logarithmic volume dependence levels off as the tumor grows bigger, presumably because of insufficient oxygen and nutrition supply via penetrating blood vessels [12].

Once recognized, there are many possibilities for attacking cancer cells by the immune system. Nonetheless, some cancer cells make it to distant locations, settle in and transform normal cells into cancer cells. Unfortunately, the biochemical processes behind these disastrous stages of cancer development – although sought after by many laboratories worldwide – are still unknown. As long as the genetic origin of cancer development is not completely understood and drugs for impeding cancerous cell division are not available yet, we still need to apply rather crude methods of fighting cancer via radiotherapy, chemotherapy, and surgery. Hope-fully, with the help of fast sequencing techniques, key mutations can be identified, and targeted drugs may be synthesized that stop uncontrolled cell divisions. At present, the leaky and porous blood vessels in the tumor volume are the main target points for pharmaceuticals and nanoparticles. Specific and targeted nanoparticles can be excited by magnetic or optical means to produce local heating, causing hy-pothermia, which is further discussed in Chapter 7 on nanomedical applications.

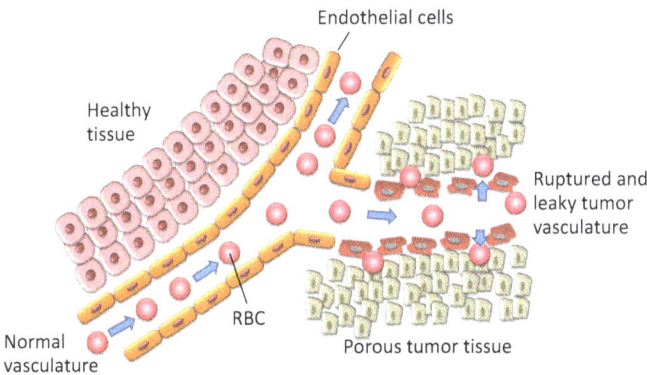

Fig. 1.4: Normal vasculature with blood tight vessels supply healthy tissue with oxygen and nutrients. In contrast, ruptured and leaky vasculature penetrate porous tumor tissue.

Healthy versus tumorous vasculature is shown schematically in Fig. 1.4. Endothelial cells line the entire circulatory system up to the smallest capillaries, including the heart and brain. This monolayer of cells has an important function as a barrier be-tween vessel lumen and surrounding tissue and for controlling the passage of gases, liquids, particles, and chemicals across the boundary. In the brain, the endo-thelium is particularly tight, allowing only the passage of water, some gases (O_2, CO_2), hydrophobic lipid-soluble molecules, and a small number of molecules such as glucose and amino acids that are essential for neural activity. It does not allow the diffusion of microscopic objects, such as bacteria or nanoparticles, or larger hy-drophilic molecules like some contrast agents to enter the cerebrospinal fluid. This tight endothelium in the brain is known as *blood–brain barrier*. In tumorous tissue, the vasculature is ruptured and leaky. Ruptured blood vessels enable contrast agents and even nontargeted nanoparticles to percolate across the wall, where they help identify tumor volumes and assist in treatment procedures.

The uncontrolled cell division is the most characteristic feature of all types of tumor cells. Un-derstanding the different forms and developments of cancer is important for early detection and possible cure.

1.4 Radiation response of cells

Radiotherapy aims at either tumor growth control and/or tumor cell destruction. In the latter case, radiation treatment intends to cause either apoptotic or mitotic cell death, a cell death that occurs during mitosis [13]. Alternatively, radiation is supposed to alter the molecular signaling pathways and block the G1/S and G2/M checkpoints to inhibit further cell division.

Before applying any radiation, a number of issues need to be considered. What kind of damage will be caused by radiation? Which radiation is best for the treatment of specific tumors: low LET[5] radiation (x-rays, γ,β protons) or high LET radiation (α-particles, neutrons)? What is the appropriate dose? Which is more effective: a single lethal dose or fractionated dose spread over several time intervals? How can a healthy tissue be protected from radiation damage?

1.4.1 LQ model

We start these considerations with the first question. Normal cells and cancer cells react in the same way to radiation. Both types of cells can self-repair if the dose is not lethal. But normal cells require a higher dose of radiation for complete apoptosis than cancer cells. We return to this critical point further below. The sensitivity to radiation varies during the life cycle of a cell. In Fig. 1.5 is plotted the single-cell survival rate as a function of x-ray radiation dose applied during different phases of the cell life cycle. The semilog plot shows that the radiosensitivity is highest during the M-phase, whereas the late S-phase is the most radiation-resistant phase. The G1-phase takes an intermediate position. Note that the initial radiosensitivity of the S-phase to low-dose radiation is little, expressed by an extended shoulder before the linear term (in the semilog plot) takes over. The shape of the survival curve $S(D)$ (not to be confused with the S-phase) as a function of dose D is usually described by the following expression [14]:

$$S(D) = \exp\left(-\alpha D - \beta D^2\right). \tag{1.2}$$

The empirical expression $S(D)$ is known as the *linear-quadratic* (LQ) model of cell survival. The first exponential term describes the linear dependence in the semilog plot, whereas the second term defines the shoulder, signaling delayed dose–response. The shoulder is typical for low LET radiation; for high LET radiation, only a linear dependence is observed. The reason becomes clear later on. The prefactor $\alpha = 1/D_0$ characterizes the slope of the survival curve, which distinguishes different phases of the cell

---

5 For a definition of LET, see eq. (6.37 in Volume II).

cycle and different types of cancer cells and their oxygen content (oxicity), as we will recognize in the following.

The open uncoiled structure of the DNA and lack of compactness may explain the radioresistance in the S-phase to low LET radiation. In contrast, the lack of re-pair mechanisms may contribute to the high radiosensitivity in the M-phase. The difference is as big as a factor of 2.5 according to the plot in Fig. 1.5.

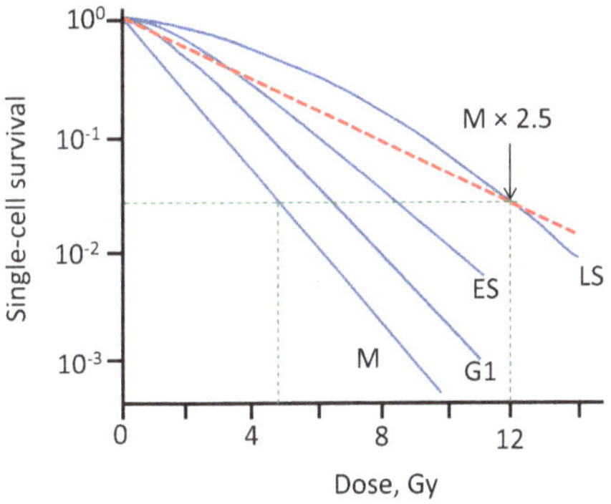

Fig. 1.5: Single-cell survival fraction versus dose upon high-energy x-ray radiation. ES (LS) = early (late) stage of S-phase. The dotted red line indicates that the radiosensitivity of the M-phase is by a factor of 2.5 higher than for the LS-phase.

1.4.2 Dose fractionation

Since the M-phase covers only a short time span in the cells' life cycle, the probabil-ity to hit the M-phase is higher if the total dose is broken down into smaller portions administered in a sequence of time intervals. This is called *fractionation*. Each frac-tionated dose is not sufficient to kill tumor cells, but the accumulated total dose will finally succeed. Furthermore, after first radiation exposure, cells readjust their cell clock with respect to their cell cycle to be more in phase with one another. How-ever, during the interval time, they also have a chance to self-repair the damage. Therefore, the total dose has to be increased compared to single-dose exposure. The LQ model (eq. (1.2)) is a guidance for fractionation planning. The higher the param-eters α and β are, the more sensitive the cells are to radiation. The ratio α/β, in turn, is a measure of the fractionation sensitivity of cells. A lower α/β ratio indicates a higher sensitivity to the sparing effect of cells via fractionation [15].

Figure 1.6 is a plot of the survival fraction of cells versus dose, typically 2 Gy per fraction. The example shows x-ray radiation with low LET compared to high LET radiation. The difference defines the *relative biological effectiveness* (RBE), in-troduced later. The survival rate per fraction is important because the fractions are repeated 30–40 times during a complete treatment. If the survival rate per fraction is 0.8, after 30 fractions, the survival rate is $(0.8)^{30} = 10^{-3}$; however, if the survival rate for a single fraction is 0.7, the final survival is reduced by two orders of magnitude to 2×10^{-5}. So the dose has to be finely tuned to have the best outcome.

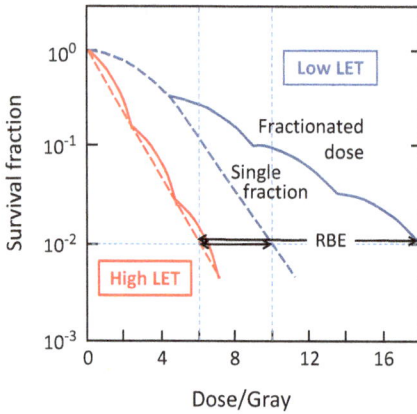

Fig. 1.6: Survival fraction as a function of dose for a single fraction (dashed lines) and fractionated dose application (solid lines) for both low and high LET radiation.

1.4.3 Biological effect of radiation

The sensitivity of cells to radiation is due to cutting chemical bonds of chromosomes. We distinguish between *direct* and *indirect actions* [16]. In the case of *direct action*, photons or secondary photoelectrons break bonds in strands of the DNA, as sketched in Fig. 1.7. *Indirect action* occurs when the radiation generates free water radicals such as H^+, OH^-, or H_2O^+ in cellular water, which increases the toxicity in a cell leading to bond breaking in the DNA.

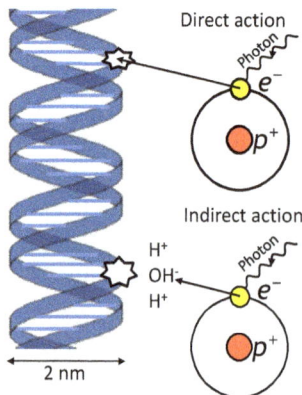

Fig. 1.7: Comparison of direct action on the DNA by the photoelectric effect and indirect action through formation of free radicals (adapted from [16]).

Radiation damage is most effective when not only one strand breaks (*single-strand break*, SSB) but when both strands of the DNA (*double-strand break*, DSB) are hit simultaneously. Then the proper repair of the DNA is impeded, and the survival chance of the cell is drastically reduced. The strands may still be partially repaired but lose their ability to reproduce. These cells often pass from the G2 to the M–phase but experience a mitotic catastrophe in the M-phase. Cancer cells are considered dead if they have lost their ability to reproduce. This may not immediately occur but may require a few more cell divisions. Malfunctioning cancer cells can be recognized in histological cuts by cell morphology changes, particularly by shrinkage and bubble formation (blebbing) of cell membranes.

Cancer cells have less ability than healthy cells for DNA repair between fractionated doses. With each subsequent mitosis, the cumulative effects of unrepaired DNA result in apoptosis of these tumor cells. Then, the original cell, together with fragments of DNA, breaks into smaller parts, while the cell membrane and organelles inside remain intact. These apoptotic cell fragments are finally removed by phagocytosis [17]. Phagocytosis is a process by which scavenger cells, such as white blood cells, ingest debris of apoptotic cell material for destruction and removal. More on phagocytosis can be found in Section 7.2.1.

1.4.4 Reaction distance

For x-ray radiation, direct or indirect actions occur more frequently with increasing photon energy. According to Fig. 6.11 in Volume 2, the absorption cross-section via photoelectric effect decreases with increasing photon energy. Beyond 100 keV, the photoelectric effect is irrelevant, and beyond 4 MeV the photon energy in matter is completely converted to Compton electrons and pair production within a short attenuation distance. This results in an increasing chance for DSB, reaching a maximum probability at a linear energy transfer (LET) of about 100 keV/μm (100 eV/nm). At this energy, the reaction distance between two hits corresponds to the diameter of a DNA molecule or about 2 nm (Fig. 1.8). For lower energies, the reaction distance is larger, causing mainly SSB, and for

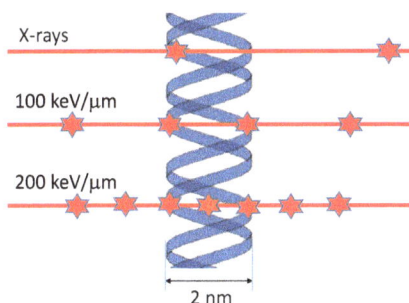

Fig. 1.8: Primary damage to the DNA for radiation with different LETs.

higher energies, the short reaction distance reduces the effectiveness of radiation damage. Therefore, the most effective energy loss per distance is 100 eV/nm. This reaction distance cannot be reached with x-ray radiation but requires heavy particles like protons, neutrons or α-particles that feature higher LET values. An overview on the LET for different particles and energies is given in Tab. 1.1.

Tab. 1.1: Linear energy transfer for different types of radiation according to [16].

Radiation	LET (keV/µm)
1.33 MeV ^{60}Co-γ-rays	0.2
250 kVp x-rays	2.0
10 MeV protons	4.7
150 MeV protons	0.5
14 MeV neutrons	12
2.5 MeV α-particles	166

> **!** Cancer cells are affected by radiation through direct action (bond breaking) and indirect action (increased toxicity). Normal cells are less sensitive to radiation than cancer cells. The tumor volume decreases upon irradiation but grows again after some time delay, if fractionated irradiation is not completed.

1.4.5 Relative biological effectiveness (RBE)

Now we reconsider the relationship between LET (LET = $-dE/dx$) and RBE. The RBE of radiation can directly be determined from Fig. 1.6 comparing survival fractions as a function of dose for different radiation sources. The RBE is defined as the ratio:

$$\text{RBE} = \frac{D(\text{photons}, V, S)}{D(\text{particles}, V, S)}, \tag{1.3}$$

where $D(\text{photons}, V, S)$ is the dose deposited by a photon source (x- or γ-rays) into a volume V with survival fraction S, and $D(\text{particles}, V, S)$ is the dose delivered by particles into the same volume and for the same survival fraction. RBE is a dose conversion factor. It converts the dose of ionizing particle radiation to the dose by photons (x-rays and γ's) for the same survival fraction and for the same cancer volume. By definition, the RBE of x-rays is 1.

In Fig. 1.9 is plotted the RBE as a function of LET. The RBE equals 1 for low LET radiation and increases as the LET reaches values higher than 50 keV/µm. The RBE goes through a maximum at 100 keV/µm, and decreases again to one for 1000 keV/µm. The maximum can be reached with a few MeV α-particles, but not with x-rays or

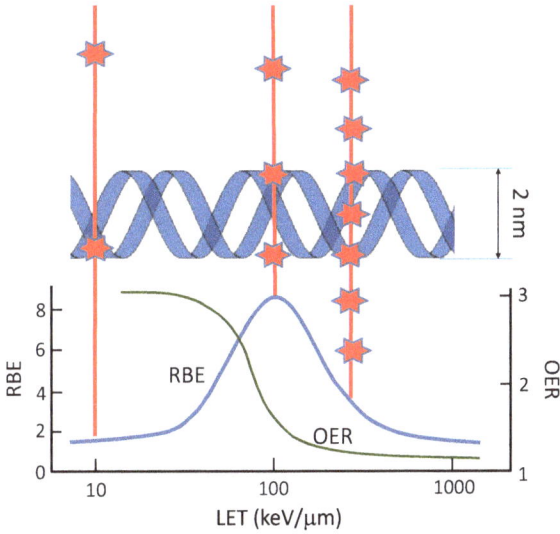

Fig. 1.9: RBE (left scale) and OER (right-hand scale) versus LET. LET is plotted on a logarithmic scale. The stars symbolize interactions of the radiation with DNA strands along their track (red lines).

γ-radiation. The RBE of the latter radiation always stays on the left side of the peak. At the maximum, the probability is high for DSBs in the DNA.

As an example we compare in Fig. 1.6 the dose required in a single fraction for a survival probability of 1%, indicated by dashed lines. The required x-ray dose is 10 Gy, whereas the dose is 6 Gy for high LET particles. The ratio yields an RBE of 1.7. Next, we compare the fractionated low LET dose and fractionated high LET dose, for the same survival probability of 1%. According to Fig. 1.6, we obtain a ratio of 18/6 = 3. This comparison also shows that the RBE depends on how the radiation is fractionated. For low LET, fractionation makes a big difference, whereas for high LET radiation, the difference is negligible. Nevertheless, dose fractionation is essential for both types of radiation because healthy tissue needs recovery between fractions.

1.4.6 Oxygen enhancement ratio

In Fig. 1.9, the oxygen enhancement ratio (OER) is also plotted against LET. OER is defined as the ratio:

$$OER = \frac{D(\text{no } O_2, V, S)}{D(O_2, V, S)}, \tag{1.4}$$

where $D(\text{no } O_2, V, S)$ is the dose to produce a survival effect without oxygen supply, and $D(\text{no } O_2, V, S)$ is the dose to produce the same effect with oxygen supply.

We notice that for low LET radiation, the OER is almost 3, whereas for high LET radiation, the OER drops to 1. This implies that for low LET radiation, the oxygenation of cancer cells is a very important parameter to obtain a positive treatment effect, whereas for high LET radiation, the oxygenation level of cancer cells has no effect on the outcome. Hypoxic and even anoxic cancer cells can be treated with high LET radiation without additional oxygen conditioning. For low LET radiation, oxygen is important to increase the indirect action effect of radiation. The discussion of the oxygen effect is continued in Section 1.5.2.

Returning to eq. (1.2), the LQ model of cell survival upon radiation exposure was used for empirically fitting experimental observations. However, later it was shown by Chadwick and Leenhouts that the two terms in the exponent are related to single-strand DNA break (quadratic term) and double-strand DNA break (linear term) [18]. High LET/RBE radiation has a much higher probability for DSB. Each hit leads to immediate cell death. Low LET/RBE radiation is more likely to cause an SSB. Therefore, cell survival is prolonged. The current status of the LQ discussion is reviewed in [19].

1.5 Radiation therapies

1.5.1 Dose control

Next we discuss typical radiotherapies of tumors in clinics. Whatever the method chosen is, the goal is to expose malign tumor cells to a critical dose level of radiation to stop their reproduction while sparing surrounding healthy cells from radiation exposure. Usually, a dose of 2 Gy per session is administered 5 times per week for 6–8 weeks. Thus, the total dose is 60–80 Gy. Between fractions, cancerous cells have a chance of:
- *self-repairing* DNA damage;
- *reoxygenation* of tumor cells;
- *redistribution* within the cell cycle;
- *repopulation* of cells by cell replication.

Those are the four "R's" in radiation biology and oncology that need to be considered before planning any radiation treatment [20, 21]. Most important in this planning is the fact that normal cells circumstantially affected by radiation also have time to recover between fractions. At a low dose of a single session, radiotherapy has the effect of growth delay but not permanent tumor recession or cure. Figure 1.10 shows the tumor growth as a function of time on a logarithmic scale. After initial exponential growth, the curve for untreated tumor volume bends over to a slower growth rate. The response to a radiation dose is a volume recession over some time. After some

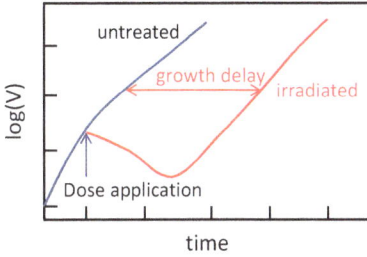

Fig. 1.10: Tumor volume V on a log scale is plotted versus time. At the onset of irradiation, tumor volume recesses but can grow again after some time delay. The growth delay–time depends on the dose.

time delay, the tumor starts growing again. This shows that the dose was too low to hinder repopulation of tumor cells.

The growth delay increases with increasing dose. Finally, cancer cells lose their ability to survive. Figure 1.11(a) shows the tumor control fraction as a function of dose. At a fraction of 50%, half of the cancer cells have died. This point is reached after an accumulated dose of about 50–60 Gy. Below this dose, cancer cells have a chance of self-repair and repopulation as already mentioned. With increasing dose, the chances are high to hit the stem cancer cells, as those are responsible for metastasizing normal cells.

Referring to previous discussions, normal cells have a higher radiation resistance than cancer cells. Permanent damage of normal cells occurs at higher dose, as indicated by the red survival curve in Fig. 1.11(a). Under optimal conditions of a radiation treatment plan, on average more than 90% of cancer cells are killed with a 5% hit rate of normal cells causing severe side effects.

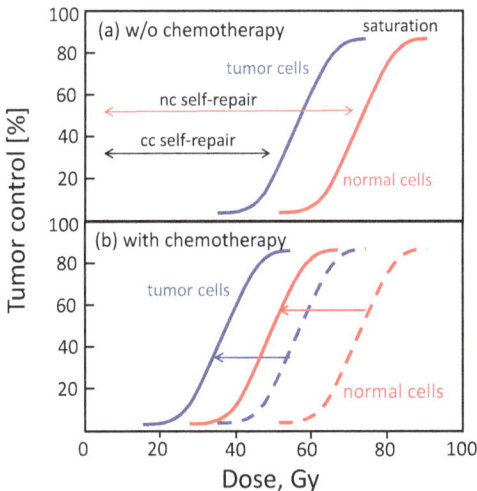

Fig. 1.11: (a) Tumor control without chemotherapy versus dose for normal cells (nc) and cancer cells (cc). (b) With chemotherapy, the radiation sensitivity increases, and the tumor control shifts to lower dose (adapted from [20]).

1.5.2 Hypoxia and chemotherapy

Blood cells infiltrate lumps of cancer cells in order to nurture and oxygenate them. However, penetration of blood vessels during angiogenesis is not a well-organized process. The oxygen supply is intermittent, and the periphery remains low in oxygen, as shown schematically in Fig. 1.12. These low-oxygen, hypoxic, or even anoxic cancer cells prove to be more radiation-resistant than cells with an adequate oxygen supply [21]. Since the early detection of this effect, there has been much speculation about its cause. Research has now confirmed that the presence of oxygen during irradiation increases the concentration of free and toxic radicals due to indirect radiation effects (see Fig. 1.7) [22]. For example, photoelectrons can combine with O_2 to form O_2^-, a strong oxidizing agent with high reactivity toward DNA.

The combined effect of SSB with a lower LET through direct radiation exposure and oxygen reactivity through indirect radiation exposure reduces the tumor survival fraction. The difference is considerable: a survival fraction of 0.01 requires 10 Gy in oxic cells, but 28 Gy in hypoxic cells, which is almost a factor of 3 in dose. During the fractionated radiation exposure, some hypoxic cells can transform into oxic cells, so that in the end, all cells are killed. This is an important point that temporarily changes the shape of the survival fraction versus dose and determines the outcome of treatment with radiation therapy. In contrast, tumor treatment with high LET radiation is less dependent on the oxygen concentration, because high LET causes predominantly DSBs without the help of indirect radiation effects.

Before applying radiotherapy, cancer tissue can be exposed to higher oxygen pressures and additional radiation sensitizers. One of these is chemotherapy, which increases the sensitivity of hypoxic cells to radiation and shifts tumor control to a lower dose, as indicated in Fig. 1.11(b). This shows that the OER is high for low LET radiation. In contrast, high LET radiation shows no visible enhancement effect via chemotherapy and therefore the OER = 1. An overview on hypoxia and radiation therapy is provided in [22].

The combination of radiotherapy and chemotherapy shows a number of other synergetic effects. Repopulation and recovery from the sublethal dose is reduced (Fig. 1.11(b)). Also noticed is an accumulation of tumor cells in the more radiation-sensitive G2 and M phases [23]. Chemotherapy improves blood circulation and the perfusion of other chemosensitizers, thereby improving the overall sensitivity to radiation. This enables a wider therapeutic range of fractionated and total doses.

Radiation therapy is like the fire brigade arriving after the house has already burned. Current and future developments focus on the early detection of cancer cells at the cellular level to enable firefighters to arrive before the house goes up in flames.

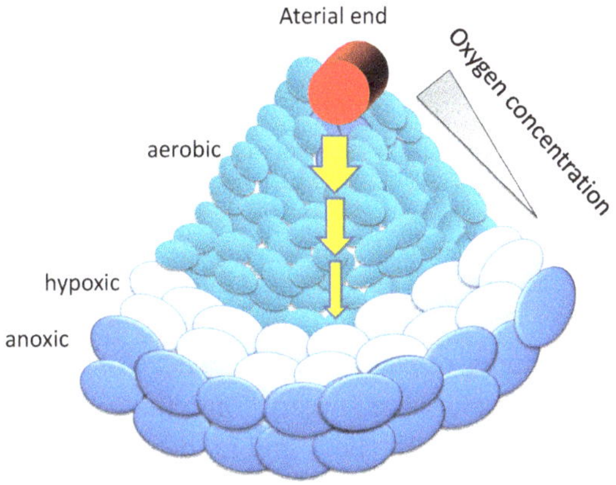

Fig. 1.12: Oxygen distribution in a lump of malignant tissue. The oxygen supply decreases with distance from blood vessels.

1.5.3 Types of radiotherapies

We distinguish between *external beam radiation therapy* (EBRT) and *internal radiation therapy* (IRT). EBRT uses x-ray, electron, proton, or neutron beams. Internal RT uses encapsulated radioisotopes that are implanted directly into the tumor mass. Boron neutron capture therapy is positioned between EBRT and IRT as it requires an external neutron beam to excite internal isotopes. Further details on these methods are given in Chapters 2–6. Another distinction of RTs is between therapies with charged particles (electrons and protons) and uncharged particles (photons and neutrons). Which RT is applied depends on the tumor type, the tumor stage, the availability of certain treatment techniques, and ultimately on the decision of the treating oncologist. For example, x-ray RT is applicable to almost any confined tumor mass, while electron beam RT can only be used to treat skin cancer or superficial tumors. Proton RT requires an extensive infrastructure, while neutron RT is used rarely and only for special cases. All methods are discussed in detail in Chapters 2–6, including an assessment of advantages and disadvantages. An overview on the various treatment methods is shown in the flowchart of Fig. 1.13. Some of the terms in the flowchart are explained in later chapters.

1.5.4 Radiation treatment planning overview

Any cancer treatment requires careful and precise planning beforehand. In the first step, a multidisciplinary team consisting of oncologists, surgeons, and radiotherapists work out a treatment protocol after the tumor has been localized and characterized. This also includes an assessment of the cancer stage (see next section) and

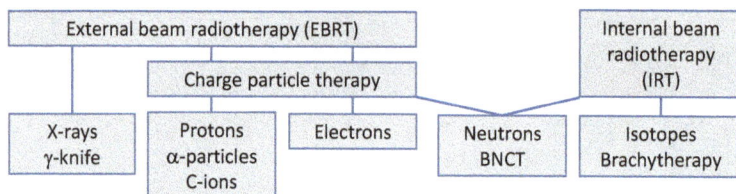

Fig. 1.13: Flowchart of radiation treatment methods used for radiotherapies of cancer.

whether the treatment is intended to be curative or palliative. Also, a decision has to be made whether RT should be applied before or after surgery. Then the patient is called again for precise localization of the tumor in the body and for determining its volume, shape, and electron density by one or several imaging modalities: MRI, CT, PET, or US. CT is the preferred imaging modality as it provides accurate pictures with a gray scale that depends on the electron density, which is the required input parameter for dosimetry calculations. MRI is more sensitive to soft tissue than CT, and US provides additional information on superficial tumor masses not detected by the other imaging methods. PET is a powerful but expensive imaging modality that is particularly sensitive to malignant cells. If combined with CT, tumors can be precisely identified with respect to location and shape, thus improving the definition of the tumor delineation, i.e., the gross tumor volume as shown in Fig. 2.7. Input data from all imaging modalities are integrated into a computer simulation program and used for calculating a 3D image of the tumor volume in the coordinate system of the body.

The information from 3D imaging of the cancer volume is used for defining a radiation plan by considering the tumor depth and the dose to be delivered. Once the dose to volume is accurately calculated, the patient goes to a simulator to determine the location of the tumor with respect to the laboratory frame. The simulator is a setup identical in dimensions to the irradiation gantry used for XRT but furnished only with x-ray imaging equipment. The coordinates are determined by taking a series of x-ray radiographs to ensure that the tumor is in the correct position relative to the treatment couch and gantry after the patient has been immobilized. The third step is the actual radiation treatment according to the radiation plan. Quality assurance (QA) during irradiation is required to avoid systematic errors, such as tumor movement. After irradiation, the tumor size is checked again with different imaging modalities. In case of local recurrence, a radiation boost is administered within a decreased target volume. Further follow-up checks conclude the treatment.

The more precise the tumor volume can be localized, the more important it is to control the positioning of the patient during radiation. Hybrid systems have been developed for improving the positioning control, combining linacs and CT scanners. Today, image-guided radiation therapy monitored by x-rays and laser-tracking systems has become the standard. As the tumor volume and surroundings change during radiation treatment, monitoring also includes dose adjustment between fractionated radiation treatments.

A flowchart of the procedures applied is shown in Fig. 1.14. Note that this describes only solid types of tumors. Hematological cancer types affecting blood, bone marrow, and lymph nodes require different treatment plans.

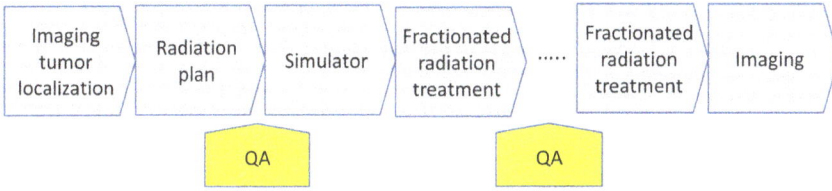

Fig. 1.14: Flowchart for radiation treatment of solid tumors. QA = quality assessment.

1.5.5 Staging

Staging cancer is a central topic of any treatment plan. Here we do not explain how staging is performed but describe only the systems used for staging and what they mean.

There are a number of staging systems that oncologists use to characterize the development of tumors. Some systems are more general for all types of tumors, and some are more specific.

The staging system TNM belongs to the more general systems.

The TNM system specifies:

T   Tumor size and extent of the tumor. TX: tumor size cannot be determined; T0: tumor cannot be found; otherwise, tumor size is graded on the scale T1–4.

N   Tumor proximity to the nearby lymph nodes that have cancer. NX: Cancer in the nearby lymph nodes cannot be measured. N0: no cancer in the nearby lymph nodes. N1, N2, N3: number and location of lymph nodes that contain cancer.

M   Distant metastasis (M): MX: Metastasis cannot be measured. M0: Cancer has not spread to other parts of the body. M1: Cancer has spread to other parts of the body.

For example, the stage T3N1M0 implies a large cancer size, a single nearby cancerous lymph node, and no spread.

Another widely used staging system is as follows:

Stage 0:   Abnormal cells are present but have not spread to the nearby tissue. This is also called *carcinoma in situ* (CIS). CIS is not cancer, but it may become cancer.

Stage I, II, III: Cancer is present. The higher the number, the larger the cancer tumor and the more it has spread into the nearby tissues.

Stage IV: The cancer has spread to distant parts of the body.

More information on tumor stages and implications can be found in [24].

1.6 Summary

S1.1   The life cycle of a cell can be subdivided into interphase and mitosis.

S1.2   Cell division takes place during mitosis.

S1.3   The mitosis is subdivided into prophase, metaphase, anaphase, and telophase.

S1.4   Tumors have six stages of development: 1. mutation, 2. extension, 3. attack, 4. penetration, 5. angiogenesis, and 6. invasion and distribution.

S1.5   Cancer is defined as an uncontrolled cell division and unlimited cell life.

S1.6   Tumor is an abnormal mass of tissue that arises from cells of preexistent tissue, and serves no useful purpose.

S1.7   Normal cells require a higher dose of radiation for complete apoptosis than cancer cells.

S1.8   Radiosensitivity is highest during the M-phase, whereas the late S-phase is the most radiation-resistant phase. The survival probabilities are different by a factor of 2.5.

S1.9   Fractionation is a radiation treatment plan with the total dose broken down into smaller portions administered in a sequence of time intervals.

S1.10 Cells are affected by radiation through direct action (bond breaking) and indirect action (increased toxicity).

S1.11 The most effective LET is reached with heavy particles like protons, neutrons, or α-particles.

S1.12 The four "R's" in radiation biology are: Repairing, Reoxygenation, Redistribution, and Repopulation.

S1.13 Oxygen-deficient hypoxic and anoxic cancer cells are more radiation-resistant than cells with sufficient oxygen supply.

S1.14 Chemotherapy enhances the radiation sensitivity of hypoxic cells and shifts the tumor control to lower doses.

S1.15 Chemotherapy allows a broader therapeutic bandwidth of fractionated and total dose.

S1.16 The oxygen enhancement ratio (OER) is defined as the radiation dose required for a certain survival fraction without the presence of oxygen and the dose required for the same survival fraction with oxygen.

S1.17 Radiation treatment with high LET radiation is independent of the oxygen level of cancer cells. For high LET radiation, the OER is one.

S1.18 Any cancer treatment requires careful and precise planning beforehand.

S1.19 QA during irradiation is required to avoid systematic errors.

S1.20 Staging specifies the size and spread of tumors in the body.

Questions

Q1.1   What are the two main phases of the cell cycle?
Q1.2   When does cell replication take place?
Q1.3   Which protein is known to be a tumor suppressor?
Q1.4   Tumors develop in six stages. Name those six stages.
Q1.5   At what stage does tumor development become noticeable by imaging modalities?
Q1.6   When does a tumor turn from a benign to a malign tumor?
Q1.7   How is the survival curve S defined as a function of the dose D?
Q1.8   When are cells most sensitive to radiation, and when are they least sensitive to radiation?
Q1.9   Why is radiation treatment fractionated?
Q1.10   What are the main effects of radiation to a cell?
Q1.11   How is the RBE defined?
Q1.12   When is the RBE highest?
Q1.13   Which types of radiation are high LET radiation?
Q1.14   Which radiation is low LET radiation?
Q1.15   What is OER?
Q1.16   For which type of radiation is OER important or not important?
Q1.17   What is the reason for the OER importance?
Q1.18   What is the effect of oxygen for tumor control?
Q1.19   What is the effect of low-dose versus high-dose application?
Q1.20   What is the effect of chemotherapy?
Q1.21   What are the four R's in radiation biology?
Q1.22   Why is QA important?
Q1.23   What does TNM stand for?

Attained competence

	+	0	–
I know the different phases of a cell's life cycle.			
I know in which phase cell replication occurs.			
I can name the different phases of cancer cells.			
I understand the origin of different slopes in the survival versus dose curve.			
I can distinguish between high LET and low LET radiation.			
I know which consequences SSBs and DSBs have.			
I appreciate the difference between direct and indirect DNA damage.			
I know which radiation has high RBE and which has low RBE.			
I am aware of the importance of oxygenation.			
I know what OER refers to.			
I can name the four R's of radiation biology.			
I know how to characterize tumors via the TNM code.			

Exercises

E1.1 **Base sequence:** What is the complementary DNA sequence to TACGACT?

E1.2 **Cell survival:** The following graph shows the two parts of eq. (1.2) independently on a semilog plot (for convenience we use here a base of 10 instead of e): $S_1(D) = 10(-\alpha D)$ and $S_2(D) = 10(-\beta D^2)$.

a. Which graph is $S_1(D)$ and which one is $S_2(D)$?

b. Determine the coefficients α and β and the ratio α/β.

c. Which survival curve is representative for x-ray irradiation and which one for neutrons?

d. What would the curve $S(D) = 10(-\alpha D - \beta D^2)$ look like if $\alpha = \beta = 0.5$

e. From the plots, determine the RBE at 0.1% survival chance and for 10% survival chance.

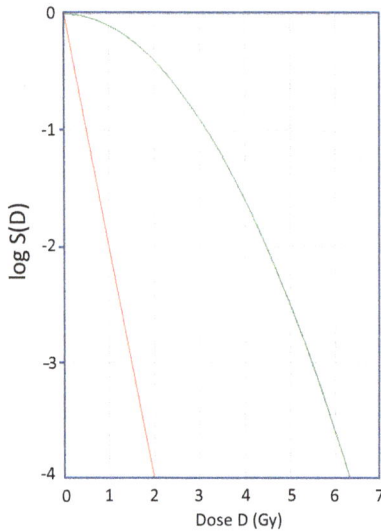

E1.3 **DNA damage:** Explain the difference of SSB and DSB and how this affects the cell survival.

E1.4 **Lethal dose:** A dose of 5–8 Gy is usually lethal. During radiation treatment, in a single fraction already a dose of 5–8 Gy is administered and much more in subsequent fractions. Why is this dose not lethal to the patient?

E1.5 **Tumor imaging:**

a. Which imaging modalities are sensitive to cancer volume?

b. What is the smallest tumor volume that can be detected?

References

[1]   Campbell NA, Reece JB. Biology. 11th edition, New York: Pearson; 2011; 2009.
[2]   Allison JP. Immune checkpoint blockade in cancer therapy. Nobel lecture; 2018.
[3]   Blackburn EH. Structure and function of telomeres. Nature. 1991; 350: 569–573.
[4]   Hayflick L, Moorhead PS. The serial cultivation of human diploid cell strains. Exp Cell Res. 1961; 25(3): 585–621.
[5]   Blackburn EH, Epel ES, Lin J. Human telomere biology: A contributory and interactive factor in aging, disease risks, and protection. Science. 2015; 350: 1193–1198.
[6]   Richard D, Chevalet P, Soubaya T. Biologie in Farbtafeln. Berlin Heidelberg: Springer Verlag; 2013.
[7]   Kumar V, Robbins S, Zhai Q, Chen J. Textbook of pathology. Peking University Medical Press; 2009.
[8]   Edlund K, Larsson O, Ameur A, Bunikis I, Gyllensten U, Leroy B, Sundström M, Micke P, Botling J, Soussi T. Data-driven unbiased curation of the TP53 tumor suppressor gene mutation database and validation by ultradeep sequencing of human tumors. 2012; 109: 9551–9556.
[9]   Sabapathy K, Lane DP. Understanding p53 functions through p53 antibodies [correction appeared in J Mol Cell Biol. 2019;11:1105]. J Mol Cell Biol. 2019; 11: 317–329.
[10]   Lewandowska AM, Rudzki M, Rudzki S, Lewandowski T, Laskowska B. Environmental risk factors for cancer – Review paper. Ann Agric Environ Med. 2019; 26: 1–7.
[11]   Ao P, Galas D, Hood L, Zhu XM. Cancer as robust intrinsic state of endogenous molecular-cellular network shaped by evolution. Med Hypotheses. 2008; 70: 678–684.
[12]   West GB, Brown JH, Enquist BJ. A general model for ontogenetic growth. Nature. 2001; 413: 628–631.
[13]   Vakifahmetoglu H, Olsson M, Zhivotovsky B. Death through a tragedy: Mitotic catastrophe. Rev Cell Death Differ. 2008; 15: 1153–1162.
[14]   McMahon SJ. The linear quadratic model: Usage, interpretation and challenges. Top Rev Phys Med Biol. 2019; 64: 01TR01, 1–24.
[15]   van Leeuwen OAL, Crezee J, Bel A, Franken NAP, Stalpers LJ, Kok HP. The alfa and beta of tumours: a review of parameters of the linear-quadratic model, derived from clinical radiotherapy studies. Radiation Oncol. 2018; 13: 1–11.
[16]   Hall EJ, Giaccia AJ. Radiobiology for the radiologist. 7th. Philadelphia, Baltimore, New York, London: Lippincott Williams & Wilkins, Wolters Kluwer; 2012.
[17]   Poon IKH, Hulett MD, Parish CR. Molecular mechanisms of late apoptotic/necrotic cell clearance. Cell Death Differ. 2010; 17: 381–397.
[18]   Chadwick KH, Leenhouts HP. A molecular theory of cell survival. Phys Med Biol. 1973; 18: 78–87.
[19]   McMahon SJ, Prise KM. Mechanistic modelling of radiation responses. Cancers 2019; 11: 205, p. 1–23.
[20]   Podgorsak EB. technical editor. Radiation oncology physics: A handbook for teachers and students. Vienna: International Atomic Energy Agency; 2005.
[21]   Radiation biology: A handbook for teachers and students. Vienna: International Atomic Energy Agency; 2010.
[22]   Rockwell S, Dobrucki IT, Kim EY, Marrison ST, Vu VT. Hypoxia and radiation therapy: Past history, ongoing research, and future promise. Curr Mol Med. 2009; 9: 442–458.
[23]   Sørensen BS, Horsman MR. Tumor hypoxia: Impact on radiation therapy and molecular pathways. Rev Art Front Oncol. 2020; fonc.2020.00562, p. 1–11.
[24]   Rosen RD, Sapra A TNM classification. In: StatPearls [Internet]. Treasure Island (FL): StatPearls Publishing; 2022. Available from: https://www.ncbi.nlm.nih.gov/books/NBK553187/

Further reading

Alberts BE, Johnson A, Lewis J, Morgan D, Raff M, Roberts K, Walter P. Molecular biology of the cell. 6th. New York: Garland Science; 2014.

International Atomic Energy Agency (IAEA). Radiation biology: A handbook for teachers and students. Training Course Series No. 42. Vienna: International Atomic Energy Agency (IAEA); 2010.

Weinberg RA. The biology of cancer. 2nd. New York: Garland Science, Taylor & Francis Group; 2013.

Kerr DJ, Haller DG, van de Velde CJH, Baumann M editors, Oxford textbook of oncology. 3rd. Oxford, New York, Athens: Oxford University Press; 2016.

Podgorsak EB technical editor. Radiation oncology physics: A handbook for teachers and students. Vienna: International Atomic Energy Agency; 2005.

Turner JE. Atoms, radiation, and radiation protection. New York, London, Sydney, Toronto: Wiley & Sons. 3rd. 2007.

Chadwick KH, Leenhouts HP. The molecular theory of radiation biology. Monographs on Theoretical and applied genetics, volume 5. Berlin, Heidelberg: Springer Berlin; 2012.

Useful websites

https://www.cancerquest.org/
http://wiki.cancer.org.au/oncologyformedicalstudents/Clinical_Oncology_for_Medical_Students#_ga=1.264882507.554164266.1468659718
http://aboutcancer.com/introduction_page.htm
https://en.wikipedia.org/wiki/DNA_replication

2 X-ray radiotherapy

Important parameters

10 MeV photons, interaction with matter	77% Compton, 23% pair production
^{60}Co γ-emitter photon energy	1.3 MeV
1 MU	0.01 Gy = 1 cGy
Peak maximum of percent depth dose	2–4 cm below the skin
Linac accelerating energy	4–25 MeV
RBE of x-rays	1 (by definition)
Electron–photon converter plate	Tungsten (W)

Acronyms used in this chapter

CK	cyber knife
CS	Compton scattering
CT	computer tomography
CTV	clinical tumor volume
FF	flattering filter
FFF	flattening filter free
GTV	gross tumor volume
ICF	iInhomogeneity correction factor
ICRU	International Commission on Radiation Units and Measurements
IGRT	image-guided radiotherapy
IMRT	intensity-modulated radiation therapy
LET	linear energy transfer
MC	Monte Carlo simulation
MLC	multileaf collimation
MRgRT	magnetic resonance-guided radiation therapy
MRI	magnetic resonance imaging
MU	monitor unit
OAR	organ at risk
OER	oxygen enhancement ratio
PDD	percentage depth dose
PE	photoeffect
PP	pair production
PTV	planning target volume
RBE	radiobiological effectiveness
SAD	source-axis distance
SSD	source-surface distance
TCE	transient charged particle equilibrium
TERMA	total energy released per unit mass
TV	target volume
US	ultrasound
XRT	x-ray radiotherapy

https://doi.org/10.1515/9783111168739-002

2.1 Introduction

Radiotherapy is intended to deposit a lethal dose of radiation to a confined volume of cancerous tissue in the body while keeping radiation damage to surrounding healthy tissue as minimal as possible. How this is achieved with x-rays is the central topic of this chapter.

The treatment of tumors is always performed by a multimodality management plan, including radiotherapy, medication treatment (chemotherapy), and eventually surgery before or after radiation exposure. Moreover, the management plan has to consider the specific organ and its functionality that is to be preserved as far as possible, while surrounding organs and, in particular, the spinal cord has to be protected from radiation. These medical and oncological aspects are not considered here any further. We concentrate our discussion on the physics of radiotherapy.

X-ray radiotherapy (XRT) has a long history. After the first excitement about the imaging capabilities of penetrating x-rays, discovered by Röntgen[1] in 1895, it soon became apparent that radiation can also heal by causing damage. Radiation was used as early as 1899 to treat successfully squamous cell carcinoma of the nose and cheek. But radiotherapy as a cancer treatment modality has taken off only during the last 50 years. The early applications of XRT were performed with x-ray tubes and with radioactive sources. Linear accelerators (linac) were developed for high-energy particle physics in the 1930s and became available for XRT in the 1960s of the last century to reduce complications that occurred with the use of ^{60}Co γ-sources. A modern version of such a linac-based cancer treatment gantry, boosting up the x-ray energies into the MeV region, is shown schematically in Fig. 2.16.

In the 1970s, lead blocks were used to define a treatment window and reduce the dose to normal tissue. In the 1980s, multileaf collimators (MLCs) were invented, leading to a 3D conformal radation exposure. Later in the 1990s, CT imaging was combined with increasing computer power for establishing 3D treatment plans. This includes computerized intensity-modulated radiation therapy (IMRT), an RT that delivers variable beam intensity throughout the irradiated volume from multiple beam angles. Concurrently, new imaging technologies (MRI, CT, US) improved the localization and definition of tumor volumes. More recent developments of XRT advance to smaller subfields of the tumor volume and to higher-resolution IMRT. The combination of robotics, artificial intelligence (AI), advanced accelerator technologies, and patient-tracking systems for high-precision RT with reduced side effects are the hallmarks of present day XRT. All these advancements are discussed later in Section 2.5. References [1, 2] give concise overviews of the historical development of XRT, and Fig. 2.1 shows a generic layout of a linac-based system for XRT. A high-energy electron beam hits a target that converts electrons to photons. The high-intensity x-ray beam is

---

1 Wilhelm Conrad Röntgen (1845–1923), German physicist and Nobel laureate in Physics 1901.

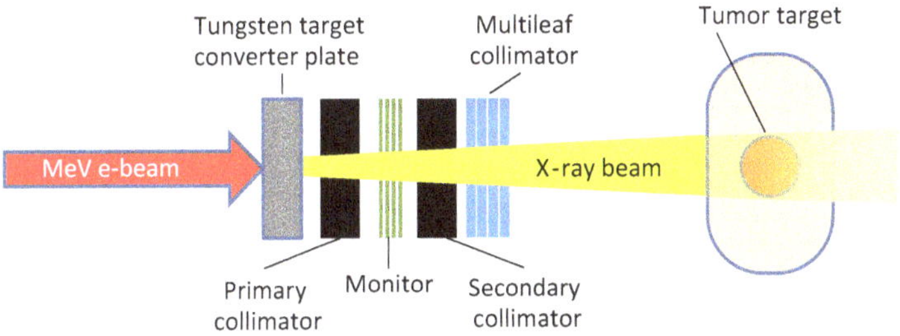

Fig. 2.1: Generic layout of beams and collimators for x-ray beam radiotherapy.

collimated and shaped to conform to the cancer volume. The x-ray beam is attenuated in the body and delivers a dose that piles up at the tumor location if hit from multiple angles.

2.2 Absorbed dose by high-energy photon beam

Before describing cancer treatment via XRT, we first need to understand the physical processes that occur when high-energy photons hit a soft matter target. More precisely, we need to know the x-ray interaction with soft matter and the dose deposited as a function of depth, that is, the dose-depth profile. Part of this issue has been discussed already for lower energy x-ray radiography in Chapters 4–7 of Volume 2. Here we focus on very hard MeV x-rays and γ-radiation used for radiotherapy.

Table 2.1 lists the three contributions of photon interaction with matter as a function of energy: photoelectric effect (PE), Compton scattering (CS), and pair production (pairing) (PP). For a discussion of these effects, we refer to Chapter 6.2 and in particular to Fig. 6.11 in Volume 2. At high photon energies beyond 4 MeV, which are relevant for XRT, CS and PP are the dominant interaction processes. Through these processes, the photon energy is converted to kinetic energy of electrons in the tissue. The high energy or "hot" electrons then is the dose-relevant radiation that causes damage to tumor cells. It is important to realize that it is not x-ray photons that are responsible for LET and RBE of tumor cells in XRT, but hot electrons. X-ray photons are merely used as energy carriers for the local generation of CS and PP electrons, which could never penetrate so deeply into the tumor volume on their own.

At 10 MeV incident photon energy, 77% of the scattering events are due to CS. Assuming that the photon scattering angle is on the order of 10°, the energy transfer to Compton electrons is about 23% or 2.3 MeV. According to Fig. 2.3, these electrons have a range of about 1 cm in water. For a photon energy of 24 MeV, the energy transfer to Compton electrons under the same condition increases to 50%, producing 10.6 MeV electrons with a range of about 5 cm in water.

Tab. 2.1: Relative weight of interactions for photons penetrating water from low energies up to high energy incident photon energies.

Photon energy (MeV)	Relative weight of interactions (%)		
	PE	CS	PP
0.01	95	5	0
0.025	50	50	0
0.06	7	93	0
0.150	0	100	0
4	0	94	6
10	0	77	23
24	0	50	50

Note: PE = photoelectric, CS = Compton scattering, PP = pair production. (Data reproduced from [3].)

2.2.1 Percent depth dose for electrons

Range and dose of electrons can be measured using an electron accelerator with tuneable energy and a water tank as a phantom target [4, 5]. Inside the water tank, detectors (dosimeters) can be positioned and shifted parallel and perpendicular to the beam to determine the local dose and the beam profile as function of depth. The setup is sketched in Fig. 2.2(a). A flat, plane-parallel ionization chamber is used as a dosimeter. The local dose $D(z)$ is then plotted versus z and normalized by the maximum dose D_{max}. The relative dose is called *percent depth dose* (PDD(z)), defined by:

$$\text{PDD}(z) = \frac{D(z)}{D_{max}} \times 100\% \qquad (2.1)$$

The body has an electron density similar to water. Therefore simulations with water reflect rather well conditions of soft tissue. The exit slit of the accelerator is placed at a standard 100 cm *source-to-surface distance* (SSD), which is a typical distance between x-ray source and patient used in clinics. Nevertheless, simulation of electron depth-dose profile with this experimental arrangement is different from the real situation, where hot Compton electrons are generated within the tissue and without crossing an air gap. On the other hand, the stopping power of air is rather low compared to water so that the 100 cm air gap causes little error in the estimate of Compton electron dose and range in tissue.

According to Fig. 2.3, the depth-dose profile of electrons in water features a rapid increase of dose below the surface, an extended plateau region, and a rapid drop off, defining a range. The plateau extension and range depend approximately linearly on the incident electron energy:

$$\langle R_{50}[\text{cm}]\rangle = \frac{1}{\rho}(0.53E[\text{MeV}] - 0.106[\text{MeV}]), \tag{2.2}$$

where ρ is the mass density and $R_{50}[\text{cm}]$ is the depth at which the primary intensity has dropped to 50%. See also Chapter 6 of Volume 2 for a discussion of range of charged particles.

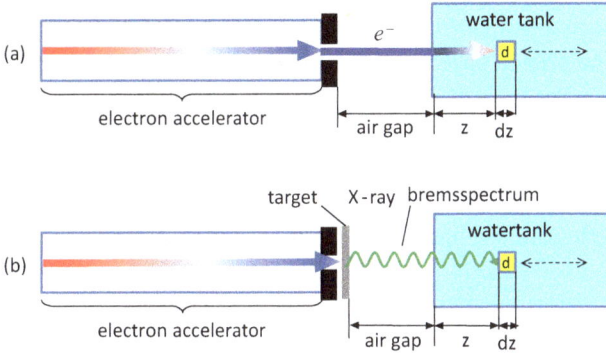

Fig. 2.2: (a) Experimental arrangement for determining the dose versus depth z for electrons in water and (b) for x-ray photons in water. Under standard source-to-surface distance (SSD) conditions, the air gap is 100 cm. The detector d indicates a plane-parallel ionization chamber as a dosimeter. The plane separation dz is also the depth resolution in the z-direction.

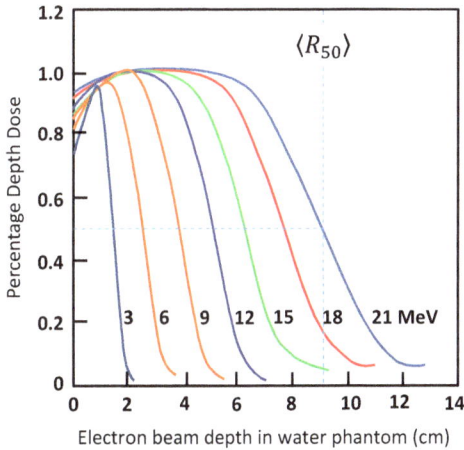

Fig. 2.3: Percent depth dose (PDD) as a function of penetration depth of high energy electron beam in water at a standard 100 cm source-to-surface distance (SSD). (Adapted from References [4, 5]).

2.2.2 Percent depth dose for photons

Next we consider the depth-dose relationship for a photon beam. The experimental setup is similar to the one for electrons, sketched in Fig. 2.2(b). Electrons are first accelerated to high energies, as in the previous case. Then the electrons are converted to x-rays by bombarding a target that generates a bremsstrahlung spectrum. Since the bremsstrahlung spectrum has a broad photon energy distribution, not the photon energy is quoted in the respective graphs but the accelerating potential which determines the cutoff energy of the photon spectrum. The ionization chamber detector in the water tank measures again the dose of charged particles generated by CS and PP conversion of x-rays in water. In Fig. 2.4(a), the PDD(z) versus depth z is plotted for two bremsstrahlung spectra generated at 6 and 15 MV. The PDD curves are characterized by a steep buildup of dose below the surface, a maximum depending only weakly on the energy, and a slow almost linear drop off beyond the maximum.

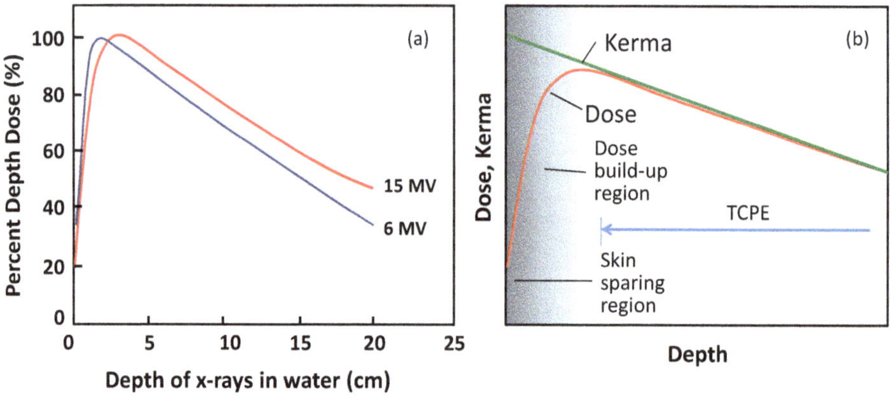

Fig. 2.4: Percent depth dose curve for photon beams in water under standard conditions of 100 cm source to surface distance. Note that the photon energy is given in terms of the accelerating potential (MV), which determines the cutoff energy of the bremsstrahlung spectrum. (b): The dose buildup region is enlarged and the kerma is plotted for the whole range.

From Fig. 2.4(a) we notice that the maximum dose is reached at a depth of 0.5 to 3 cm below the surface, depending on the x-ray energy. Near the surface, the dose is very low, which is known as the *skin-sparing effect*. Beyond the maximum, the dose drops off almost linearly. The initial buildup of dose below the surface can be explained as follows. When the x-ray beam enters the phantom or the body, it starts generating Compton electrons and electrons/positrons from pair production. Those particles travel a certain distance in the forward direction before being stopped and depositing their energy by producing further secondary photoelectrons. For instance, a 15 MV x-ray bremsstrahlung spectrum has an average photon energy two-third of the maximum energy, or 10 MeV. We have already noticed that 10 MeV photons produce 2.3 MeV Compton electrons at a scattering angle of 10°, which are stopped within a range of 1 cm. So, the maximum dose naturally occurs at the respective depth below the surface. Beyond the maximum, the dose drops off because the x-ray beam's intensity decreases and, consequently, the number of Compton electrons produced per distance traveled. Dose buildup and skin-sparing region is shown again on an enlarged scale in panel (b) of Fig. 2.4.

Figure 2.4(b) also shows the kerma as function of depth. Kerma is defined as the energy fluence of an x-ray beam converted into kinetic energy of charged particles (Section 7.3.5, Volume 2):

$$K = \Psi_{ph} \left(\frac{\mu_{tr}}{\varrho} \right) \tag{2.3}$$

According to this definition, kerma starts immediately as x-ray photons interact with matter. Then photon energy is transferred to Compton electrons and converted to electron-positron pairs. As explained before, those hot electrons and positrons travel a certain distance before depositing their energy and contributing to the dose. Hence, kerma is high from the beginning without buildup region, followed by a continuous dropoff with diminishing x-ray intensity. Within the buildup region, the ratio:

$$\beta = \frac{K(z)}{D(z)} > 1, \tag{2.4}$$

whereas beyond the maximum, the ratio:

$$\beta = \frac{K(z)}{D(z)} = 1, \tag{2.5}$$

The region where $\beta = 1$ holds is called the region of *transient charged particle equilibrium* (TCPE) region [6]. In this region, dose and kerma decrease simultaneously while their ratio remains constant. This is because the average energy of the generated secondary electrons and their range is roughly constant with depth in the medium.

2.2.3 Reference dosimetry and beam quality correction factor

We return to the question of local dose and what it means to measure the dose using a dosimeter. Usually, the dose from radiation is measured with a dosimeter in air. The measured dose at a certain distance from a standard source of constant and spatially homogeneous radiation is shown in the following equation [6]:

$$D_{air} = \frac{Q}{m_{air}} \left(\frac{W_{air}}{e} \right) \tag{2.6}$$

A standard radiation source for calibration is a ^{60}Co γ-source. In eq. (2.6), Q is the total charge produced by the radiation in a defined volume, m_{air} is the air mass of that volume, and W_{air} is the medium ionization energy of molecules in the air, which has been determined to be on average 33.97 eV/ion pair = 33.97 J/C. The true dose at the same position in air and distance from the source but with a removed dosimeter is:

$$D_{air,0} = \frac{Q_0}{m_{air}} \left(\frac{W_{air}}{e} \right).$$ (2.7)

Clearly, the true dose $D_{air,0}$, which equals the air kerma, cannot be measured. But we can make assumptions about how much the presence of a dosimeter changes the true value $D_{air,0}$. The difference is due to wall absorption, scattering in the small dosimeter cavity, and recombination effects. A correction factor k_0 collects all these factors and relates both quantities so that the true value can be estimated from the measured dose:

$$D_{air,meas} = D_{air,0} k_0 = K_{air},$$ (2.8)

where K_{air} is the air kerma. Next, we define a cavity air calibration coefficient $N_{D,air}$ by the expression:

$$N_{D,air} = \frac{D_{air}}{M_q}$$ (2.9)

where M_q is the reading of the dosimeter and $N_{D,air}$ relates the reading to the local dose in air.

$$D_{air,meas} = M_q N_{D,air}$$ (2.10)

In eq. (2.9), the subscript q stands for the beam quality.[2] The calibration beam quality of a laboratory specialized to prepare standards is designated as q_0, and the beam in a clinical setting has the quality q.

Ultimately, we need to know the absorbed dose measured in a water phantom at the reference depth z:

$$D_{w,q_0}(z) = M_{q_0} N_{D,w,q_0}$$ (2.11)

The subscript q_0 indicates a reference beam in the absence of the dosimeter and M_{q_0} is the fully corrected dosimeter reading under the reference conditions used in a standards laboratory. N_{D,w,q_0} is the calibration coefficient in terms of the absorbed

---

2 In the literature, the beam quality factor usually carries a capital subscript letter Q. However, in order to avoid confusion with Q standing for charge in eqs. (2.6) and (2.7), the lower case letter q is chosen to symbolize the beam quality.

dose to water. In case the dosimeter is used in a different beam environment characterized by a beam quality q instead of q_0, the absorbed dose to water is given by:

$$D_{w,q}(z) = M_{q_0} N_{D,w,q_0} k_{q,q_0} \qquad (2.12)$$

Here

$$k_{q,q_0} = \frac{D_{w,q}}{D_{w,q_0}} \qquad (2.13)$$

is the *beam quality correction factor* for the difference between a standard reference beam and the actual user beam in a gantry.

As dosimeter, the use of plane-parallel ionization chambers is recommended. The separation of the planes dz and the area A define the integration volume $A dz$ and the depth resolution dz in the z-direction. The integration volume should be small as not to disturb the photon flux, but large enough for sufficient sensitivity. As the ionization chamber has walls separating the chamber from water and since it is filled with air instead of water, a correction factor k_{q,q_0} is introduced, as discussed above, for calculating the true dose to water. The walls are made of light material (PMMA or carbon), such that the beam perturbation is kept at a low level. The correction factor is the product of two contributions for the air-filled cavity and for the cavity walls:

$$k_{q,q_0} = k^c_{q,q_0} \times k^w_{q,q_0}. \qquad (2.14)$$

Without going through further lengthy calculations, it turns out that the correction factor k_{q,q_0} for electrons deviates from one by not more than 1% in an energy range from 5 to 20 MeV, which has been confirmed by Monte Carlo (MC) simulations [7].

2.2.4 Monte Carlo simulations

For radiation therapy, it is important to calculate or simulate the depth-dose profile regarding the actual machine settings, including beam energy, intensity profile, and collimation. The result of these calculations/simulations is used for the treatment plan. In the past, the calculations were performed analytically with simplifying assumptions. An early Monte Carlo (MC) simulation of fast charged particles was published by Berger in 1963 [8]. Only in recent years, with increasing computer power and speed, MC simulations have become routine.

In 2004, CERN published and made available to the public an MC simulation package for calculating high-energy particle trajectories called GEANT4 [9]. With this package, thousands of charged and uncharged particle trajectories can be calculated and their total dose determined. Since the introduction of GEANT4, numerous MC simulations have been performed for medical practice. Alternative packages are also available and offered by the manufacturers of XRT gantries (Siemens, Varian, Elekta) such as FLUKA, PENELOPE, and EGSnrc. However, GEANT4 has advantages concerning flexibility, allowing a wide variety of beam conditions and geometries.

With the help of GEANT4, it could be shown that the MC simulations for the calculation of dose and range agree very well with the experimental results for both photons and electrons [4]. The good agreement gave reasons to hope that the water phantom's relatively simple test arrangement can be reliably simulated as well as more realistic conditions relevant for radiation treatment [10]. The MC simulations have now reached a very high level of maturity. The range and dose of radiation can be calculated organ-specifically, allowing for OER, and even organ movement due to heart beat and respiration is considered [11]. MC simulations have become the "gold standard" in radiation treatment planning with x-rays and charged particles. An overview of the state-of-the-art and perspectives of future high-speed MC simulations is given in [12, 13]. A good overview on MC methods can also be found in [14], and a historical account on the development of MC in medical physics is given in [15].

In the following, we give a brief outline of the MC method as it is used for clinical treatment plans, following descriptions provided in [5, 14]. We distinguish between three categories of dose calculation algorithms:
1. Factor-based algorithm
2. Convolution-superposition algorithm
3. Monte Carlo simulation

1. Factor-based algorithm

This algorithm uses a semiempirical approach based on experimentally determined data obtained from phantom measurements. Beam profile, attenuation, and PDD are the main input data. For estimating the dose to the body, various correction factors are taken into account, such as surface contours, air gaps between organs, and variations of the mass attenuation factor. The *inhomogeneity correction factor (ICF)* is defined as:

$$\text{ICF} = \frac{\text{dose in heterogeneous medium}}{\text{dose at same point in homogeneous medium}} \tag{2.15}$$

$$= \%\text{per cm} \times \text{inhomogeneity thickness}$$

ICF is used for correcting dose calculations for radiation plans. Correction factor-based algorithms are fast but less accurate than the other two methods mentioned below.

2. Convolution-superposition algorithm

Here we first consider the photon fluence $\Psi_{ph}(\vec{r})$ emanating from the source. As the photons enter the tissue, both photons and electrons generated by the photons, contribute to the local dose. Therefore, instead of determining the kerma of the radiation, the essential parameter for the local dose is the *total energy released per unit mass*, called TERMA:

$$T(\vec{r}) = \frac{\mu}{\varrho} \Psi_{ph}(\vec{r}) \tag{2.16}$$

Next we define a convolution kernel $A(\vec{r} - \vec{r}')$ of the deposited dose. The kernel is a volume element of deposited energy by photons and electrons at the interaction site in the body. The kernel can be considered as a "dose cloud" at any point $\vec{r}$, where dose is generated. The local dose $D(\vec{r})$ at point $\vec{r}$ in the body is then the convolution integral of the TERMA and the kernel:

$$D(\vec{r}) = \int T(\vec{r}') A(\vec{r} - \vec{r}') d^3\vec{r}' \tag{2.17}$$

In the final step, we have to superimpose all kernels for all relevant energies and obtain:

$$D(\vec{r}) = \int dE' \int T(\vec{r}', E') A(\vec{r} - \vec{r}', E') d^3\vec{r}' \tag{2.18}$$

Algorithms based on the convolution-superposition principle are faster and more accurate than the factor-based algorithms, but still not as accurate as MC algorithms to be considered next.

3. Monte Carlo simulation

The types of interactions of photons and electrons with matter are well known and the cross sections are precisely determined. However, at what time a photon is Compton scattered or absorbed by PE or PP cannot be predicted. When we follow an MeV photon track in matter, it may travel a short distance undisturbed, then becomes Compton scattered and finally generates an electron-positron pair. The interaction occurring from one volume element to the next is a stochastic process. The photon may travel, scatter, or transfer its energy to another particle with certain probabilities determined by the cross sections. A second photon has a different track and a different history of events. If we add thousands or millions of photon tracks for a defined geometry, the calculated dose distribution will become increasingly accurate. Presently, MC simulation provides the most reliable radiation planning data, and with modern computers, the calculation time is reasonably short. Because of the high reliability, MC simulations are considered the "gold standard" in dose calculations, against which other methods have to be compared. Figure 2.5 shows MC simulations of the PDD and the beam profile in a water phantom at different depth z compared to experimental results. The agreement is very good and has become much better in later simulations.

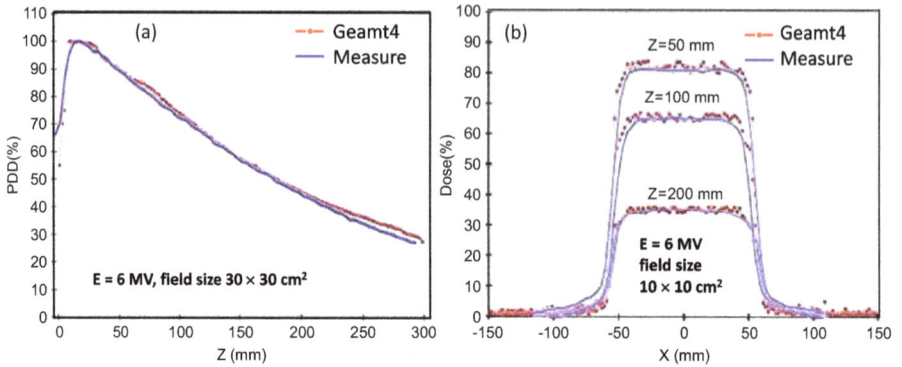

Fig. 2.5: Comparison of Monte Carlo simulations and experiments: (a) Percentage depth dose profile of photons in a water phantom; (a) and (b) beam profile. Adapted from [4].

2.2.5 Dose to isocenter

When calculating the dose to a tumor volume, two methods are commonly used, as illustrated in Fig. 2.6 [5, 6]. The first one, known as constant *source-to-surface distance* (SSD) assumes a constant distance from the source to the surface of the patient. The second method is the isocentric setup with a constant *source-axis distance* (SAD). The center of the target volume is placed at the machine isocenter such that the distance to the target point is kept constant for all beams at different angles. For radiotherapy, the SAD technique is easier to implement as no readjustment of the patient in relation to the radiation source is required. On the other hand, under SSD conditions the dose to volume is easier to simulate for establishing a treatment plan. SSD implies, however, that the isocenter is on the skin. Therefore, the treatment couch and the patient must be moved laterally and vertically with respect to the source to meet this condition. Note that:

$$SAD = SSD + d, \tag{2.19}$$

where d is the distance between surface and target.

In the exercise section, a simple example is worked out for calculating dose to the isocenter assuming SAD conditions. The dose delivered by linacs is calibrated and quoted in Gy per monitor unit (Gy/MU), where 1 MU = 1 cGy.

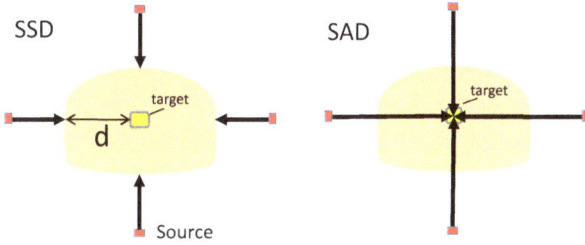

Fig. 2.6: Two different methods for planning radiotherapy. Left: constant source-surface distance (SSD), all arrows have same length. Right: constant source-axis distance (SAD).

2.3 Target volume and collimators

2.3.1 Target volume definition

The first task in radiation treatment planning is the precise localization of the tumor volume to be irradiated, while sparing organs at risk (OAR). According to the recommendations of the International Commission on Radiation Units and Measurements (ICRU) report 50, 62, and 71 [16], the following distinctions should be made, which are also illustrated in Fig. 2.7:

1. Gross tumor volume (GTV) is the macroscopic tumor volume defined by imaging methods.
2. Clinical target volume (CTV) is a tissue volume containing the GTV and/or subclinical malignancies with some chance of becoming therapeutically relevant. CTV is an anatomical and clinical assessment and not a physical measurement. Clinical assessment attempts to answer whether additional cancerous tissue may be present beyond the well-defined tumor mass, how far microscopic branches can reach, the likelihood of local recurrence, and the growth rate of known histological tumors. If the tumor was surgically removed prior to RT, no GTV can be defined and CTV is the volume of subclinical disease.
3. Planning target volume (PTV) is a geometric concept introduced for treatment planning and evaluation. It is the recommended volume for determining beam size and dose distribution to ensure that an adequate dose is delivered to all parts of the CTV. PTV also includes potential tumor movement due to organ movement within the body, patient movement on the treatment couch, and machine compatibility. PTV is defined in the treatment room frame of reference and assuming it is at the isocenter of the gantry. Usually, the PTV is taken as follows: PTV = CTV plus a shell of 1 cm thickness.

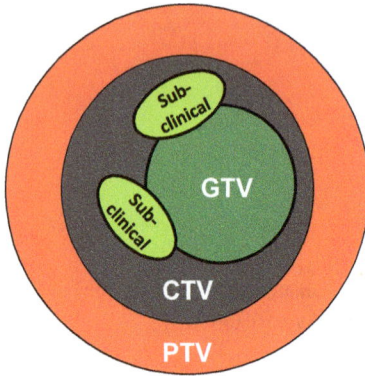

Fig. 2.7: Defining volumes according to the ICRU report 62. GTV = Gross Tumor Volume; CTV = Clinical Target Volume; PTV = Planning Target Volume.

With the help of *image-guided radiotherapy* (IGRT), the position of the tumor volume during XRT can be checked continuously and eventually corrected. Although IGRT adds extra dose to the patient when monitored by x-ray CT, the overall improvement of the geometric accuracy justifies its use during XRT and may decrease the volume of tissue being irradiated. Figure 2.8 compares schematically a CTV with and without the use of IGRT. More recent versions of IGRT use a combination of laser and x-rays (cyberknife (CK)) or hybrid MRI/XRT systems, which will be presented later.

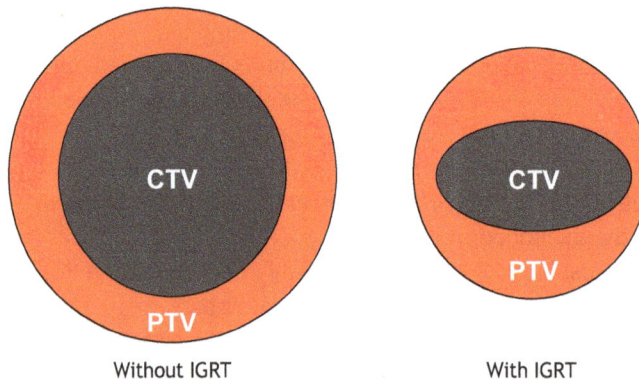

Without IGRT   With IGRT

Fig. 2.8: Image-guided radiotherapy increases the accuracy of localizing the clinical target volume and thereby helps reduce the planning target volume.

The positional accuracy of 3.5% is presently the accepted tolerance. But the required accuracy depends on the specific situation. High-dose small-volume RT, such as in brain irradiation, requires more accuracy and precision than a large or total body irradiation. Here we use the terms *precision* and *accuracy*. High precision implies hitting the same spot with minimal scatter, but the narrow Gaussian distribution may lie outside of the target area. Accuracy implies hitting the target, but eventually with a wide Gaussian distribution about the target center. High accuracy and high precision imply hitting the target's center with minimal scatter (see Chapter 8 for further illustration). A stereotactic head frame completely immobilizes the head to reach this goal in brain tumor irradiation, as presented in Section 2.6.

The contour of the GTV and CTV is currently drawn by experienced oncologists based on images taken by MRI, CT, US, and experience. In the future, the contours will be defined through the use of AI [17]. This is an algorithm that takes into account the same MRI images and, in addition, a huge database of many patients with similar diseases. Current comparisons between contours drawn by experts and those drawn by AI already show that the latter is superior. Tolerance is reduced with a better definition of the CTV, and the OAR is improved. Moreover, latest developments show that XRT can be combined with MRI during radiation exposure for a high precision PTV definition, as we will discuss further.

After defining the aim of XRT and the volumes to be treated, the next step is to prescribe the dose to the PTV per fraction, the number of fractions, and the total dose to the PTV. This is the topic of the next sections.

2.3.2 Multileaf collimator

In the past, simple rectangular collimator blades defined beam size and treated volume (TV). Figure 2.9 shows a top view of the geometric arrangement and the dose profile defined by the two pairs of adjustable blades arranged perpendicularly. The adjustable collimator is one of several collimators that defines the x-ray beam path in the gantry from the source to the PTV. A schematic side view of a standard collimator system in an XRT gantry is shown in Fig. 2.17.

According to the ICRU, the TV area should at least receive the absorbed dose determined as the minimum dose to the PTV. The schematics in Fig. 2.9 shows that the TV is naturally larger than PTV, and normal tissue is also exposed to a high dose. With such a rectangular slit system, single, double, or multifold exposures can be performed, as shown in Fig. 2.10.

Note that single field exposure implies a dose on the proximal side to be much higher than 100% of the calculated dose at the tumor site due to beam intensity attenuation in the tissue, see Fig. 2.10(a). With two exposures from anterior and posterior, the dose at the target center is reduced to 50% from either side in order to add up to 100% from both sides. Hence the dose on the proximal side of the tumor can be lowered to about 60% (Fig. 2.10(b)). Similarly with a four field exposure, the single dose from either side is only 25% to add up to 100% at the PVT (Fig. 2.10(c).

The example in Fig. 2.10 shows that an increasing number of exposures from different sides, the dose outside of the target center decreases, while at the tumor volume, the isodose contour becomes better defined. This is important for *sparing*

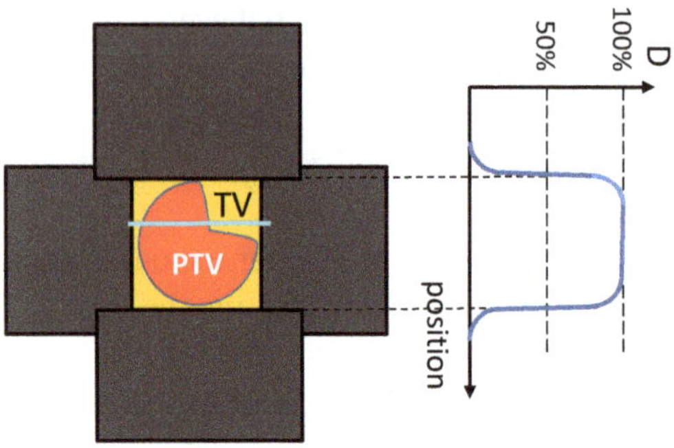

Fig. 2.9: Simple rectangular double-slit collimator defining the treated volume (TV) containing the PTV to be exposed; right panel shows a dose profile for the slice indicated by the horizontal line. Fifty percent dose level defines the beamwidth.

the healthy tissue from radiation dose and side effects. It shows that even with a linear dependence of dose versus depth, a high dose can be accumulated in a specific location at a certain depth if the exposure is administered from many different angles. *Multibeam exposure* is mandatory for XRT and unique to this method. A

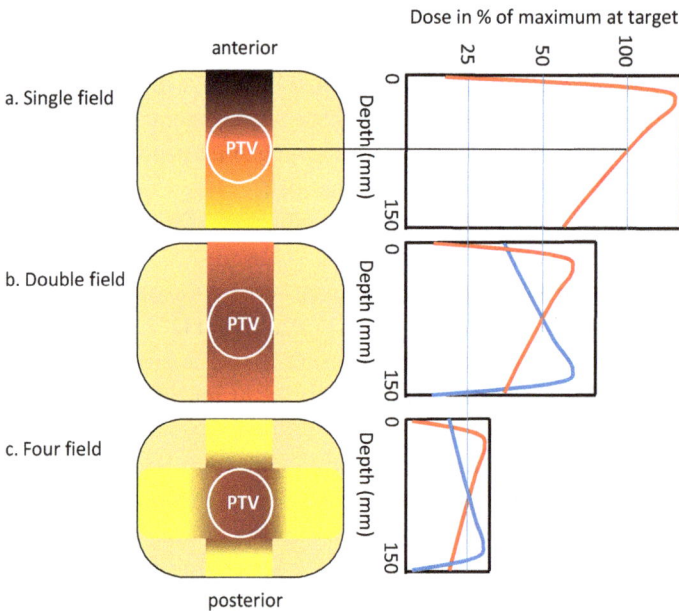

Fig. 2.10: Radiation exposure using a rectangular collimator (shown in Fig. 2.9): (a) single field exposure from anterior; (b) double field exposure from anterior and posterior; (c) four field exposure. Lines indicate isodose in percent. In case of four field exposure (anterior, posterior, left, and right) the total dose at the center of the tumor adds up to 100% with some sparing of healthy tissue at a dose level of 25%.

decisive advantage of proton beam therapy is the need of only a single beam exposure that can be precisely positioned.

Exposures using single or double-slit collimators are not sufficiently precise for
a high-quality XRT. Therefore wide open slit collimators are used only for palliative
treatments. To overcome problems occuring with a double slit, first lead blocks
were cast to conform to the individual shape of the PTV. However, one lead block
conforms only to the PTV from one side, and more lead blocks are required to conform to the other sides. This made the conformal exposure an expensive and time-
consuming procedure. In the 1980s, the concept of MLCs was introduced, consisting
of two opposing banks of moveable tungsten sheets (Fig. 2.11) [18]. The higher the
number of sheets, the more precisely the conformity can be adjusted. These slits are
pneumatically driven and computer-controlled. While the gantry moves around the
patient, the beam is stepwise opened at specific angular positions and the sliding
leaves are adapted to the projected profile of the PTV. This conformal MLC–XRT
spares the surrounding tissue and reduces radiation side effects. An example of
MLC applied to the prostate is shown in Fig. 2.12. The MLC is adapted to the shape
of the prostate and radiation treatment sparse exposure of colon and rectum.

Fig. 2.11: Multileaf collimator that conforms to the actual projected profile of the PTV.

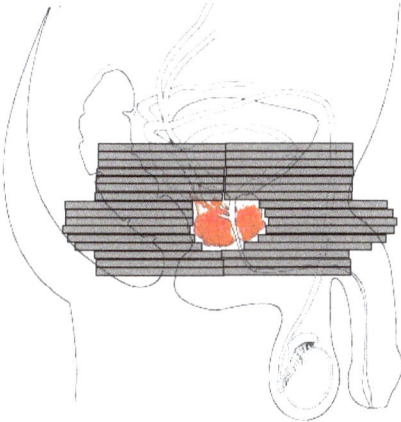

Fig. 2.12: MLC application for prostate radiotherapy with conformal adaptation that spares the colon, rectum, and bladder.

2.3.3 Intensity-modulated radiotherapy

After introducing fast-reacting high-speed moveable collimator sheets, the next invention was the implantation of *intensity-modulated radiation therapy* (IMRT, sometimes referred to as IMXT). IMRT combines MLCs to define the PTV and intensity modulation to deposit the required dose at each location. The principle is explained in Fig. 2.13. Let us first consider one fixed gantry angle. The exposure time at one fixed angle is subdivided in time intervals Δt_i. During each time interval, the collimator slits are moved by computer control to yield a laterally modulated intensity distribution. The voxel being open all the time receives the highest dose, a neighboring voxel open only for part of the time gets less radiation, etc. Thus the time modulation of the collimator slits translates into a lateral dose modulation, as indicated in Fig. 2.13 (b).

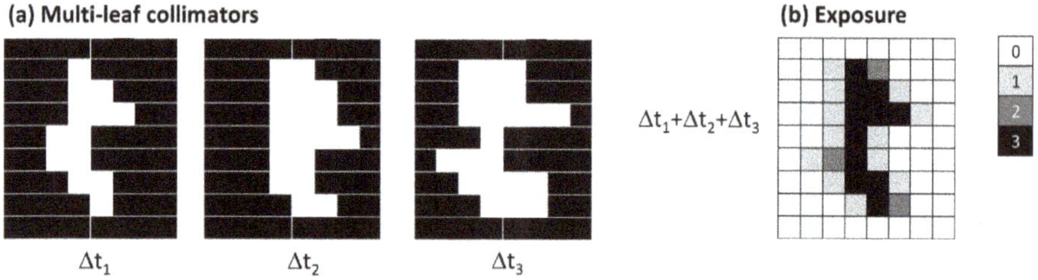

Fig. 2.13: During exposure at a fixed angle of the gantry, MLC open and close during time intervals Δt_i to yield a modulated dose distribution in the PTV: (a) white fields correspond to open slits, black fields to closed slits; (b) white fields correspond to zero exposure, black fields to highest exposure for open slits during all time intervals.

The intensity modulation procedure is repeated at all angular positions of the gantry, as exemplified in Fig. 2.14. The accumulated dose to the tumor volume corresponds precisely to the simulated radiation treatment plan. Clearly, QA during the whole procedure has to confirm the agreement between the anticipated dose and the actually delivered dose. With IGRT, the proper positioning of the patient can be controlled and eventually corrected. The latter procedure requires two beams: one at lower energies for imaging and another one at higher energies for XRT.

The treatment plan aims to deliver a homogeneous isodose to the PTV, a sharp contour at the edges, and a rapid dose dropoff beyond. With continuously improved treatment plans supported by advanced imaging methods and 3D simulation programs, this goal is achieved nowadays better than ever. Nevertheless, we should keep in mind that x-ray beams precisely delivered to the PTV create locally Compton electrons and electrons from pair production. Those have their range causing radiation damage beyond the well-defined borders of the PTV. This effect is known as the *penumbra effect*. Two contributions to the penumbra effect can be distinguished:

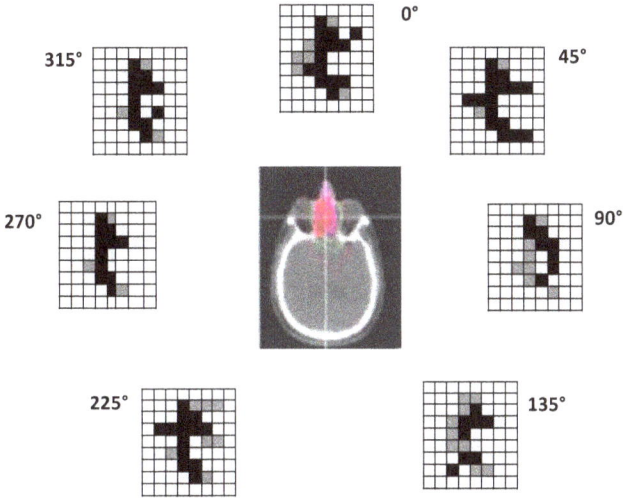

Fig. 2.14: Intensity-modulated fields for all gantry angles add up to a precisely defined dose to volume distribution of the tumor volume.

geometric and physical. The geometric penumbra is due to an extended source size and to slits which do not completely block the beam, see Fig. 8.3 in Volume 2. The physical penumbra effect occurs due to photon beam scattering and electron beam straggling.

The geometric penumbra can be reduced by well-designed collimators. The physical penumbra is intrinsic. Because of a more precisely defined range, proton radiation therapy, discussed in the next chapter, features a better-defined physical penumbra. Further information about IMRT can be found in Refs [5, 6, 19].

2.4 Linear accelerators for x-ray generation

It should be recalled that x-rays for cancer treatment are much harder (higher energies) than x-rays for imaging. For imaging, electron acceleration energies of up to 150 keV are used. For cancer treatment, a typical value in the past was 250 keV. This was the upper limit of what could be achieved with standard x-ray tubes and static electrical fields. For higher energies, γ-radiation such as a ^{60}Co-source with an average energy of 1.3 MeV was employed and is still applied for the so-called γ-knife, discussed in Section 2.8. Radiation therapy has greatly benefitted from developments that took place in high-energy physics. Elementary particles (electrons and protons) are being accelerated to very high energies (MeV to TeV) with linear accelerators (linacs) using microwave cavity technology. A spinoff of these developments are linacs for generating very hard x-rays in the MeV range for cancer treatment. The advantage of using these very high energy x-rays is a much-enhanced LET, therefore higher RBE and shorter exposure times. We refer to Section 4.3.2 in Volume 2 and Ref [20]. for more details on linacs.

For clinical applications, 0.5–1.5 m long linacs are installed, producing electron energies between 4 and 25 MeV. Figure 2.15 shows a schematic cross section of a gantry for XRT. A hot filament supplies electrons fed into evacuated cavities, where they are accelerated to high energies by microwaves with frequencies up to 3 GHz. At the end of these cavities, the electron bunches are guided by a bending magnet to an exit slit. Then the 1.5–2 mm wide electron beam hits a tungsten target. Like in a standard x-ray tube, the high-energy electrons produce a wide spectrum of hard x-ray bremsstrahlung radiation with a cutoff energy according to the maximum energy of the electrons. Main manufacturer of XRT-linac systems are Varian (Swiss), Elekta (Sweden), Hitachi (Japan), Accuray (USA), and ViewRay (USA).

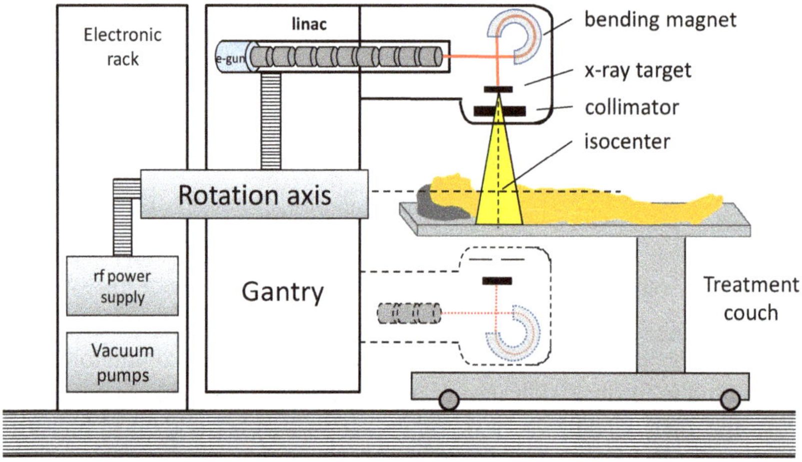

Fig. 2.15: Medical linac used for cancer treatment. The gantry houses a linac, bending magnet for beam guidance, collimators, and shutters. It circles around the isocenter where the tumor of patient is positioned by adjusting the treatment couch.

As described in the previous section, the exit x-ray beam is roughly confined by tungsten "jaws" and further refined by MLCs. An electron gun, accelerators, target, and collimators are housed in a gantry (see also Fig. 2.1). Since the gantry is rather bulky, rotation is only possible in a coplanar plane about one horizontal axis, and all beams meet at the isocenter. The patient on the treatment couch is moved such that the cancer target coincides with the isocenter. The anterior and posterior positions for radiation treatment are indicated in Fig. 2.15. Figure 2.16 shows a modern medical linac used for cancer x-ray radiation treatment.

The energy distribution of the x-ray beam is very broad, typical for a bremsstrahlung spectrum. Furthermore, the intensity is often inhomogeneously distributed in the lateral direction due to x-ray scattering processes in the tungsten target. A flattening filter (FF) is inserted behind the tungsten target (Fig. 2.17) to remove lateral beam intensity gradients. The treatment is called flattening filter free (FFF)

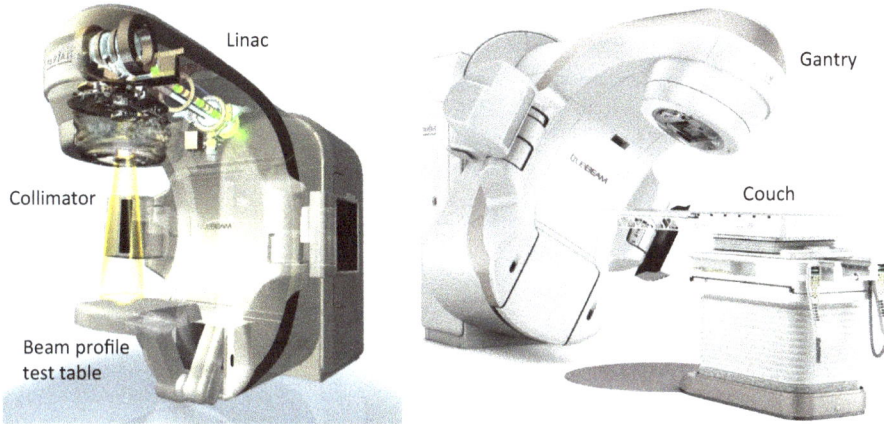

Fig. 2.16: Modern medical linac used for cancer x-ray radiation treatment. The gantry housing a linac, collimators, and shutters circles around the patient at the center for irradiating tumors with high energy x-ray photons incident from different angles. (Courtesy Varian Medical Systems International AG, Switzerland; All Rights Reserved).

exposure if the FF is removed. A dual ionization chamber monitors the x-ray intensity before the beam exits through the MLC to the target. Calibraion of the monitor is performed with a phantom of similar density as the body tissue. The dose to be applied is then quoted in monitor units (1 MU = 0.01 Gy). Alternatively, the tungsten target can be removed, and the electron beam will directly be funneled to the target for treatment. For this purpose, the 1.5–2 mm wide electron beam is widened by inserting a scattering foil instead of the tungsten target. Recall from the discussion of Fig. 2.3 that the range of electrons in tissue is much shorter than for x-rays. There is no skin-sparing effect, and the buildup of dose is minimal. Therefore, treatment with an electron beam is only justified for near surface exposure, for instance, skin cancer. However, skin cancer is usually treated either by brachytherapy (Chapter 5) or by laser exposure (Chapter 6).

In recent years it has been shown that the high-energy photons generated in clinical linear accelerators also produce photon-induced neutrons by (n,y) reactions. Whenever photons with energies higher than the binding energy of neutrons hit a target, neutrons can be released. Since the binding energy of neutrons in nuclei is on the order of 4–8 MeV, this phenomenon occurs in medical linear accelerators operating above 5 MeV. The generated neutrons either penetrate the head of the gantry or are captured by other materials. In the latter case, neurons activate the material, followed by β-excitation and prompt γ-emission. A significant neutron production rate for linacs operating at energies above 10 MeV has been observed. After switching off the linac, a *radiation afterglow* is detected, which consists of a fast neutron background and gamma spectral lines [21]. Therefore, shielding this radiation from patients and clinical staff exposure is mandatory.

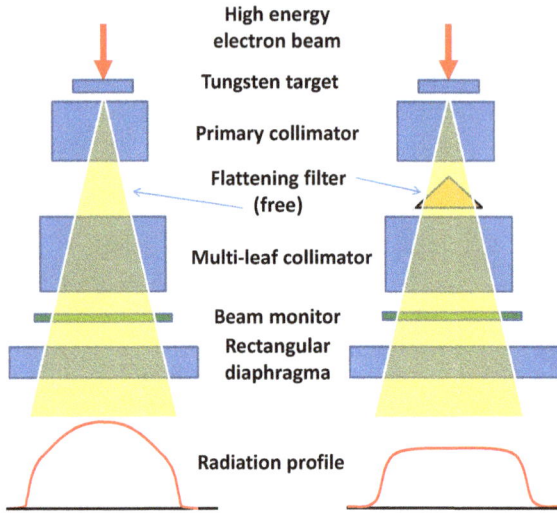

Fig. 2.17: X-ray intensity profile with and without flattening filter.

2.5 New developments

2.5.1 Cyberknife technology

In recent years, there has been a groundbreaking invention in clinical radiation therapy known as cyber knife (CK). Combining three major technology advancements, CK overcomes several obstacles of more conventional linacs for XRT. Those are 1. Precision robotics; 2. Image guidance; and 3. Breathing compensation. The high precision of the CK allows treatment with a much higher submillimeter definition of tumor areas in the brain, spine, lung, liver, kidneys, and prostate, which reduces the PTV and decreases the risk from exposure to healthy tissue [22].

CK technology is based on the same principle of high-energy x-ray photon production via a linac, as discussed in the previous section. However, the linac is operated at higher frequencies of 9.3 GHz instead of 3 GHz. This produces a much steeper field gradient, allowing shorter microwave cavities. Thus the CK linac producing 4–6 MeV electrons is shorter (0.5–1 m) and more lightweight (~150 kg) than standard linacs for XRT. Furthermore, CK uses a fine focus x-ray beam scanned across the PTV, called dose painting. Therefore, heavy slits and MLCs can optionally be spared.

The short linac is placed on a robotic arm with six degrees of freedom, as is common for industrial robots, with a positional accuracy of 0.2 mm. A second arm carries the treatment couch. The positioning of the accelerator and angle of exposure is much more flexible than in a standard gantry, which has only one coplanar

plane of rotation. Instead, the CK can deliver beams from thousands of noncopla-
nar, isocentric, or nonisocentric angles. At the same time, the lower photon energy
reduces the risk of photon–neutron production.

The image guiding system is a computer-controlled position tracking system,
using an infrared camera combined with x-ray imaging, which monitors patient
movement and compensates for small deviations from the original plan. Therefore
the patient does not need to be immobilized as it is required for standard XRT
equipment. The synchrony camera (Fig. 2.18) is an infrared camera that follows LED
markers placed on the body and compares their position with gold markers inside
the body, which are tracked by two x-ray sources. X-rays from both tubes are oriented at right angles and meet at the center of the target. X-ray images, therefore,
deliver a stereoscopic picture that is matched with the CT images taken during the
planning and simulation phase. Reference points are compared, and any deviations
are corrected for by robotic repositioning. With this tracking system, the spatial accuracy of the target location is within 0.6 mm.

Moreover, technical innovations allow the treatment of tumors that move during
respiration, such as lung tumors. Also, the liver and kidneys move up to several millimeters during respiration, which constitutes a big problem in standard XRT. Due to
the movement, the PTV has to be much larger than the GTV. However, as indicated
above, CK technology overcomes this problem by an intelligent image tracking system. The external breathing motion is recorded by an infrared camera and balanced
with the internal tumor position as seen by x-ray imaging using preimplanted gold
markers. The robot arm follows the breathing pattern and delivers the radiation

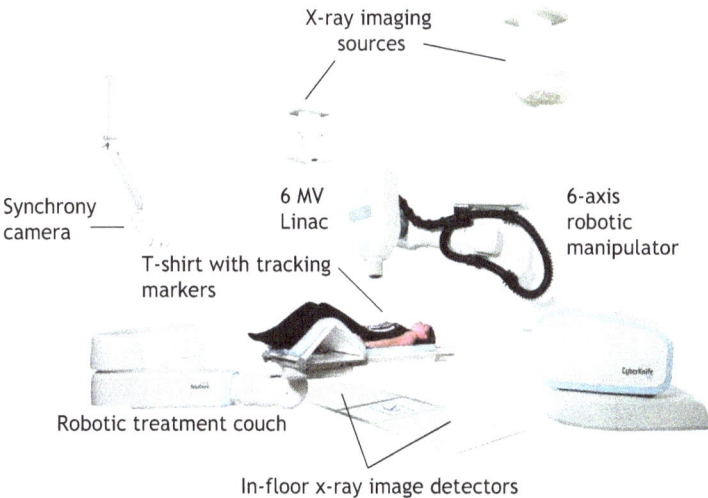

Fig. 2.18: Cyberknife treatment facility. (Adapted from ACCURAY; http://www.accuray.com/ and
Wikipedia, © creative commons).

precisely to the target. Raytracing procedures and MC simulations complement the optimized radiation treatment, which typically takes 30 to 120 min. Thus CK treatment is a form of highly focused position tracking radiotherapy named *stereotactic radiotherapy* and also known as *stereotactic radiosurgery*. Despite the term "surgery," this type of radiotherapy is noninvasive. Furthermore, robotic-guided high-precision CK technology allows treatment of inoperable and surgically complex tumors, such as those in the brain, close to the spinal cord, or at the backbone. A recent review on the CK stereotactic radiosurgery is given in [23].

Although the cyber-knife technology has many advantages compared to more traditional XRT systems, a few problems should be mentioned. Those are:

1. The distance between the x-ray source and the isocenter is not fixed due to the open frame geometry. This complicates dose calculations and hampers the use of phantoms. Thus, quality assurance (QA) is more important and difficult to confirm than in more traditional settings.
2. The electron energy of about 5–8 MeV may sometimes be too low. However, the accelerator is not large enough to generate higher energies. On the other hand, larger linacs become heavy and cannot be moved precisely by the robotic arm.
3. CK uses a finely focused x-ray beam, a so-called cone beam, in place of a MLC. This improves the dose delivery precision to the PTV but considerably extends the radiation exposure time. This technique is acceptable for small and well-defined tumors, but for larger and spread out tumors, MLC is preferable.

Meanwhile, manufacturers of more traditional XRT gantries have retrofitted their equipment with laser and x-ray tracking systems. Surface-guided radiography has now become the standard. Therefore, gated radiation exposure to compensate for organ movement has become available with gantries that feature only one axis of rotation [24].

2.5.2 MR-linac hybrid systems

MRI and CT imaging for clinical staging of tumors is a standard procedure and part of the QA. Ideally, both imaging modalities are applied since CT and MRI show different sensitivity and specificity for various types of cancers. In addition, x-ray monitoring during radiation treatment is often performed in hybrid systems such as the CK, discussed in the previous section. In contrast, MRI monitoring appeared in the past prohibitive because of the high magnetic field required for MRI that disturbs the electron beam path in the linac and electron focusing optics. Nevertheless, two companies offer MRI-linac hybrid systems: Elekta (1.5 T MRI) [25] and Viewray MRIdian (0.3 T) [26]. The hybrid system of Elekta is shown in Fig. 2.19 with the front cover taken off so that the MR tube becomes clearly visible. These systems allow live imaging during radiotherapy. The first results are convincing [27, 28]. The MR-linac hybrid system offers improved soft-tissue contrast, real-time adaptive radiotherapy, instantaneous reaction to changes in the tumor volume, and readjustment of the radiation dose. All these features contribute to a much improved and precise radiotherapy with superior OAR sparing. Simultaneously, the overall gantry design and further options such as MLC are preserved for IMRT. IGRT by using a MR-linac hybrid is often referred to as *magnetic resonance-guided radiation therapy* (MRgRT).

The stray magnetic field problem could be solved by covering the linac including electron optics with magnetic shielding material (μ-metals) [26]. The x-ray beam is funneled through the gaps of the Helmholtz coils. Because the MR-linac combination is new, protocols for dosimetry in magnetic fields are required, dose correction factors must be recalculated, and depth-dose profiles must be measured and simulated.

Compared to other IGRT systems using cone beam CT, the MR-linac shows higher sensitivity for many types of tumors but does not add more dose to the patient during monitoring. Therefore, it is, in many respects, the ideal combination for IGRT. Furthermore, MR-linacs allow more complex dose distributions in a shorter period of time, which improves the patient experience. However, the MR-linac hybrid does not provide patient tracking as is done with CK. Therefore, patient immobilization is required like in standard XRT systems. Finally, trimodality imaging using PET/CT/MRI in combination with radiotherapy has been tested and recently reported [29].

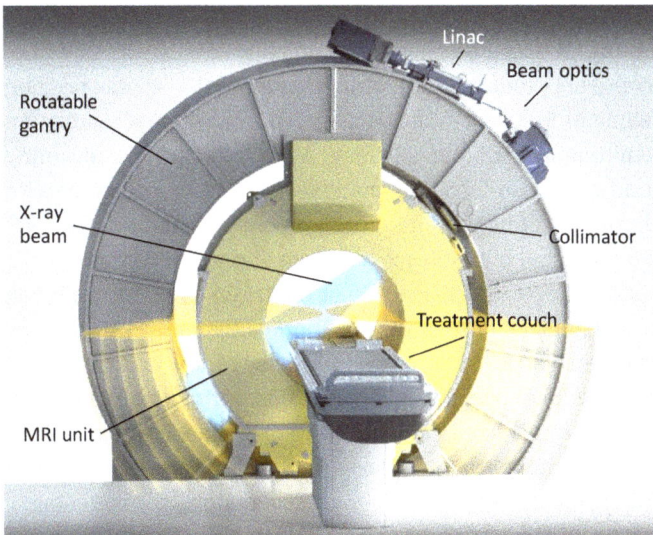

Fig. 2.19: MRI-linac system from Elekta with gantry cover taken off. The MRI unit at the center is clearly visible. The MRI coil is split in half. The top can be taken off for particular radiation angles. Image courtesy of Elekta.

2.5.3 FLASH-RT

FLASH is a radiation treatment with very high dose rate of roughly 100 Gy/s in short exposure time periods of 100 ms [30]. The high dose is equivalent to the entire RT treatment plan or a large fraction of it. The delivery time is in the subsecond regime. The short radiation time has the added advantage of minimizing motion uncertainties between fractions.

Researchers have noticed in animal studies that the application of a high dose at a high rate shows tumor control similar to conventional treatment while sparing healthy tissues from side effects. The biochemistry of FLASH is not completely understood at this point, but it is speculated that the effect is comparable with high LET radiation (see Section 11.4.3). In any case, FLASH-RT has proven to be a promising tool for obtaining similar tumor control compared to conventional radiotherapy with very good sparing effect. First patients have successfully been treated with FLASH-RT [31]. For review and historical notes see [32, 33].

From a physics point of view, the question arises of how to deliver a dose rate that is high enough to meet the FLASH conditions using a conventional x-ray gantry. With protons, much higher dose rates can be reached than with photons.

2.6 Gamma knife

A gamma knife is an RT using ^{60}Co isotopes emitting γ-radiation of about 1.3 MeV photon energy [34]. ^{60}Co isotopes are produced by neutron capture of ^{59}Co via the reaction ^{59}Co(n,γ)^{60}Co with a lifetime of 5.26 years. The decay via β$^-$ and prompt γ-radiation consists of two γ-lines emitting at 1.173 and 1.33 MeV. More information concerning ^{60}Co isotopes can be found in Section 5.2.3. Gamma knife is specifically designed for RT of tumors in the brain. It cannot be used for other purposes. Similar to XRT, γ-RT does not have a well-defined range, but the exposure from many different angles deposits the highest dose at the target center. Outside of the target edge, the radiation rapidly drops off with an insignificant dose to adjacent normal tissue. The head of the patient has to be fixed in a stereotactic frame to precisely deliver the dose to the target volume. This ensures that the head position is reproducible during treatment planning and exposure. After the head is fixed, it is covered with a stainless steel helmet containing a large number of boreholes. The holes are plugged up with cylindrical tubes accepting the ^{60}Co isotopes. Figure 2.20 shows schematically a stereotactic frame with helmet. Gamma radiation emanates from different sides to a target at the isocenter. The number of plugs filled determines the dose at the target position. The patient is moved on a hydraulic couch into a radiation unit, and irradiation can start. γ-Knife is the only RT modality where γ-rays from many different sides all hit the target simultaneously. All other irradiation procedures have to move the source sequentially from one position to the other. γ-knife RT saves some treatment time, which is beneficial for the patient arrested in an uncomfortable stereotactic head frame but is insignificant with respect to cell-cycle time.

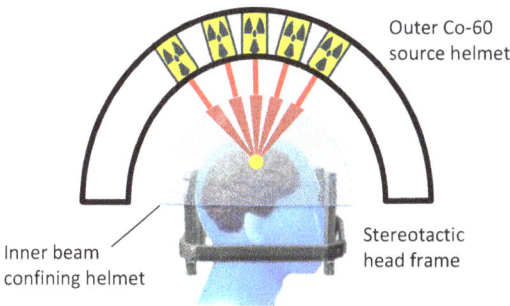

Fig. 2.20: Stereotactic head frame for gamma-knife treatment of brain tumors.

To summarize, the γ-knife RT is used for the treatment of small tumors in the brain. In the past, this method was justified because hard x-rays of about 250 kVp do not have sufficient intensity after penetrating the skull and not sufficient LET to damage tumor cells. However, this problem has been overcome by using hard x-rays from MeV linacs with photon energies higher than the 1.3 MeV γ-energy from the ^{60}Co-source. The target tracking possibility using CK technology fulfills the second requirement of precisely positioning the head and suppression of any movement. Furthermore, larger tumors can be treated better with the CK than with the γ-knife since the focus can be moved, which is not possible using a stereotactic frame. For these reasons, the γ-knife RT may not be necessary anymore in future brain radiation therapy. This would also circumvent the handling of radioactive isotopes, starting from the production of isotopes via neutron irradiation, safeguarding highly radioactive material during use, and finally depositing low radioactive nuclear waste. An overview on the use of γ-knife treatment in neurosurgery is given in Ref [35].

2.7 Summary

S2.1   Any cancer treatment requires careful and precise planning beforehand.
S2.2   Quality assurance (QA) during irradiation is required to avoid systematic errors.
S2.3   At 10 MeV incident photon energy about 80% of the scattering events are due to Compton scattering, the rest is due to pair production.
S2.4   The dose-depth profiles of electrons and photons in the soft matter are very different.
S2.5   High energy electrons feature a plateau and a range with respect to particle intensity.

S2.6   Photon beams do not have a range, but their intensity diminishes with penetration depth due to Compton scattering and pair production.

S2.7   For calculating dose to a tumor volume, two different methods are commonly used: constant *source-to-surface distance (SSD)* and a constant *source-axis distance* (SAD).

S2.8   For defining the target volume, three different volumes are distinguished: gross tumor volume (GTV), clinical target volume (CTV), and planning target volume (PTV).

S2.9   Multileaf collimators (MLCs) consist of two opposing banks of moveable tungsten sheets that can be adapted to the shape of the tumor volume.

S2.10   Intensity modulated radiation therapy (IMRT) is a combination of MLC and intensity modulation for each individual angle of radiation exposure.

S2.11   Penumbra can occur (a) for *geometric* reasons due to an extended source size and transmission through slits and (b) for *intrinsic* physical reasons due to the range and straggling of particles in tissue.

S2.12   Linacs are used to boost the energy of electrons into the MeV range, which is not possible with standard x-ray tubes.

S2.13   High energy electrons accelerated in a linac hit a tungsten target and produce very hard bremsstrahlung x-ray photons.

S2.14   The advantage of using very high energy x-rays is a much enhanced LET, resulting in higher RBE and shorter exposure times.

S2.15   To improve beam homogeneity, a FF is inserted after the tungsten target.

S2.16   CK combines three major advancements in technology and overcomes a number of problems of more conventional linacs for radiation therapy: (1) precision robotics; (2) image-guiding system; and (3) breathing compensation.

S2.17   MRI-linac hybrids combine magnetoresonance imaging with linac radiotherapy for precision localization of tumors and real-time imaging during irradiation.

S2.18   FLASH is a new radiation therapy of tumors that delivers a very high dose of about 50–100 Gy in a single session at a very high dose rate of about 100 Gy/100 ms.

S2.19   Gamma knife is a radiation therapy using γ-radiation from a ^{60}Co isotope source, emitting 1.3 MeV photons.

S2.20   Gamma knife is specifically designed for RT of neurosurgical disorders in the brain.

? Questions

Q2.1   What are typical x-ray energies for XRT?

Q2.2   How are high-energy x-ray photons generated?

Q2.3   When a 10 MeV incident photon energy hits a water target, what is the percentage of photons undergoing Compton scattering as compared to pair production?

Q2.4   The dose-depth profiles of electrons and photons in soft matter are very different. Why?

Q2.5   Why is the kerma-depth profile different from the dose-depth profile?
Q2.6   If electrons are responsible for the dose buildup, why are electron beams not directly used for radiation therapy?
Q2.7   In XRT, how far below the skin's surface does the peak of the depth dose profile lie?
Q2.8   What is the physical reason for the dose buildup below the surface?
Q2.9   What is the difference between SSD and SAD?
Q2.10 What are the advantages and disadvantages of SSD versus SAD?
Q2.11 What is the difference between GTV, CTV, and PTV?
Q2.12 What is image-guided radiation therapy good for?
Q2.13 XRT does not have a range; the depth dose profile continuously drops off. Why is it nevertheless possible to deposit a high dose at one specific target volume inside the body?
Q2.14 What is IMRT–MLC?
Q2.15 What are the essential ingredients of CK technology?
Q2.16 What does MRgRT stand for?
Q2.17 Why is FLASH a potential option for RT?
Q2.18 What is gamma-knife used for?

Attained competence   +   0   –   ⚡

I can name the main components of a medical linac for cancer treatment

I can draw the depth-dose profiles for high-energy electrons in water

I can draw the depth-dose profile for high-energy photons in water

I can distinguish between the buildup region and the transient charged particle equilibrium region

I know why the kerma is high in the buildup region

I can distinguish between GTV, CTV, and PTV

I can distinguish between physical and geometric penumbra

I know what the difference is between IMRT and IGRT

I appreciate how dose is build up during exposure from different sides

I understand what MLCs are good for

I can list the essential advantages of cyberknife technology and MIgRT

I know what the γ-knife was used for

I know what stereotactic irradiation implies

i **Exercises**

E2.1 **Dose to the isocenter:** Calculate the dose to the isocenter, assuming SAD conditions. The dose of linacs is calibrated and quoted in Gy/MU, where 1 MU = 1 cGy. Assume a three-field exposure: one from anterior and two from posterior. Using a 6 MV photon beam 100 MU are delivered from all three sides.

    a. What is the PDD at a distance of 10 cm from the skin to the isocenter? As there is no further obstruction in this field, the weighing factor is taken to be 1.

    b. After exposure of 100 MU from anterior, what is the dose at the isocenter in cGy?

    c. For the posterior field, the distance from skin to isocenter is longer and the PDD is therefore lower at 47%. An additional weighting factor of 0.7 takes into account the smaller posterior field, widening of the beam, and bones in the path with a higher absorption coefficient than soft tissue. What is the dose delivered to the isocenter from the posterior sides?

E2.2 **Boosting up the dose to the isocenter:**

    a. The treatment plan requires that 70% of the anterior side to be delivered from either posterior side. The dose from the anterior side at the isocenter is 73 cGy. What is then the dose to be delivered from the posterior sides?

    b. In order to reach this dose on the posterior side, the dose from the linac has to be boosted up. What is the boosting factor?

    c. What is now the total dose delivered?

    d. Assuming that the prescribed dose at the isocenter is 200 cGy instead of 155 cGy, all three field exposures have to be boosted up. What is the boosting factor?

    e. What is the dose to be delivered by the linac at all three fields in units of MU?

    f. Show all numbers calculated in table form.

E2.3 **Exposure time:** A patient is to be treated with a total dose of $D_{patient}$ = 5 Gy to a rectangular tumor field of A = 5 × 5 cm^2 at a depth d = 10 cm below the skin. Suppose that the 6 MeV linac delivers radiation with an intensity of 0.1 Gy/cm^{2}min. For how long is the patient to be irradiated in a single fraction?

References

[1] Gianfaldoni S, Gianfaldoni R, Wollina U, Lotti J, Tchernev G, Lotti T. An overview on radiotherapy: From its history to its current applications in dermatology. Open Access Maced J Med Sci. 2017; 5: 521–525.

[2] Hawkes N A comprehensive history of cancer treatment, 2015, online: http://raconteur.net/history-of-cancer-treatment/

[3] Johns HE, Cunningham JR. The physics of radiology. 4th. Springfield, IL: Charles Thomas; 1983.

[4] Sardari D, Maleki R, Samavat H, Esmaeeli A. Measurement of depth-dose of linear accelerator and simulation by use of Geant4 computer code. Rep Pract Oncol Radiother. 2010; 15: 64–68.

[5] Khan FM, Gibbons JP. The physics of radiation therapy. 5th. Philadelphia, Baltimore, New York, London: Wolters Kluwer; 2014.

[6] Podgorsak EB, editor. Radiation oncology physics: A handbook for teachers and students. Vienna: International Atomic Energy Agency; 2005. Available at http://www-pub.iaea.org/mtcd/publications/pdf/pub1196_web.pdf

[7] Zink K, Wulff J. Monte Carlo calculations of beam quality correction factors k_Q for electron dosimetry with a parallel-plate Roos chamber. Phys Med Biol. 2008; 53: 1595–1607.

[8] Berger MJ. Monte Carlo calculation of the penetration and diffusion of fast charged particles Methods in Computational Physics. ed Alder B, Fernbach S, Rotenberg M, New York: Academic; 1963, 1, 135–215.

[9] https://geant4.web.cern.ch/collaboration/workshops/users2002/tutorial

[10] Francescon P, Beddar S, Sariano N, Das IJ. Variation of k_Q for the small-field dosmetric parameters percentage depth dose, tissue-maximum ratio, and off-axis ratio. Med Phys. 2014; 41: 101708.

[11] Carrier JF, Archambault L, Beaulieu L. Validation of GEANT4 (an object-oriented Monte Carlo toolkit) for simulations in medical physics. Med Phys. 2004; 31: 484–492.

[12] Muraro S, Battistoni G, Kraan AC. Challengs in Monte Carlo simulations as clinical and research tool in particle therapy: A review. Front Phys. 2020; 8: 567800.

[13] Kim DW, Park K, Kim H, Kim J. History of the photon beam dose calculation algorithm in radiation treatment planning system. Prog Med Phys. 2020; 31: 54–62.

[14] Salditt T, Aspelmeier T, Aeffner S. Biomedical imaging. Principles of radiography, tomography, and medical physics. De Gruyter. Berlin, Boston: Graduate Text; 2017.

[15] Rogers DW. Fifty years of Monte Carlo simulations for medical physics. Phys Med Biol. 2006; 51: R287–301.

[16] International Commission on Radiation Units and Measurements Prescribing. Recording and reporting photon beam therapy. ICRU Reports 50, 62, 71.

[17] Bai T, Balagopal A, Dohopolski M, Morgan H, McBeth R, Tan J, Lin MH, Sher DJ, Nguyen D, Jiang SB. A proof-of-concept study of artificial intelligence assisted contour revision. Radiology: Artificial Intelligence, 2022; 4, No. 5, p. 1–5.

[18] Price P, Sikora K, Illidge T. editors, Treatment of cancer. 5th. London: Hodder Arnold Publication; 2008.

[19] Mayles P, Nahum A, Rosenwald JC editors, Handbook of radiation therapy physics, theory and practice. Boca Raton, London, New York: CRC Press, Taylor & Francis Group; 2007.

[20] Anderson R, Lamey M, MacPherson M, Carlone M. Simulation of a medical linear accelerator for teaching purposes. J Appl Clin Med Phys. 2015; 16: 1–17.

[21] Israngkul-Na-Ayuthaya I, Suriyapee S. Evaluation of equivalent dose from neutrons and activation products from a 15-MV X-ray LINAC and Phongpheath Pengvanich. J Radiat Res. 2015; 56: 919–926.

[22] Tan TJ, Siva S, Foroudi F, Gill S. Stereotactic body radiotherapy for primary prostate cancer: A systematic review. J Med Imaging Radiat Oncol. 2014; 58: 601–611.

[23] Ding C, Saw CB, Timmerman RD. Cyberknife stereotactic radiosurgery and radiation therapy treatment planning system. Med Dosim. 2018; 43: 129–140.

[24] Scobioala, S., Kittel, C., Elsayad, K. et al A treatment planning study comparing IMRT techniques and cyber knife for stereotactic body radiotherapy of low-risk prostate carcinoma. Radiat Oncol. 2019; 14: 143.

[25] Winkel D, Bol GH, Kroon PS, van Asselen B, Hackett SS, Werensteijn-Honingh AM, Intven MPW, Eppinga WSC, Tijssen RHN, Kerkmeijer LGW, de Boer HCJ, Mook S, Meijer GJ, Hes J, Willemsen-Bosman M, deGroot-van Breugel EN, Jürgenliemk-Schulz IM, Raaymakers BW. Adaptive radiotherapy: The Elekta unity MR-linac concept. Clin Transl Radiat Oncol. 2019; 18: 54–59.

[26] Klüter S. Technical design and concept of a 0.35 T MR-Linac. Clin Transl Radiat Oncol. 2019; 18: 98–101.

[27] Wegener D, Zips D, Gani C, Boeke S, Nikolaou K, Othman AE, Almansour H, Paulsen F, Müller AC. Stellenwert des 1,5-T-MRLinearbeschleunigers für die primäre Therapie des Prostatakarzinoms. Der Radiologe. 2021; 9: 839–843.

[28] Cuccia F, Alongi F, Belka C. et al., Patient positioning and immobilization procedures for hybrid MR-Linac systems. Radiat Oncol. 2021; 16: 183.

[29] Decazes P, Hinault P, Veresezan O, Thureau S, Gouel P, Vera P. Trimodality PET/CT/MRI and radiotherapy: A mini-review. Front Oncol. 2021; 10: 614008.

[30] Borghini A, Vecoli C, Labate L, Panetta D, Andreassi MG, Gizzi LA. FLASH ultra-high dose rates in radiotherapy: Preclinical and radiobiological evidence. Int J Radiat Biol. 2021; 16: 1–9.

[31] Bourhis J, Sozzi WJ, Jorge PG, Gaide O, Bailat C, Duclos F, Patin D, Ozsahin M, Bochud F, Germond JF. Treatment of a first patient with FLASH-radiotherapy. Radiother Oncol. 2019; 139: 18–22.

[32] Wilson JD, Hammond EM, Higgins GS, Kristoffer P. Ultra-high dose rate (FLASH) radiotherapy: Silver bullet or fool's gold?. Front Oncol. 2020; 9: 1–12.

[33] Binwei L, Feng G, Yiwei Y, Dai W, Yu Z, Gang F, Tangzhi D, Xiaobo D. FLASH radiotherapy: History and future. Front Oncol. 2021; 11: 644400.

[34] Sanders J, Nordström H, Sheehan J, Schlesinger D. Gamma Knife radiosurgery: Scenarios and support for re-irradiation. Phys Med. 2019; 68: 75–82.

[35] Ganz JC. Gamma knife neurosurgery. Berlin, Heidelberg, New York: Springer Verlag; 2011.

Further reading

Williams JR, Thwaites D. editors. Radiotherapy physics in practice. 2nd edition. Oxford, New York, Athens: Oxford University Press; 2000.

Hall EJ, Giaccia AJ. Radiology for the radiologist. 7th edition. Philadelphia, Baltimore, New York, London: Wolters Kluwer, Lippincott Williams & Wilkins; 2012.

Podgorsak EB. editor. Radiation oncology physics: A handbook for teachers and students. Vienna: International Atomic Energy Agency; 2005. Available at http://www-pub.iaea.org/mtcd/publications/pdf/pub1196_web.pdf

Khan FM, Gibbons JP. The physics of radiation therapy. 5th edition. Philadelphia, Baltimore, New York, London: Wolters Kluwer; 2014.

Useful websites on cyberknife technology

https://www.accuray.com/cyberknife/
https://www.elekta.com/radiotherapy/
https://viewray.com

3 Charged particle radiotherapy

Acronyms and physical parameters of	
PRT	proton radiotherapy
PBT	proton beam therapy
OER	oxygen enhancement ratio
OAR	organs at risk
RBE	relative biological effectiveness
CIRT	carbon ion radiotherapy
SOBP	spread-out Bragg peak
NTD	nontarget dose
Charge particles for tumor therapy	protons, carbon ions
Range monitoring in tissue	PET, prompt γ-emission
Skin sparing	no
LET of 160 MeV proton beam	6 keV/cm in Bragg peak
Range of 110 MeV protons in water	10 cm

3.1 Introduction

Proton beam therapy (PRT) dates back to a 1946 publication by Wilson.[1] In his seminal paper he proposed the use of newly available particle beams from cyclotrons for medical treatment and particularly for cancer irradiation [1]. Clinical applications of proton beams started in the early 1990s of the last century. According to the Mayo Clinic in Minnesota, *proton radiotherapy* (PRT)[2] has by now been used with great success for many types of tumors, including brain, breast, esophageal, eye, gastrointestinal, gynecological, head and neck, liver, lung, lymphoma, prostate, soft tissue, spine, and many pediatric cancers. As the dose can precisely be deposited into the tumor volume without distal fall-off, children with cancer benefit the most from PRT. Their organs are still developing, and those at risk can be spared from radiation using PRT. For adults, the benefit of PRT in comparison to more conventional and cheaper radiation treatments is discussed, as long-term comparative studies are still lacking. The same arguments apply to heavy-ion beam radiation such as carbon ion beams.

The higher the energy of charged particle beams, the higher the manufacturing costs of these facilities and the higher the treatment costs for patients. XRT requires a 0.5–2 m long electron accelerator producing 4–20 MeV electrons at the cost of

---

1 Robert R. Wilson (1914–2000), US-American physicist and first director of the Fermilab.
2 In the literature one can find different acronyms. The most frequent ones are: PBT (proton beam therapy), CPT (charge particle therapy), PT (proton therapy), and PRT (proton radiation therapy). Here we chose to use the acronym PRT to distinguish it from XRT.

https://doi.org/10.1515/9783111168739-003

about 1–5 M\$. PRT requires a 130–250 MeV accelerator at the cost of 200 M\$ and up. This high capital investment and consequently high treatment cost warrants a careful justification of why PRT is needed and for which type of cancer it is most effective. Eventually, as demand for new facilities grows and advanced smaller accelerators are developed, investment capital may decline, making PRT more viable. Compared to XRT, there are much fewer PRT centers around the globe (~100 at the time of writing) and even fewer carbon ion beam therapy centers (~10). Reference [2] gives an overview of existing and planned PRT centers. In this short chapter, we will first present the main properties of charged particle beams and their interactions with tissues. Then we continue discussing the implementation of proton beams, required facilities, and practical issues.

3.2 Proton beam therapy: overview

We recall from Chapter 6.3 in Volume 2 that charged particles (protons, α-particles, carbon ions) in matter have a well-defined range. The range depends on the initial energy and the atomic number Z of the target material. At the end of their track, the particles deposit their remaining kinetic energy and then stop within a short straggling distance. The stopping power, i.e., the energy loss per distance traveled, is low and constant for 95% of the track but rises sharply toward the end and forms a so-called Bragg peak.[3] The dose to be delivered to the cancer volume is contained in this Bragg peak. There is no radiation on the distal side of the peak.

The characteristic Bragg peak defining a range for protons in matter, in contrast to x-ray radiation, is the main incentive for *proton radiation therapy* (PRT). Within the region of the Bragg peak, the energy loss per distance traveled is highest, and so is the dose as well as the *relative biological effectiveness* (RBE) for destroying cancerous cells. In Fig. 3.1(a), the range of protons and x-rays is compared. The x-ray dose goes through a maximum within 5 cm below the skin and then drops off continuously without a defined range, as is expected from the standard Lambert–Beer law (see also discussions in Chapter 2, Fig. 2.4). In contrast, the proton beam is characterized by an almost constant fluence (particles/area), implying that only a few protons get lost along the track. Close to the end of the track, the fluence drops rapidly and the dose rises steeply due to the increase in energy loss to the environment. The width of the Bragg peak is determined by the range straggling.

With PRT, the Bragg peak delivering the highest dose can be placed right into the cancer volume by tuning the proton beam's initial energy. By stacking up Bragg peaks with varying incident proton energy and beam intensity, a Bragg-peak plateau

---

3 William Henry Bragg (1862–1942), British physicist and Nobel laureate in Physics 1915 (see footnote #7 in Chapter 6).

is formed that covers the entire cancer volume (Fig. 3.1(b)). The plateau region is de-noted as *spread-out Bragg peak* (SOBP). The dose in the proximal region is low and zero in the distal region. Therefore, PRT does not require a multifield exposure to pile up the dose at the target center. In many cases, a single field is sufficient, as we will see in the next sections. Technically, beam tuning is achieved by inserting an attenu-ator wedge into the primary proton beam, which controls both beam energy and intensity.

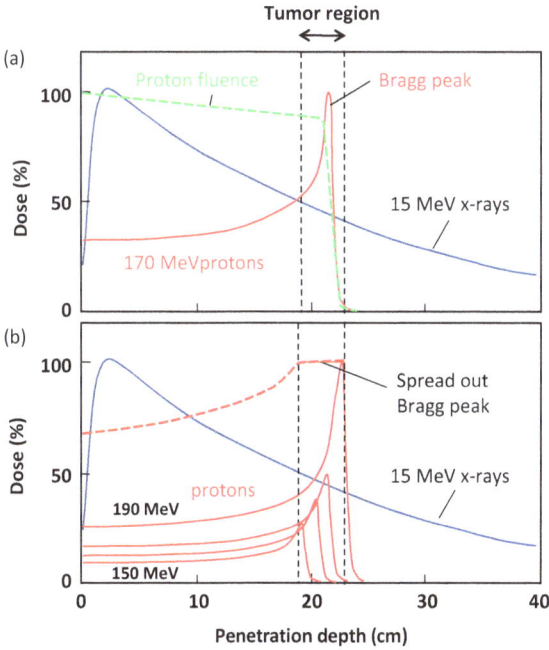

Fig. 3.1: (a) Comparison of dose versus depth for x-rays and protons. Plots are normalized to the maximum dose. Protons feature a long track with low dose that ends in a peak of high dose. The Bragg peak with the highest linear energy transfer (LET) and RBE can be placed at the location of the tumor volume by tuning the incident energy. In contrast to protons, x-rays do not have a well-defined range. (b) A dose plateau can be formed (spread-out Bragg peak) that conforms to the tumor volume by varying the incident energy and intensity of the proton beam.

For comparison, Fig. 3.2 shows the dose build-up for photons and protons in a water phantom for a single field and multifield exposure. The water tank has a pro-jected area of 30×30 cm^2. The target is at the center, stretching from 10 cm below the surface to 20 cm. The primary energy is chosen such that the photon and proton beams reach 100% of the specified dose at the target center [3].

The top panels (a) and (b) relate to single field exposure with one photon and one proton beam entering from the left side. The dose of the 15 MV photon beam rises steeply up to 200% within a few millimeters below the water surface and then continuously drops off, reaching 100% at the center of the target. The single field photon beam features two characteristic properties: the dose on the proximal side is much higher than needed at the target center; and second, the dose on the distal side shows attenuation but no range. The proton beam energy varies from 150 to 170 MeV to cover the target extending from 10 to 20 cm below the water surface. The dose is only about 70% on the proximal side, increasing to 100% at the target center and dropping off to zero on the distal side. This implies that a single field proton exposure does not affect critical tissues in the distal area. Such a condition is only possible with charged particle beams but never with photons.

Now we focus on the double field exposures in panels (c) and (d) of Fig. 3.2. As the x-ray beam dose adds up at the target center, the dose of the x-ray beams on both anterior sides can be lowered to a maximum of about 120%. The same effect but more pronounced is seen for the proton beams. Here the dose on the proximal sides drops from 70% to about 50%, adding up to 100% at the target center.

Finally, considering the quadrupolar exposure in panels (e) and (f), the trend seen in panels (c) and (d) continues. Aiming for 100% dose in the target area, we need a maximum dose of 50% on the proximal side of all four x-ray beam directions, dropping to $4 \times 25\% = 100\%$ in the target area. In contrast, for protons, the same dose is reached at the center, while the dose on each proximal side is lowered to 20%. This example clearly shows that PRT is more effective than XRT in sparing the proximal part of the tissue in all cases, single, double, or quadrupole exposure fields. This sparing effect helps to reduce healthy tissue acute side effects, such as inflammation and nausea. However, in the case of multiple field exposure, proton beams do not protect the skin as much as photons do, where the buildup of the dose reaches a maximum after penetrating about 2 cm below the skin (Fig. 3.1).

In PRT, the absence of dose on the distal side and the lower dose on the proximal side of the target center results in an overall lower *nontarget dose* (NTD) compared to XRT. This provides more flexibility for selecting the number and directions of proton beams and distributing NTD over healthy tissue. The reduction of NTD can be significant and has been estimated to be about 50% for prostate cancer [4]. Whatever is said here about PRT also applies even more to heavy charged particle therapy, such as *carbon ion radiotherapy* (CIRT). Most of this chapter describes PRT, but the last section is a brief account on CIRT.

It is evident that proton radiation has a distinct advantage over photon beam irradiation as concerns NTD and organs at risk (OAR). However, this advantage is credited to a considerably more complex and expensive facility, which is to be discussed later. The emerging technology committee of the American Society of Radiation Oncology judged in their 2012 published report: Intensity-modulated radiation therapy (IMRT) produces excellent local control rates and genitourinary and gastrointestinal toxicity are present but manageable in most patients with low rates of long term dysfunction. Therefore, the bar is set high for a new technique such as

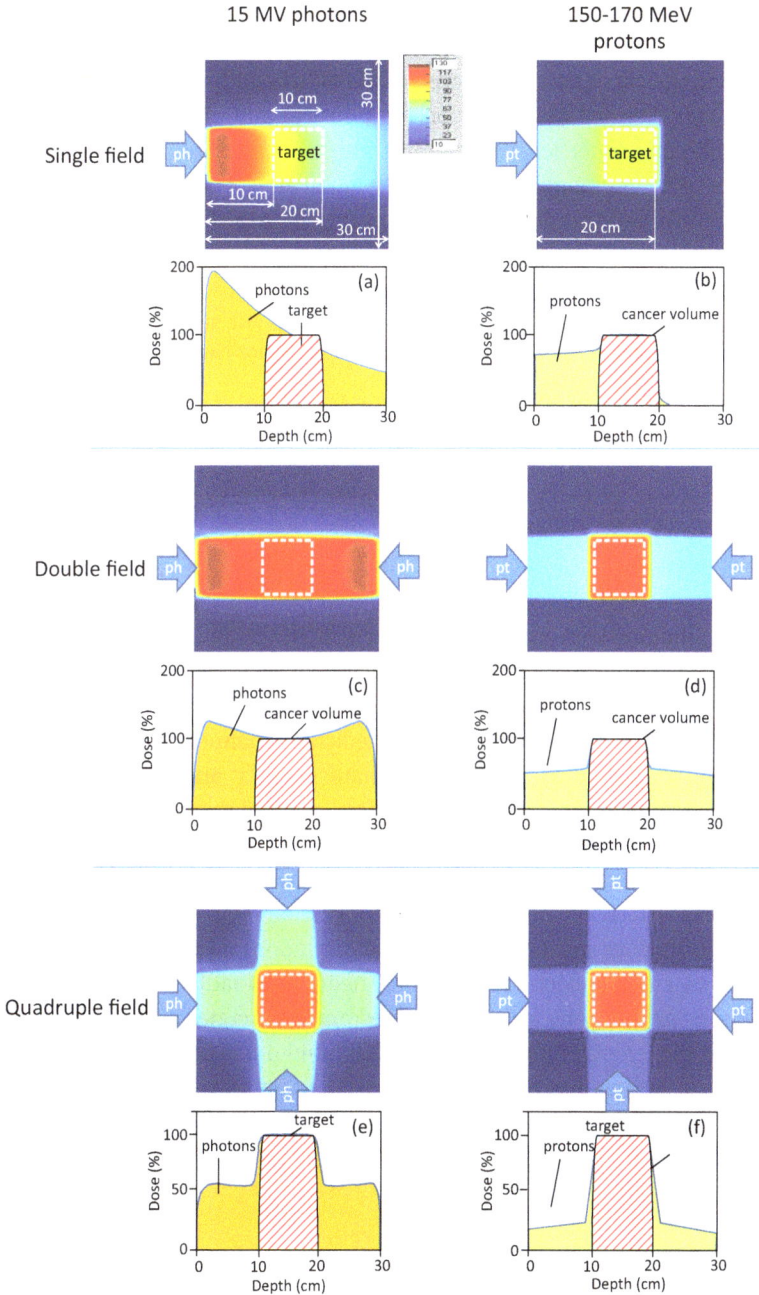

Fig. 3.2: Comparison of photon and proton radiation therapy for (a) single field exposure, (b) double opposing field exposure, and (c) quadrupole field exposure of a target volume at the center of a water phantom. The dose of the radiation is schematically color-coded and plotted in the adjoining graphs. (Adapted from Roelf Slopsema with permission, CRC handbook) [3].

proton radiotherapy (PRT) to deliver either improved tumor control or reduced toxicity over IMRT" [5].

Having pointed out the main characteristics of PRT, the differences to XRT, and the main advantages, we will now focus on the physical and biophysical aspects. They include the beam properties, collimators, and beam steering, range control, dosimetry, accelerator, and gantries. The purely medical aspects can be found in the relevant literature. Informative overviews on the physics of proton therapy is provided in [6, 7].

> **!** Protons are not exponentially attenuated in matter like x-rays. Protons lose most of their energy in the Bragg peak at the end of their track and then stop. Therefore, protons have a range, which depends on their initial energy.

3.3 Characteristics of proton beams

3.3.1 Interaction channels

The interaction of photons with matter is a rather straightforward matter. Only two types of cross sections are of any importance for photon energies beyond 1 MeV: Compton scattering and pair production. In comparison, the interaction of protons with matter is complex [8]. The most relevant interaction channels are sketched in Fig. 3.3. Those can be subdivided in electronic interactions (a) and nuclear interactions (b–d). The electronic interaction consists of ionization of atoms and transfer of kinetic energy to electrons, similar to the Compton effect. The nuclear interactions can be either Coulomb scattering at the nucleus, also known as Rutherford scattering (b), small-angle elastic recoil scattering of nuclei (c), or large angle inelastic recoil. Inelastic scattering includes nuclear reactions and excitations comprising a number of reaction products, such as electrons, positrons, neutrons, γ-radiation, and eventually further hadrons and mesons (d). PRT requires proton energies of 70 MeV up to about 250 MeV. In this energy range, the processes (a–c) dominate. Process (d), although a minor contribution, has recently become important for particle beam monitoring, as we will discuss in Section 3.6.

3.3.2 Proton beam range

As we recognized in the previous section, at the start of the path through the body, the energy loss of protons is low and consists mainly of ionization according to interaction channel (a) in Fig. 3.3. In water and tissue, mainly the K-shell electrons are excited. The scattering range of knocked-out electrons is on the order of 1 mm,

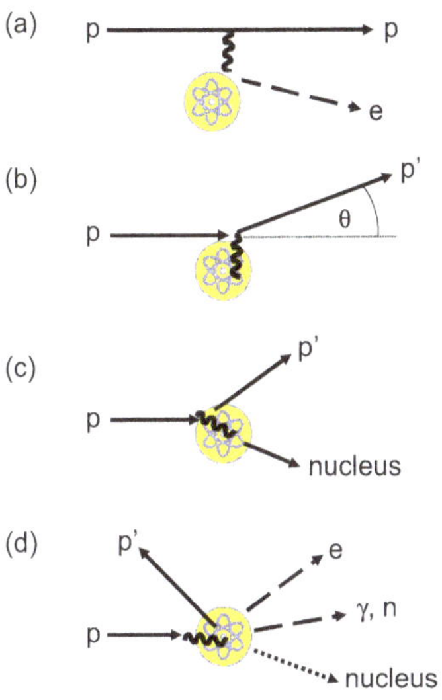

(a)

(b)

(c)

(d)

Fig. 3.3: Interaction channels of high energy protons with nuclei in the target: (a) ionization interaction; (b) Rutherford scattering; (c) elastic recoil scattering; (d) nuclear reactions and excitations. Adapted from [6].

while the proton path continues straightforward. In this region, the LET is low as well as the dose to the body. Recalling from Chapter 6 in Volume 2, LET (or stopping power) is the average energy loss $\langle -dE/dx \rangle$ per distance x traveled (see eq. (6.37)):

$$\text{LET} = \left\langle -\frac{dE}{dx} \right\rangle. \tag{3.1}$$

Assuming that the electronic interaction dominates the stopping power, the energy loss is described approximately by the Bethe–Bloch equation (see eq. (6.38) in Volume 2 for further information):

$$-\frac{dE}{dx} = \frac{4\pi k_B Z_{\text{eff}}^2 e^4 \rho_e}{mc^2 \beta^2} \ln\left(\frac{2mc^2\beta^2}{I(1-\beta^2)} - \beta^2 \right), \tag{3.2}$$

where ρ_e is the electron density. After successive scattering events, the protons will slow down, and the LET increases rapidly and reaches a maximum. Here, the dose varies almost inversely with the remaining proton energy. With some more straggling, the protons come to a complete halt. The peak in the LET as a function of depth is known as *Bragg peak*, shown again for a water phantom in Fig. 3.4. In this example of a 160 MeV proton beam entering from the left and traveling to the right, an LET peak is formed shortly before the beam stops at a depth of 17.5 cm. At the peak, the LET is 60 MeV/cm = 6 keV/μm or merely 6 eV/nm. Assuming a mean ionization energy of 30 eV, ionizing collisions occur on the average every 30 nm close to the end of the track.

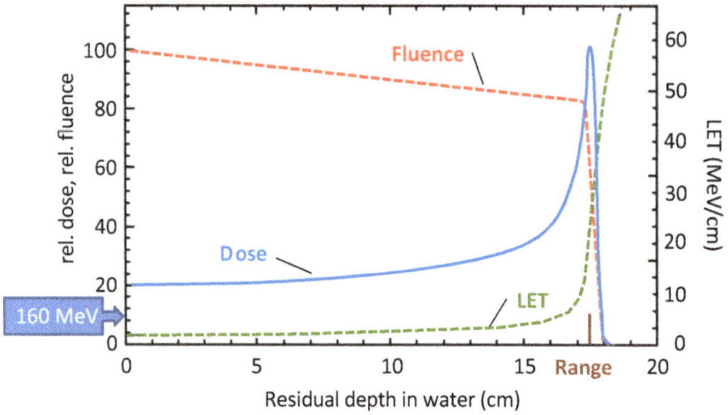

Fig. 3.4: Dose, fluence, and linear energy transfer (LET) of 160 MeV protons, entering in a watertank from the left side. The range is 17.5 cm.

For a homogeneous material like water, the range is determined by integrating the stopping power from start to stop (see Chapter 6, eq. (6.35) in volume 2):

$$R(E) = \int_E^0 \frac{1}{-\frac{dE'}{dx}} dE' = \int_0^E \left(\frac{dE'}{dx}\right)^{-1} dE'. \tag{3.3}$$

It turns out that the average range of protons in water $\langle R_{\text{water}} \rangle$ follows a simple power law and can be estimated with a semiempirical equation [1, 9]:

$$\langle R_{\text{water}} \rangle (\text{cm}) = 2 \times 10^{-3} \left(E_p(\text{MeV})\right)^{1.8}, \tag{3.4}$$

which covers the energy range of PRT. Here E_p is the kinetic energy of protons just before entering a water phantom in units of MeV and R is in units of centimeters.

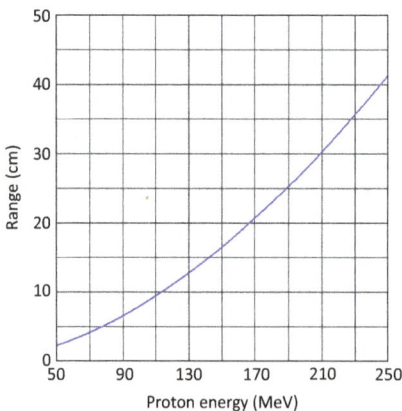

Fig. 3.5: Range of protons in water as a function of initial proton kinetic energy at the point of entry [9].

This function is plotted in Fig. 3.5 and gives guidance on initial proton energies needed for reaching deeper parts of the body.

According to eq. (3.2), the stopping power is proportional to the electron density ρ_e along the path traveled. Since the electron density and mass density of tissues vary, the range is usually quoted in reference to the density of water:

$$R_{water}(cm) \times \rho_{water}(g/cm^3) = \tilde{R}_{water}\left(\frac{g}{cm^2}\right) \tag{3.5}$$

The actual range in tissue is then:

$$R_{tissue}(cm) = \tilde{R}_{water}\left(\frac{g}{cm^2}\right)/\rho_{tissue}(g/cm^3). \tag{3.6}$$

In an inhomogeneous medium with densities ρ_i extending over thicknesses d_i, the range is the weighted sum over the individual thicknesses:

$$R_{medium} = \tilde{R}_{water}\left(\frac{1}{\rho_1}\frac{d_1}{d} + \frac{1}{\rho_2}\frac{d_2}{d} + \ldots\right) = \tilde{R}_{water}\frac{1}{d}\sum\frac{d_i}{\rho_i}, \tag{3.7}$$

where d is the total thickness, as illustrated in Fig. 13.6.

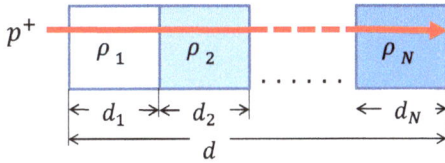

Fig. 3.6: Areas of different densities in the body add up for estimating a total range of protons.

It is essential to calculate the range of protons in tissues along the selected exposure field in order to spare OAR from toxic radiation. Just a few millimeters off can have dramatic consequences. The range can be estimated according to eq. (3.6) together with (3.3). However, the actual electron density along the path is needed for an improved range calculation. This information can be gained from CT images and the conversion of gray scales to Hounsfield units. Though the best range calculation is performed by Monte Carlo simulations of the stopping power, which has become the gold standard for establishing radiation treatment plans. In addition, recently in vivo proton beam monitoring has been introduced, which is the ultimate control and safeguarding of patients. Because of its importance, Section 3.4 is devoted to the discussion of range monitoring.

In practice, a maximum range of protons in the body of $R_{max} = 32\,cm$ turns out to be sufficient to reach all parts of the body. According to Fig. 3.5, a range of 32 cm requires a maximum beam energy of 230 MeV. Most cyclotrons in operation for PRT are specified for this maximum beam energy. Only accelerators used for uveal melanoma have lower energy of 70 MeV because of the short proton track in the eye.

3.3.3 Beam profile

The monoenergetic proton beam delivered by the accelerator typically has a lateral cross-section of 1×1 cm^2. When this beam is funneled into a water phantom, one can determine the profile with the help of a dosimeter or a semiconductor detector. The longitudinal beam range is well defined, as already discussed and shown again in Fig. 3.7(a). Typical values are a dose dropoff from 90% to 10% within 0.6 mm. The lateral beam profile plotted in panel (b) has a flat top and a narrow penumbra with an 80–20% dropoff within 2–3 mm. With increasing length of the trajectory, the beam widens only slightly. But at the end of the track, the beam spreads by about 50% within the Bragg-peak area. For specific clinical applications, such a beam may be too wide or narrow. If the beam is too wide, the beam cross section can be reduced with a set of collimators. Narrow beams are used for pencil beam irradiation, discussed in the next section. If the beam is too narrow, it can be widened by using a scattering plate, schematically shown in panel (c) of Fig. 3.7. Scattering plates may be parallel and flat or others may be laterally modulated to shape the beam direction and intensity.

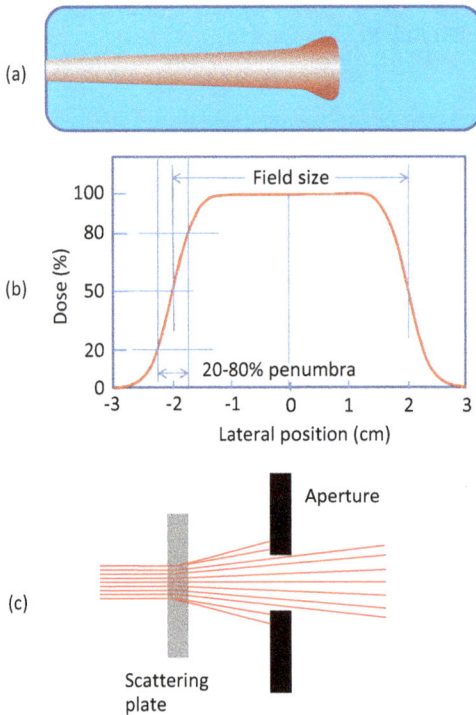

Fig. 3.7: (a) Proton beam in a water phantom. The beam's cross section is well defined and widens only slightly toward the end of the track; (b) a flat top characterizes the beam profile with a narrow penumbra; (c) the proton beam cross section as delivered by the accelerator can be widened using a scattering plate. An aperture defines the final beam cross section.

A diaphragm limits the beam cross section exactly to the tumor volume's lateral (transverse) size. However, apertures do not affect the longitudinal beam profile. For this purpose, the collimator can be combined with a scattering plate. The scattering plate modulates the beam energy to conform the proton range to the longitudinal tumor volume. Such an *intensity-modulated proton beam therapy* (IMPT) is the analog to the intensity-modulated radiation therapy (IMRT) already discussed in Chapter 2 [10]. The modulation plate is made of either some wax or plastic material with a density close to water. Figure 3.8 shows MC simulations of the lateral and longitudinal beam profile with and without modulator plate. It is evident that the conformity of the beam profile to the tumor volume is much improved with the use of a modulator plate. The disadvantage of this method is that the modulator plate causes scattering of the proton beam, increasing the penumbra and contributing to range uncertainty. Alternative approaches use a laterally modulated SOBP or a pencil beam that can be scanned across the tumor volume (see below). The next two sections present both options.

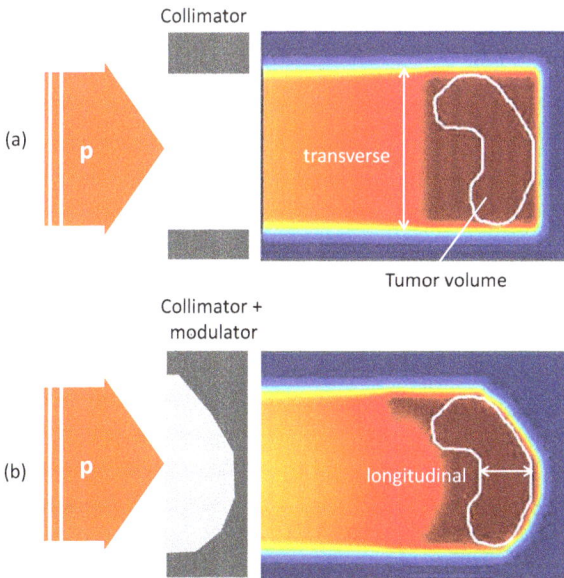

Fig. 3.8: Beam profile in the region of the Bragg peak: (a) with the use of a collimator, the transverse beam profile conforms with the lateral extension of the tumor volume; (b) full conformity of the beam profile in the lateral and longitudinal direction is achieved by using a collimator and a modulator plate. Adapted from [11].

The proton beam profile shows a narrow penumbra in the lateral and longitudinal direction, which broadens slightly towards the end of the track. The beam shape can be accomodated with the use of modulators.

3.3.4 Formation of SOBP

The *spread-out Bragg peak* (SOBP) is one of the hallmarks of PRT. SOBP allows to precisely tune the dose across the tumor volume. The SOBP can be described and modeled analytically to find the optimal weight factors for the contributing single Bragg peaks [3, 10]. Assuming that we know the shape of the Bragg peaks via phantom experiments, their range as a function of energy can be modeled. Then we represent the SOBP as the sum of weighted single Bragg peaks:

$$\text{SOBP}(R, d) = \sum_{i=1}^{N} W_i \times \text{PP}(R_i(E)). \tag{3.8}$$

Here $\text{PP}(R_i(E))$ is the pristine peak depth-dose with range R_i and E is the proton beam energy at the entrance to the phantom or tissue. The weighting factor W_1 of the first PP_1 is usually taken to be 80% of the top maximum, and R_1 is the maximum range. The other PP_i and their weighting factors W_i must be determined numerically or via MC simulations. As illustrated in Fig. 3.9, the sum yields the SOBP as a function of range R and distance d to the surface (skin).

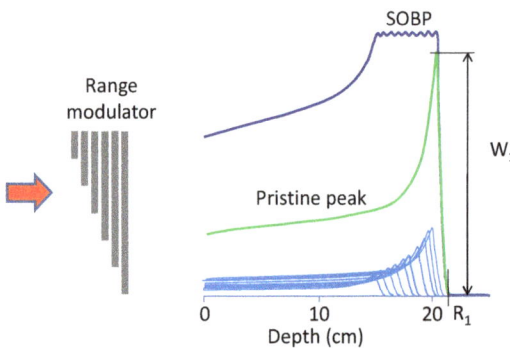

Fig. 3.9: Generation of SOBP by the weighted sum of pristine Bragg peaks.

In order to generate an SOBP, both the proton range (energy) and the beam flux have to be controlled. There are two methods to tune the proton energy and beam flux either by changing the accelerator energy or by inserting attenuators in the beam. Both methods have advantages and disadvantages. Tuning the accelerator energy does not disturb the beam quality but is rather slow and requires alterations of the machine settings. Using range tuning allows faster beam variation but widens the beam through scattering and causes a lateral penumbra. Moreover, modulators enhance the background via nuclear reactions and lower the beam quality. Despite these adverse effects, clinics prefer beam tuning via range modulation.[4] There are several designs of range modulators either in the form of wheels with stacks of attenuators (energy stack), wedges, or ridge filters [3, 10].

3.3.5 Beam delivery and scanning

Next issue to be discussed is beam delivery. There are several options. The main decision is between a wide beam and a pencil beam. A wide beam must be shaped both laterally and longitudinally and also in terms of exposure time to conform to the tumor volume. This so-called IMPT requires sophisticated fabrication of scatter plates, modulators, and collimators as we have already seen before.

Using a narrow pencil beam requires two pairs of crossed electromagnets for steering and scanning the proton beam across the tumor volume (Fig. 3.10). The scanning follows a particular point-by-point pattern, line by line, or stitching forward–backward. At each point, time and energy of the beam can be varied to control the dose and depth locally. The scanning can also be performed layer-by-layer, starting from the back and changing energy to reach the front.

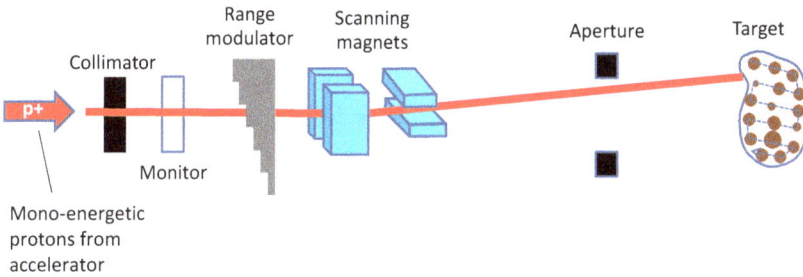

Fig. 3.10: Schematic layout of a beam delivery system for scanning a proton pencil beam. Collimators limit the transverse beam size and range modulator control the longitudinal beam shape. Scanning electromagnets steer the beam across the tumor volume. Scanning is performed in a stop-and-go fashion. At each point the dose can be tuned according to the treatment plan.

With such a pencil proton beam, the entire planned target volume (PTV) is covered point by point, essentially "dose painting" the radiation into the PTV. Pencil beam PRT effectively treats the most complex tumors with curved contours, like those in the prostate, brain, and eye, while leaving healthy tissue and other critical OAR intact. The pencil beam PRT is a very powerful tool that could eventually outdate multileaf collimators (MLC) and replace IMPT techniques. The pencil beam irradiation

---

4 This preference is due to the fact that most PRT centers operate with cyclotrons and the extraction energy of the proton beam from a cyclotron is fixed.

features a high flexibility and conformity but takes more time and is therefore more demanding on the patient's patience than the wide beam option [11, 12].

> **!** Proton beams can be delivered to the target area by a conformal beam shape or by rasterizing a pencil beam across the tumor volume.

3.3.6 Dose requirements

The dose deposited in the tumor region leads to direct and indirect bond breaking in the DNA (see Fig. 11.7), where double-strand breaks are most effective in reducing the survival chance of cancer cells. The dependence of the survival fraction on the dose D is plotted in Fig. 3.11 for photons, protons, and carbon ions. According to the definition, the RBE for photons is 1. At a survival fraction of 0.1, the RBE for carbon ions is $RBE_{carbon} = D_{photon}/D_{carbon} = 2.4$, whereas for a survival fraction of 0.01, the $RBE_{carbon} = 2$. In all cases, the RBE of carbon ions is significantly higher than for photons. An RBE of 1.1 ± 0.1 has been determined for protons independent of the dose, which is only 10% higher than for photons. The OER for protons was determined to be about 2.7 ± 0.1 [13]. The higher RBE of protons and carbon ions entails a lower total dose to be administered for a complete treatment. In the case of hard x-rays, a total dose of 60–80 Gy is usually administered in fractions of about 2 Gy per session. According to the higher RBE of protons and carbon ions, the total dose and the number of sessions can be reduced accordingly while the dose per session is kept constant.

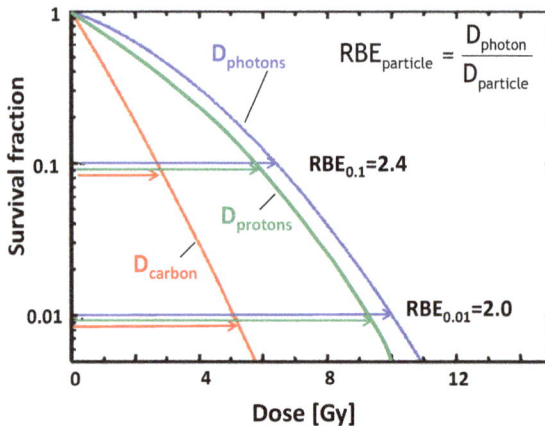

Fig. 3.11: Comparison of survival fraction of cancer cells as function of dose for photons, protons, and carbon ions.

The shape of the survival fraction curve S as a function of dose D can be described by the empirical linear-quadratic (LQ) model , as we have already seen in Chapter 1, eq. (1.2):

$$S(D) = \exp(-\alpha D - \beta D^2).\tag{3.9}$$

The first exponential term describes the linear dependence in the semilog plot, whereas the second term defines the shoulder, signaling delayed dose response. The shoulder is typical for low LET radiation, such as XRT and PRT, whereas for higher LET radiation with carbon ions the linear dependence dominates.

Protons beams are low LET radiation. The RBE for PRT is only 10% higher than for XRT !

3.3.7 Charge particle fluence

In PRT, a total dose of about 50–70 Gy is usually administered in 10–20 fractions. Each fraction of 2–7 Gy takes about 1 min of exposure time. We want to estimate what proton fluence and flux is required for such a dose rate. The result is important for the design specifications of the proton accelerator, delivering the proton beam.

First, we recap some definitions in Chapters 5 and 7 in Volume 2. The particle fluence is defined as the number of particles N per area a

$$\Phi = \frac{dN}{da}\tag{3.10}$$

The particle flux F is defined as the number of particles per time t:

$$F = \frac{dN}{dt}\tag{3.11}$$

The intensity I is the number of particles per area a and time t.:

$$I = \frac{d^2N}{da\,dt} = \frac{d\Phi}{dt} = \frac{dF}{da}.\tag{3.12}$$

The beam current is defined similarly to the electric current as the number of charges per time or flux times the elementary electric charge:

$$I_q = q_e\frac{dN}{dt} = q_eF\tag{3.13}$$

The dose D_{CP} of charged particles is given by (see eq. (7.25) in Volume 2):

$$D_{CP} = \frac{1}{\rho} \sum_i (\Phi_i \times LET_i \times RBE_i) \tag{3.14}$$

And for completeness, we also express the charge particle dose rate:

$$\dot{D}_{CP} = \frac{1}{\rho} \sum_i (\dot{\Phi}_i \times LET_i \times RBE_i) \tag{3.15}$$

where $\dot{\Phi}_i$ is the particle intensity. Assuming that the particle fluence is predominantly due to protons, then the fluence is determined by:

$$\Phi = \frac{\rho D}{RBE \times LET}. \tag{3.16}$$

To estimate the required proton fluence for a single session, we assume that an accumulated dose of 5 Gy is deployed. Then, with RBE = 1.1, and a stopping power in water of about 40 MeV/cm (4 keV/μm) according to Fig. 3.4, we find a fluence:

$$\Phi = \frac{\rho D}{RBE \times LET} = \frac{10^3 kg/m^3 \times 5J/kg}{1.1 \times 40 \times 10^8 eV/m \times 1.6 \times 10^{-19} J/eV} = 7 \times 10^{12} \frac{Protons}{m^2}$$

For a beam cross section of 1×1 cm^2, the total number of protons required for one session is:

$$N_p = \Phi \times a = 7 \times 10^8 \text{ protons.}$$

Finally, we determine the proton current required for this radiation treatment if the dose of 5 Gy is delivered within 60 s:

$$I_q = q_e \frac{dN}{dt} = \frac{7 \times 10^8}{60s} 1.6 \times 10^{-19} As = 1.8 \times 10^{-12} A = 1.8 pA.$$

A beam current density of 1.8 nA/m^2 and a total current of 1.8 pA can easily be provided by proton beam accelerators [14]. We return to this point in Section 3.6 on accelerator technologies.

3.4 Positioning and range monitoring

3.4.1 Treatment plan and positioning

We have already pointed out the main advantages of charged particle (p, ^{12}C) therapy over XRT: finite beam range and conformal energy deposition in the PTV that spares surrounding healthy tissues and OAR. Both the dose delivery to the PTV and sparing OAR are due to the characteristic LET as a function of distance in tissue featuring a Bragg peak and a sharp drop-off at the distal end of the particle track.

These advantages also constitute a considerable challenge since PRT and CIRT are only useful if the beam profile precisely conforms to the PTV. Uncertainties of the particle range by only a few millimeters can lead to underdosage in the PTV or over-dosage in the OAR. Therefore, an elaborate treatment plan is essential before applying charged particle RT to patients. The treatment plan consists of:

1. localization of the PTV by CT with respect to patient coordinates;
2. converting Hounsfield values to electron densities for evaluating the stopping power in different tissues;
3. simulation of the particle beam range;
4. determining the total dose and fractionation;
5. analysis of the particle beam shape, straggling and widening in tissue; and
6. optimization of the beam delivery concerning the patient's anatomy.

Two main tools are used to support the treatment plan:

a. Monte Carlo (MC) simulations for calculating the particle track and range as a function of initial particle energy and material density; for simulations the package GEANT4 provided by CERN is most widely used [15].
b. Phantoms for modeling as realistically as possible patient settings, including shapes and electron densities representing bones, muscles, and tissues.

An additional challenge arises in the gantry room. While the application of XRT is almost routine, the corresponding machines are at a very high industrial level, including a variety of patient position controls and feedback systems, proton and carbon particle RT is not routine at all and there are no standards to follow so far. Each gantry is constructed based on the customer's specifications, with many options to choose from. The first basic decision is between a fixed particle jet nozzle or a rotating nozzle [16]. The fixed nozzle allows for a smaller gantry but requires an extensive patient motion and control system supported by laser systems and CT. The rotating nozzle entails a 5 m diameter gantry weighing 100–200 tons with the patient at the isocenter. In Fig. 3.12 a rotatable gantry is shown schematically together with the cyclotron and the beam delivery system. The patient at the isocenter appears tiny compared to the rest of the construction. Here, too, a comprehensive control system ensures that the patient is positioned precisely according to the treatment plan. Medical physicists are in great demand for the planning, implementation, and operation of these extensive facilities.

One more piece of information is still missing. This concerns the in vivo control of the particle track in the body. This is important to remove range uncertainties due to varying RBE in the body. Therefore, monitoring the beam expansion and exact range to match with MC simulations will optimize the beam delivery. The physical principles of such monitoring have been established for some time [17] and, after many years of research, are now also being implemented in the clinical environment [18]. The monitors use either γ-radiation after positron annihilation or prompt γ-emission, both of which result from nuclear reactions and excitation of charged particle beams along their trajectory in the body (see Fig. 3.3(c) and (d)). The main characteristics of both methods are outlined in the following two sections.

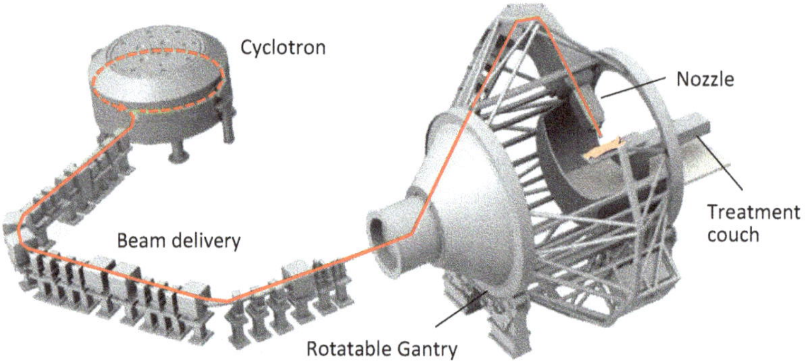

Fig. 3.12: Schematic outline of the beam delivery system from the cyclotron to the gantry. The gantry features a nozzle, which rotates about the patient at the isocenter. (Adapted from the Israelian Proton Therapy Center).

3.4.2 PET monitoring

In Section 3.3 we have argued that charged particles initially lose energy in matter mainly by electronic excitations. In addition to this dominating process, some particles strike nuclei in the target, producing radioisotopes with very short lifetimes. Most of them are β^+-emitters. The most prominent nuclear reactions that produce β^+-emitters in tissues are listed in Tab. 3.1, together with reaction threshold energies, half-life, and positron energies. The initial energy of a proton beam of some 100–200 MeV is high enough to excite any of the reactions listed in Tab. 3.1. As the proton beam reaches the end of the track, the particle energy is no longer sufficient to excite even those reactions with the lowest reaction threshold energy. For this reason, we expect that the positron production rate ceases before the proton beam comes to a stop. This scenario is shown schematically in Fig. 3.13 and confirmed by experiments using phantoms. First, there is a slight increase in positron yield along the proton track, which suddenly drops to zero as we reach the average range of the proton beam. Positrons are scattered in tissue and straggle before they come to rest and annihilate together with electrons, emitting two γ-quanta flying apart in opposite directions: $\beta^+-\beta^- \rightarrow \gamma\gamma'$. Although the total straggling path length depends on the initial positron energy, in the end, the annihilation process takes place within a distance of about 1–2 mm of its origin. Hence, detecting γ-emission via coincidence counting as described in Chapter 10 of Volume 2 should be a good monitor for tracking proton beams and, in particular, their range.

Tab. 3.1: Nuclear reactions producing positron emitters via fast protons in tissue [19, 20].

Nuclear reactions via proton impact	Reaction threshold energy (MeV)	Half-life (min)	Positron energy (MeV)
$^{16}O(p,pn)^{15}O$	16.8	2.04	1.72
$^{16}O(p,\alpha)^{13}N$	5.7	9.96	1.19
$^{14}N(p,pn)^{13}N$	11.4	9.96	1.19
$^{12}C(p,pn)^{11}C$	20.6	20.4	0.96
$^{14}N(p,\alpha)^{11}C$	3.2	20.4	0.96
$^{16}O(p,\alpha pn)^{11}C$	59.6	20.4	0.96
$^{12}C(p,p2n)^{10}C$	34.5	0.32	1.87

The practicability of monitoring charged particle tracks via β^+-decay/γ-emission can be tested experimentally by using phantoms on the one hand and by simulations on the other hand. Phantoms consisting of PMMA (chemical sum formula $C_5H_8O_2$, density of 1.18 g/cm³) are more suitable than water phantoms as they contain carbon ions constituting an important source for positron production.

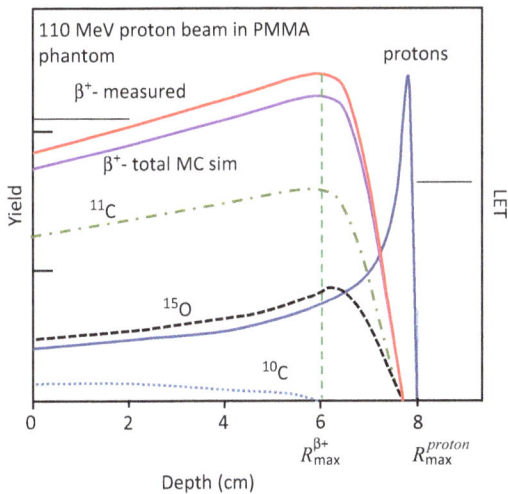

Fig. 3.13: Schematic representation of positron yield in relation to the proton linear energy transfer (LET) and range in tissue for three different positron emitters: ^{11}C, ^{15}O, and ^{10}C. Adapted with permission from [21].

Figure 3.13 shows schematically the results of a PMMA phantom experiment using a pencil-like proton beam with initial beam energy of 110 MeV [21]. The proton beam produces the β^+-emitting radioisotopes ^{11}C, ^{10}C, and ^{15}O via (p,n) or (p,2n) reactions of ^{12}C and ^{16}O nuclei in the target. The coincidence detector records all 0.511 MeV γ-emissions without discriminating between the individual contributions, as all β^+-emitters decay via the same annihilation channel. Additional MC simulations determine the total yield of all positron-emitting nuclei, which is then compared to the experimental results. We note that the β^+-emitter ^{11}C has the largest contribution to the β^+-activity, followed by ^{15}O, whereas the contribution of ^{10}C is negligible. Most importantly, the maximum of the Bragg peak and the delivered dose is a few millimeters distal to the maximum β^+-activity distribution. This is to be expected because the proton beam no longer has sufficient energy for nuclear excitations at the end of the track. However, for PET monitoring to be helpful in clinical environments, this distance should be precisely known and as small as possible [22].

In practice, two methods are used to monitor the position and range of charged particle beams via β^+-decay/γ-emission: (a) online monitoring with a PET system mounted on the rotatable proton gantry; (b) offline monitoring with a PET/CT scanner. Both options, schematically illustrated in Fig. 3.14, have advantages and disadvantages.

Online monitoring, sketched in panel (a) of Fig. 3.14, has the advantage that little time passes between particle irradiation and PET monitoring. The short half-life, diffusional escape, and biological washout demand a time gap as short as possible. However, the PET scanner has a limited field of view because of geometric restrictions in the gantry environment. Furthermore, it cannot be combined with a CT scanner, which otherwise would provide the most useful information.

Offline monitoring features all advantages which are missing in the online option. After RT, the patient is moved to the CT/PET machines. This will cost some time, resulting in lower count rates and a higher probability of biological washout. However, these disadvantages are compensated by a higher image quality taken

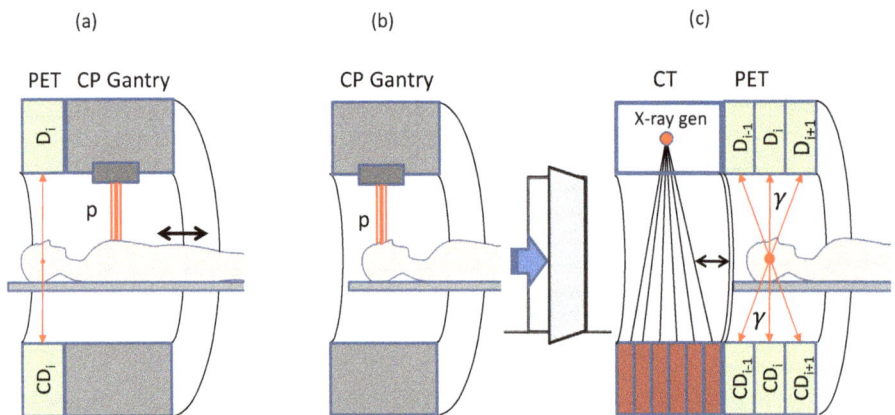

Fig. 3.14: Illustration of (a) online monitoring and (b,c) offline monitoring of positron decay with a PET scanner after irradiation with charged particles (CP) delivered through a rotatable gantry. Double arrows indicate required patient movement between irradiation nozzle and PET scanner (left) and between CT and PET scanner (right).

with a full and uncompromised PET scanner, lower background radiation, and an optional CT scan. In most radiation clinics, these two scanners are already combined in one unit.

Figure 3.15 shows results from a patient with pituitary adenoma, which is a tumor of the hypophysis [23]. The treatment plan reproduced in the top left panel of Fig. 3.15 requires a dose of 0.9 GyE (GyE = Gy × RBE) per square field delivered by two orthogonal proton fields from lateral and from posterior–anterior, confirmed by MC dose simulation in the top right panel. After irradiation, the prescribed dose was reviewed by offline monitoring. The PET and MC – PET maps in the left and

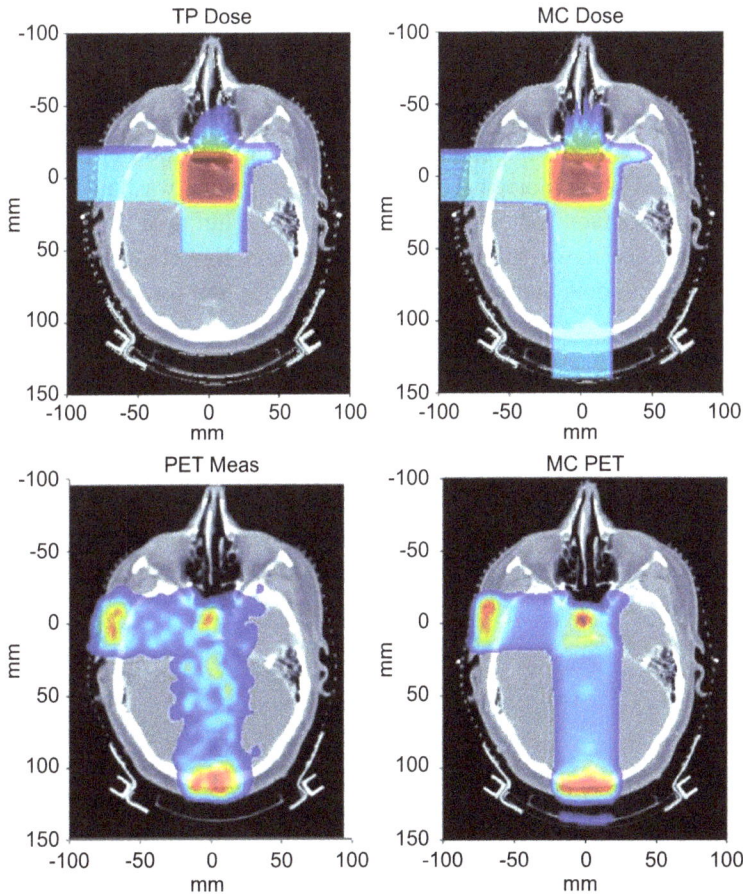

Fig. 3.15: Top panels: Treatment plan (TP) and Monte Carlo (MC) dose calculation for a patient with pituitary adenoma receiving two orthogonal fields: first from lateral and second from posterior–anterior. Bottom panels: measured PET (left) and Monte Carlo simulated PET and overlayed images. Delay times between the first and second-field applications and the beginning of imaging were about 26 min (lateral) and 18 min (posterior-anterior), respectively. Color range goes from blue (minimum activity) to red (maximum activity) (Reprinted from [23] with permission from Elsevier Limited).

right lower panels, respectively, were taken after 26 min and 18 min delay since first and second-field applications respectively. With a delay time of 20 min and more, the PET activity is solely due to the decay of ^{11}C since ^{15}O has a half-live ten times shorter than ^{11}C. This helps in calculating and interpreting recorded PET maps. The PET maps in the lower panels of Fig. 3.15 indicate that the positron activity was washed out, causing enhanced activity in the entrance fat and bone as well as in the distal part of the tumor volume. According to the authors of this investigation, the enhanced activity is due to a higher composition in carbon (19.4%) in the brainstem white matter than in the surrounding gray matter (9.5%) and cerebrospinal fluid (0%) that is not properly taken into account by the MC calculations.

These studies show that PET/CT imaging is a helpful and promising tool for monitoring charged particle tracks in the body. However, various uncertainties need to be overcome to make it a routine and reliable clinical procedure. The uncertainties are mainly due to realistic models for body composition and density along the particle track, more accurate cross sections for nuclear activation of positron emitters in the body and better biological models for perfusion and washout effects. At present, the range accuracy has been reduced to 3–5 mm, but an accuracy of at least 2 mm is required. So there is still room for improvement. Nevertheless, PET/CT monitoring of charged particle therapy remains a supportive tool for in vivo range verification and is presently the only clinically viable one [24].

3.4.3 Prompt gamma monitoring

Alternative to recording delayed-emission from β^+-decay for range verification of proton beams, it has been proposed to use the prompt γ-emission following nuclear excitations for real-time monitoring [25–27]. While the proton beam travels through

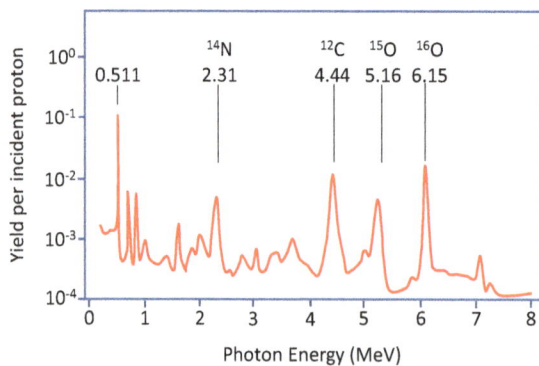

Fig. 3.16: Prompt γ-emission spectrum from protons hitting a phantom filled with water. (Data adapted and combined from references [26, 27]).

the tissue, nuclei are excited to emit prompt γ-rays. "Prompt" refers to a timescale of nanoseconds. Because of the short time scale, only online monitoring is feasible.

The spectrum of promt γ-emission lines covers a wide energy range, as the spectral analysis in Fig. 3.16 shows [26, 27]. The line at 0.511 MeV is due to positron annihilation and is not of interest in this context. All other peaks are prompt γ-emission lines, which are either due to nuclear excitations denoted by p,p' in Fig. 3.3 or due to partial fragmentation denoted by p,x ($x = \alpha,d,n$). Which prompt γ-lines actually can be detected depends on the chemical composition of the target. The line at 2.31 MeV is due to the reaction $^{16}O(p, d\gamma_{2.31})^{14}N$, the line at 4.44 MeV is dominantly from the reactions $^{16}O(p, \alpha\gamma_{4.44})^{12}C$ and $^{12}C(p, p'\gamma_{4.44})^{12}C$, the line at 5.16 MeV is mainly due to the reaction $^{16}O(p, n\gamma_{5.18})^{15}O$, and the one at 6.15 MeV is from the reaction $^{16}O(p, p'\gamma_{6.13})^{16}O$. In case that the target contains carbon but no oxygen, the last two emission lines are missing, whereas the line at 4.44 MeV is still present. From the intensity of the spectral lines, the concentration of oxygen and carbon can be estimated. For prompt γ-emission proton beam monitoring the lines at 4.44, 5.16, and 6.15 MeV are most useful.

Next, we explore whether any of the γ-lines precisely indicates the range of the proton beam. Of crucial interest for clinical applications is the exact position of the Bragg peak and, in particular, the distal fall-off. These questions can be tested with an experimental setup illustrated schematically in Fig. 3.17 [26, 27]. A proton pencil beam of variable energy enters a phantom from the left. The proton beam produced

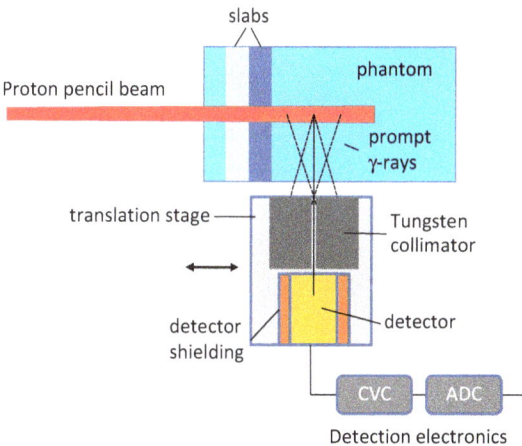

Fig. 3.17: Experimental arrangement for measuring the prompt γ-emission yield along the proton path in a water phantom. Slabs of different materials are inserted to modify the γ-emission spectrum. The detector output current is converted to a voltage pulse by a charge-voltage converter (CVC), where a discriminator sets an energy window for the spectral line to be inspected. The signal is fed into an analog – digital converter (ADC) and further processed by the data acquisition system (adapted from [26, 27].).

by a linac has a pulsed time structure with pulses about 10 ns apart. Monitoring the pulsed beam and triggering the γ-detector to a delay time shorter than 10 ns assures that detected photons are due to prompt γ-decay rather than from a more continuous background. The phantom contains mainly water, but slabs of different material densities and compositions can be inserted that mimic bones or lungs. The detector on a translation stage has a long bulky tungsten slit in the front that accepts γ-rays from a narrow angular width along the proton beam path. The detector output is connected to charge-voltage converter, pulse height analyzer, and analog-digital converter.

Figure 3.18 shows the results of experiments measuring the prompt γ-emission along the proton beam path in a water phantom with incident proton energy of 110 MeV [25]. The black line shows the dose deposited by the protons, which is given as an integral depth-dose per area in units of Gy cm². The plot shows that the 4.44 MeV γ-emission line conforms rather well to the deposited dose, whereas the 5.2 MeV γ-emission line is less useful. But the total γ-spectrum integrated from 3.0 to 7.0 MeV also serves as a very good indicator for the range. This is important as recording the total intensity does not require an energy discriminator and is therefore more adaptable to in vivo clinical practice. From these experiments and anticipated further refinements, prompt γ-emission monitoring may, in the long run, be even more promising than PET monitoring as artifacts due to washout can be avoided. Furthermore, those spectral lines can be used with energy discrimination

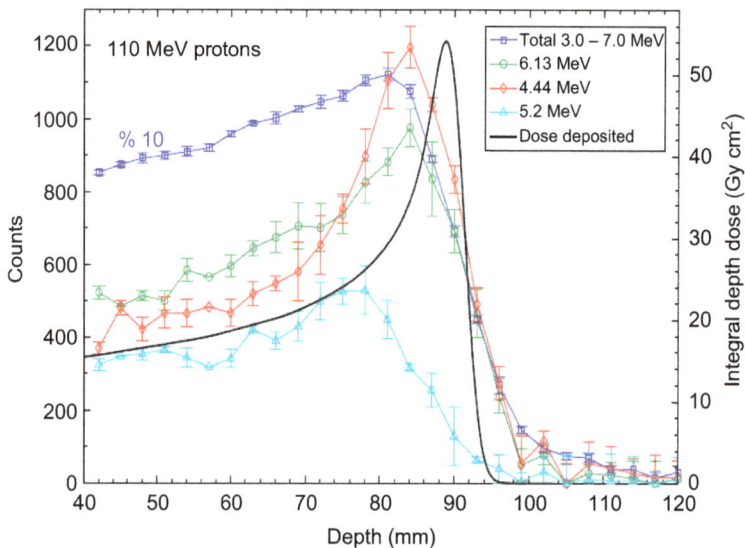

Fig. 3.18: Gamma emission count rates as a function of depth along a proton path in a water phantom. For comparison the dose deposited over the same range is also plotted. (Reproduced with permission from [25]).

as fingerprints of the oxygen concentration in tumorous tissue. As we realize, the OER is most important for tumor survival and, therefore, essential information.

> In vivo proton range monitoring can be performed by PET scanning or by prompt gamma emission recording. **!**

3.4.4 Patient monitoring

There is one more monitoring that needs to be done. This is the monitoring of the patient's movement as a whole or of particular organs. There are two strategies to cope with movement: stereotactic fixation and/or monitoring. Head fixation and eye movement monitoring is the only option for choroidal melanoma PRT. In general, patient monitoring for PRT follows the same procedures as for XRT. Laser and x-ray tracking are also used here, although x-ray monitoring with a C-arm and cone beam is bulky and cumbersome and increases the total radiation dose to the patient [28]. For these reasons, alternatives have been proposed. As we have seen in XRT, the combination of MRI and PRT would also constitute the ideal image-guided RT. MRI is fast enough to follow the respiratory cycle and eventually fast enough for cardiac cycle gating. However, this assessment is based on the premise that all other connected facilities operate with similar high speed and sufficient computer power for image processing and beam control.

Unfortunately, combination of MRI–PRT faces even more serious implementation problems than MRI–XRT due to the proximity of bending magnets in the beam delivery nozzle. Electromagnetic field interaction between the MRI coils and the proton delivery system can degrade beam quality and contribute to range uncertainties. MRI–PRT hybrid systems were tested in a pilot study [29, 30], and technical challenges were identified despite the mentioned problems. The overall results are promising and clinical implementation can be expected within the next 5–10 years.

3.5 Examples for proton beam therapy

3.5.1 Posterior fossa tumor

There are many similarities between IMRT and IMPT. But there is also an important difference to be noted. By virtue of the Bragg peak, a real three-dimensional (3D) intensity modulation can be achieved with IMPT in contrast to 2D IMRT. An example is shown in Fig. 3.19 where a comparison is made of dose distributions for treatment of the posterior fossa with protons, photons, and IMRT. Posterior fossa is a brain tumor located near the bottom of the skull causing medulloblastoma in patients of young

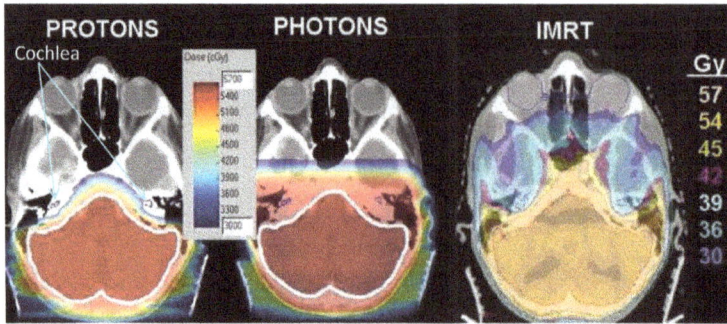

Fig. 3.19: A comparison of dose distributions for treatment of the posterior fossa in a patient with medulloblastoma, using protons, standard photons, and IMRT. Substantial sparing of the cochlea, outlined in blue and indicated by cyan-colored arrows, is only possible with protons. Note the different color coding for PRT, XRT, and for IMRT. (Reproduced with permission from [31]).

age. The radiation treatment covers the brain and also critical organs such as the co-chlea. A dose of 100% of the cochlea with standard slit XRT (middle panel) can be reduced to 90% with the use of MLC XRT, to 33% with IMRT (right panel), and to merely 2% using protons via IMPT (left panel) [31].

3.5.2 Prostate tumor

The prostate is a well-defined local gland the size of a walnut located below the male bladder. In the case of prostate cancer, radiation therapy plays an important role as nonsurgical and curative treatment [32]. In most cases, x-ray IMRT with MLC was used in the past. More recently, cyberknife technology combined with pencil beam and image guidance is employed. In any case, the lateral area exposed to XRT is well defined, while the radiation depth is unrestrained. Therefore, most radiation energy in patients receiving x-ray RT is deposited in nontargeted tissue in the pelvic space. with IMRT, In particular, the OAR are the rectum and the bladder. In con-trast, most radiation energy with proton RT is deposited inside the targeted area. Clinic studies have yet to show whether this basic difference in radiation dose depo-sition between x-rays and protons leads to an improved clinical outcome in the short and long term for patients of different age groups [33].

Figure 3.20 shows the exposure fields used in XRT for prostate radiation treat-ment in comparison to PRT. The top panels (a) and (b) show the simulated radiation levels from the treatment plans overlaid on CT images of the transversal plain con-taining the prostate. The bottom panels (c) and (d) outline the organs exposed. X-ray RT uses eight exposure fields for building up the required dose at the target center; PRT uses only two exposure fields. It is evident that PRT results in a superior defined target volume with sharper radiation dropoff on the proximal sides while sparing bladder and rectum from radiation.

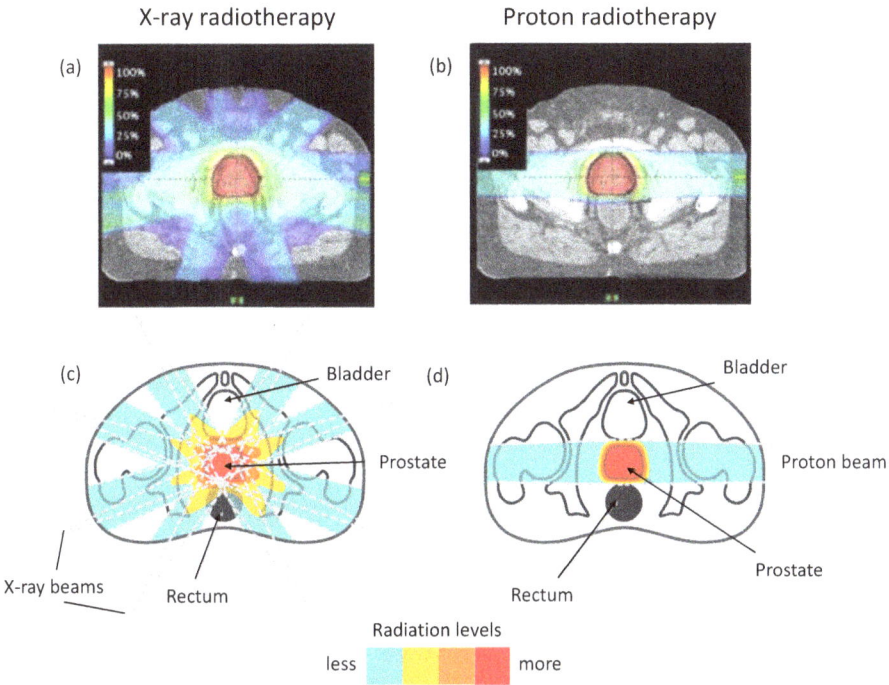

Fig. 3.20: (a) Eight radiation exposure fields for x-ray irradiation of the prostate cancer volume at the center. The radiation level is color-coded from red (high dose) to blue (low dose). (b) Same for the proton irradiation of the cancer volume, using two exposure fields. Panels (c) and (d) outline the organs in the transverse plane containing the prostate with respect to the exposure fields. Graphs are adapted from Maryland Proton Treatment Center and Seattle Proton Center.

3.5.3 Uveal melanoma

Because of the exceptionally high local control and precision, PRT has successfully been applied for treating uveal melanoma [34, 35]. Uveal melanoma is the most common ocular tumor occurring on the retina close to the optic nerve. Surgery is critical and almost impossible. IMRT does not provide the required local control nor does it spare the nearby optic nerve. In contrast, PRT is the treatment modality of choice for this type of cancer and has successfully been applied for more than 20 years. Proton therapy for uveal melanoma requires a ∼ 70 MeV proton beam for the depth of the eye (∼24 mm), which is moderate compared to PRT requirements of other cancer types. A schematic outline of the treatment setup is depicted in Fig. 3.21. Similar to the γ-knife therapy, the patient's head has to be fixed stereotactically during irradiation.

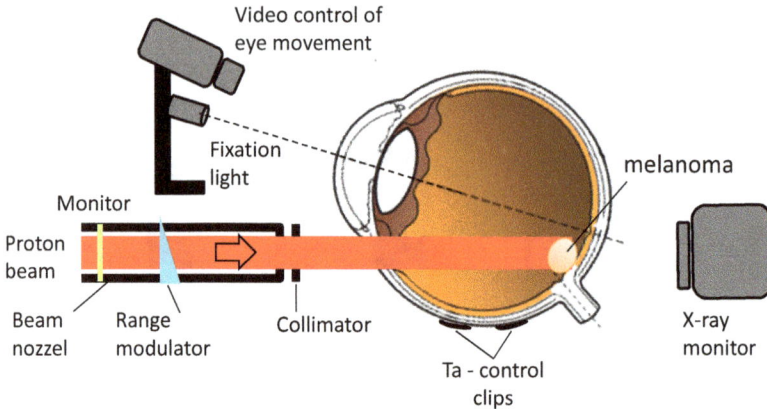

Fig. 3.21: Proton beam therapy of uveal melanoma. During RT the head is fixed and the eye should not move. Ta–clips on the backside of the eye ball define individual eye coordinates. A wedge-shaped range modulator determines the position of the Bragg peak.

Ta-markers implanted beforehand provide individual coordinates of the eye necessary for the treatment plan and irradiation simulation. Eye movement is monitored by a fixation light and by video recording. Individually shaped collimators are adapted to the tumor size and location. A plastic range monitor serves for additional range control. An update and comparison of five different facilities offering uveal melanoma via PRT is published in [36].

More examples of PRT and excellent comprehensive reviews of the physics of proton therapy can be found in [6, 37], including further topical references. Another excellent source of information is the CRC book on *Proton Therapy Physics*, edited by H Paganetti and listed under further reading.

3.6 Accelerators and gantries for proton beam therapy

PRT requires large-scale facilities that are beyond the scope of local hospitals. Usually, those facilities are part of a regional or national health care center. A general layout of a PRT facility is sketched in Fig. 3.22(a). It consists of four main parts: 1. Ion source and a linear accelerator for ramping up the energy of proton bunches to about 5 MeV; 2. Synchrotron to store the protons on a circular track guided by bending magnets; 3. Beam extraction and delivery system to the gantries; 4. Gantries for proton delivery to patients in treatment rooms. The number of gantries determines the number of patients that can be treated simultaneously. Usually, one gantry has a fixed horizontal beam with lower energy for treating uveal melanoma, and the other gantries are equipped with rotatable nozzles. A modern version that includes the capability of gated-beam

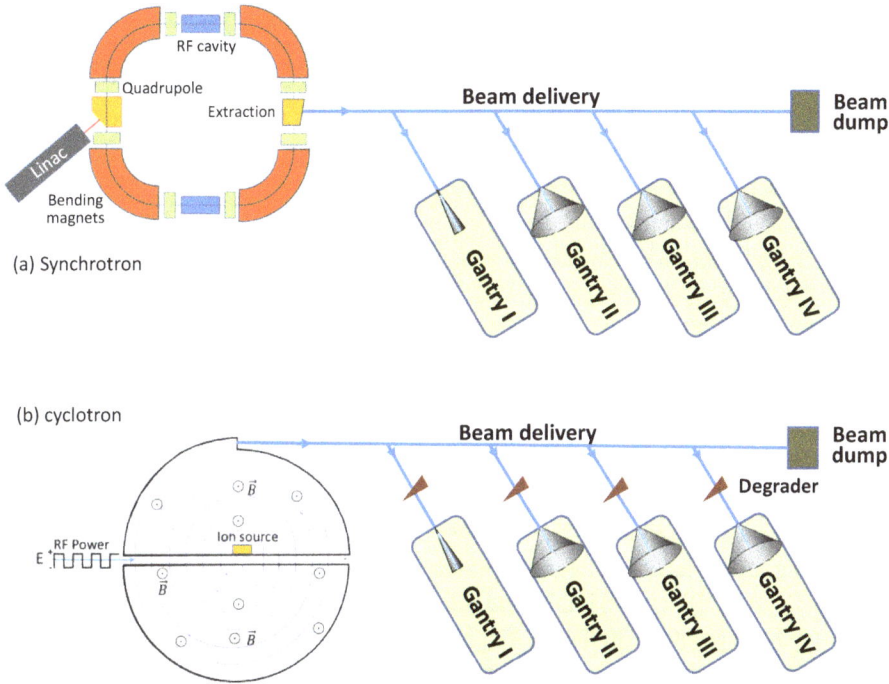

Fig. 3.22: (a) Basic layout of proton synchrotron radiation facilities consisting of a linac, storage ring with bending magnets and RF coils, beam delivery lines, and a set of gantries for patient treatment. (b) Alternative layout replacing linac and synchrotron by a cyclotron. Gantries for fixed and rotatable beams are indicated.

irradiation synchronized with the patient's respiratory cycle and spot scanning capabilities is described in [38].

Alternatively, the proton linac and synchrotron may be replaced by a cyclotron or by a synchrocyclotron; the schematics is shown in Fig. 3.22(b). Synchrocyclotrons have a design similar to conventional cyclotrons but with a varying RF field to compensate for relativistic effects as protons approach the speed of light [39]. In contrast to synchrotrons, cyclotrons have an ion source but no linac. In cyclotrons, protons are accelerated up to the highest energy and then extracted into a continuous beam (aside from microbunches). The proton extraction energy is fixed by the central magnetic field and the cyclotron's radius. Variation of energy for forming an SOBP is only possible by using a wedge-type beam degrader.

Synchrocyclotrons can be much reduced in size by using superconducting coils. Recently systems have been developed that are considerably smaller and lighter than accelerators used so far, allowing alternative strategies for beam delivery [40]. In one concept, a 20 ton synchrocyclotron is designed to rotate around the patient for PRT, similar to an x-ray gantry [41]. Another concept proposes the use of a dielectric wall, which is a 2 m long accelerator for reaching 200 MeV protons [42, 43]. The dielectric wall uses a high voltage gradient insulator to handle high electric field gradients without being short-circuited, a concept which is promoted by the Lawrence Livermore Laboratory in the USA. Although this technology is not yet ripe for application in PRT, there is a clear challenge for finding cost-efficient and space-saving alternatives to presently available facilities. Development of novel accelerator concepts can be expected in the coming years.

All accelerators have advantages and disadvantages. Synchrotrons are the most flexible machines, cyclotrons are the most cost-effective ones. The main differences are as follows: synchrotrons deliver a pulsed beam, which is very flexible with respect to proton energy and time structure, and adaptable to the respiratory cycle. Whenever the costing protons in the storage ring are needed, they are adiabatically condensed into bunches and accelerated further in RF cavities to the maximum specified energy; Moreover, synchrotrons can operate with different ions, such as H, ^{4}He, ^{12}C, and ^{20}Ar, which is impossible with cyclotrons. Synchrotrons, though, require more space and are more delicate to operate. Cyclotrons, in contrast, deliver a fixed energy and a continuous beam of higher intensity than a synchrotron. However, energy change and time-gated operation is more difficult to implement and leads to degradation of the beam quality. Which type of system is finally chosen must be carefully considered by the clinical staff. Table 3.2 compares the pros and cons of both accelerator types.

Tab. 3.2: Comparison of synchrotrons and cyclorons.

Comparison	Synchrotron	Cyclotron
Orbit	Fixed	Increases with ramping energy
Magnetic field	Variable	Fixed
Time structure	Pulsed proton bunches, gated	Continuous
Max Energy	Variable	Fixed
Beam stability	High	Moderate to high
Beam current	Lower	Higher
cost	Higher	Lower
Foot print	Larger	Compact
Size	Medium	Smaller and lighter
Ions	Variable (p, ^{12}C)	Fixed (p)
Operation	Flexible	Constraint

! High energy proton beams are provided by particle accelerators like synchrotrons or cyclotrons

3.7 Carbon ion beam therapy

3.7.1 LET and OER

Protons have a superior range property compared to x-rays. But the RBE of 1.1 is not much higher than for x-rays. PRT is therefore characterized as a high conformity – low LET radiation therapy. Irradiation with much heavier carbon ions features similar advantages as proton irradiation does in terms of position control. In fact, the Bragg peak is even sharper than for protons, as shown in Fig. 3.23 by the red line. Moreover, because of the heavier mass of carbon ions, multiple scattering and range straggling is about three times less than for protons, resulting in a sharper edge in the lateral direction (perpendicular to the beam) and, with some limitations to be discussed below, in the longitudinal direction (parallel to the beam). Edge sharpness is an important factor for deep-seated tumors, where beam widening due to scattering may blur the PTV contour, known as penumbra effect.

The LET for carbon ion RT is about a factor of 10 higher than for the lighter protons:

$$\text{LET}_C = \left\langle -\frac{dE}{dx_C} \right\rangle = 60\,\frac{\text{keV}}{\mu\text{m}} - 100\,\frac{\text{keV}}{\mu\text{m}}. \tag{3.17}$$

For this LET value, the RBE of carbon ions is at least a factor of two higher than for protons (compare Fig. 1.9 and Fig. 3.24). Therefore, carbon ions are classified as high LET radiation, featuring enhanced therapeutic benefits in treating radiation-resistant tumors. Furthermore, they have a low oxygen enhancement ratio (OER), meaning that they effectively treat hypoxic cells, which otherwise resist XRT requiring high OER. And CIRT is less dependent on the cell cycle, implying that cell death occurs (necrosis) independent of the cell-cycle phase [44]. For these reasons, CIRT has been proposed as an alternative to PRT and XRT. Historical notes on the implementation of CIRT can be found in [45, 46].

The dose–depth profile for all three types of radiation is compared in Figs. 3.23 and 3.24. The particle energy is chosen such that protons and carbon ions have the same SOBP range at the same depth below the surface. The dose is given in effective units and scaled such that photon and proton dose have the same value at the tumor center. The difference between proton and carbon ion doses within the tumor region is due to their different RBE.

Although the dose on the proximal side is nearly the same for protons and carbon ions, there is a notable difference on the distal side. The tail in the dose distribution of Carbon ions is assigned to their nuclear fragmentation. The fragments consist of H, He, Li, Be, and B ions, as has been identified in studies using a water phantom [47]. The lighter fragments have their own and more extended range and straggle forward, but their overall dose contribution is considered tolerable compared to the

Fig. 3.23: Comparison of the dose as function of depth for XRT, PRT, and CIRT. The carbon ion energy is given in atomic mass units. The particle energies are chosen such that the Bragg peaks occur at the same distance in water. (Adapted from [6, 44]).

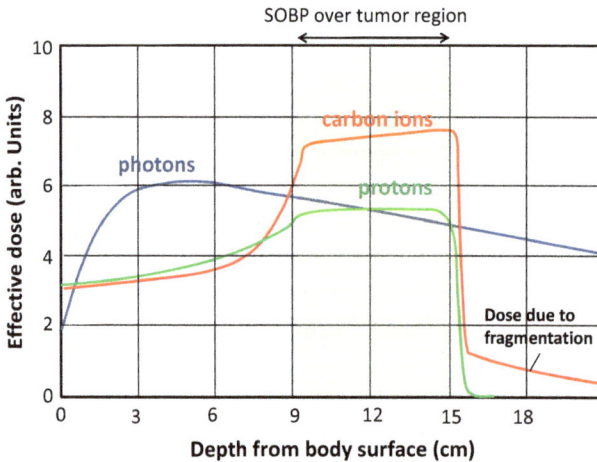

Fig. 3.24: Dose distribution as a function of depth from the body surface for photons, protons, and carbon ions. The dose is given in effective units and scaled such that photon and proton dose have the same value at the tumor center. The difference between proton and carbon ion doses within the tumor region is due to their different RBE. The carbon ion dose tails off due to fragmenting carbon nuclei before coming to rest. SOBP = spread-out Bragg peak. (Adapted from [6]).

other benefits that CIRT offers [48]. On the other hand, if a crucial organ is located in the tail region, which cannot tolerate the dose, CIRT is not applicable.

3.7.2 Implementation

CIRT requires much higher beam energy to reach the same point in the tissue than PRT. Referring to Fig. 3.23, a 150 MeV proton beam has a Bragg peak at 14.5 cm below the water surface. In order to reach the same depth, a carbon energy of 270 MeV/u is required, which is a factor of 1.8 higher than the mass difference would suggest. This makes the application of CIRT even more complex and costlier than PRT. While it is much more difficult to stir proton beams around patients than an electron beam, the heavy machinery required for CIRT appears to make beam stirring almost impossible. Nevertheless, there exists one facility for CIRT with a rotatable gantry located at the University of Heidelberg, Germany. A picture of the gigantic facility is shown in Fig. 3.25. For further discussions on the present status of

Fig. 3.25: Gantry for carbon ion radiation treatment at the Heidelberg Ion Beam Therapy Center is a 360° rotating beam delivery system. It weighs 670 tons, is 25 m long, and 13 m in diameter spanning three stories. (Photo reproduced from Heidelberg University Hospital; https://www.klini kum.uni-heidelberg.de/Gantry, image-2012.129328.0.html).

carbon beam therapy and a decade of carbon beam therapy experience at the Heidelberg Ion Beam Therapy Center, we refer to the reviews published in [49, 50].

In conclusion, the advantages of CIRT include a more narrow Bragg peak than a proton beam, less lateral scattering and therefore a narrow penumbra, and an RBE which is a factor 10 higher than for protons. The disadvantages are a distal tail due to nuclear fragmentation and the immense investment cost for a CIRT facility.

> **!** Carbon ion RT is classified as a high LET, high RBE, and low OER treatment. CIRT is characterized by a narrow Bragg peak, high-precision and high-accuracy beam delivery, but a long dose tail beyond the Bragg peak due to nuclear fragmentation.

3.7.3 Concluding remarks

Although proton therapy has clear advantages over other radiotherapies as outlined in this chapter, there is not yet sufficient clinical evidence from long-term studies on the benefits of PRT over more conventional radiotherapies. For a status report, we refer to [51]. In the case of uveal melanoma, PRT has a clear edge [52]. Furthermore, for treating children with malignant central nervous system tumors, PRT is preferable because it lessens the chance of harming healthy, developing tissue [53]. Children may receive proton therapy for cancers of the brain, the spinal cord, and the eye. For other types of cancer, the main advantage is a better sparing of surrounding tissues, thereby reducing the side effects. The essential question of whether or when survival or cure can be improved by applying PRT can presently not be answered due to lack of long-term experience. However, it is a topic of current medical investigations. As more proton radiation facilities come online worldwide, the experience will considerably increase soon.

3.8 Summary

S3.1   Interaction of protons with matter is dominated by electronic interactions, Coulomb interaction with the nucleus, and small angle recoil of nuclei.

S3.2   For PRT, protons of energies 130–200 MeV are used.

S3.3   Protons have a well-defined range in matter.

S3.4   The dose per depth shows a peak at the end of the proton track known as Bragg peak.

S3.5   The position of the Bragg peak can be tuned by varying the incident energy of the proton beam.

S3.6   RBE of protons is 10% higher than for photons and is highest in the Bragg peak.

S3.7   LET of a proton beam is low in the beginning and increases sharply in the Bragg peak.

S3.8   PRT is low in LET.

S3.9   The proximal dose is lower than for XRT and is zero on the distal side of the target volume.

S3.10   PRT has better conformity to cancer volume than other radiation treatment methods.

S3.11   Pencil-type PRT has the ability to paint the entire volume of cancer tissue.

S3.12   PRT is useful for local well-defined tumors and for tumors near-sensitive parts of the body such as near the optic nerve, cochlea, brain, and spinal cord.

S3.13   PRT is more effective than XRT in sparing the proximal part of tissue.

S3.14   PRT does not spare the skin as much as x-rays do.

S3.15   By tuning the energy, a spread-out Bragg peak can be generated that covers the tumor volume.

S3.16   Carbon ions have an RBE of 2.2–2.4 and are therefore classified as high LET radiation.

S3.17   Carbon ion radiation treatment has a low OER.

S3.18   Protons are accelerated by synchrotrons or cyclotrons.

S3.19   Synchrotrons operate with a pulsed proton beam, cyclotrons with a continuous beam

S3.20   Two types of gantries are commonly used: fixed beam nozzle and rotatable nozzle.

Questions

Q3.1   What are the main interaction channels of protons with matter?
Q3.2   When is the LET highest for protons, at the beginning or at the end of its track?
Q3.3   Do protons have a range?
Q3.4   What is the maximum proton energy and range needed to reach all parts in the body?
Q3.5   What is the RBE for protons?
Q3.6   Is PRT considered to be low or high RBE treatment?
Q3.7   Is PRT high or low LET radiation?
Q3.8   How can a spread out Bragg be formed?
Q3.9   What is the main advantage of PRT compared to XRT?
Q3.10   How can proton beams be monitored?
Q3.11   The dose-depth profile for carbon ions exhibits a tail beyond the Bragg peak. What is the reason for the tail?
Q3.12   What are the main advantages of CIRT?
Q3.13   What are the main disadvantages of CIRT?
Q3.14   What are the four essential parts of a conventional proton radiation facility?

Attained competence checker   +   0   –

If asked by a patient, I can tell the differences between XRT and PRT

I know why protons have a finite range in tissues

I know how a spread-out Bragg peak can be formed

In clinics, the transverse and the longitudinal shape of the proton beam must be adapted to the tumor volume. I know how to achieve this.

I can distinguish between a wide beam and a pencil beam and their respective advantages/disadvantages

I know what the LET and RBE values for PRT are

I know why carbon radiotherapy can be of interest

I understand that PRT requires a large facility of four main components

I can name the four essential parts of a proton radiation facility

I know the difference between a cyclotron and a synchrotron

I appreciate the consequences that a fixed versus a rotatable nozzle has.

I know why CIRT is considered as an alternative to PRT

Exercises

E3.1 **Cyclotron magnetic field:** A cancer volume is located at 27.5 cm below the surface and shall be treated with a proton beam. A cyclotron is available for delivering a proton beam of sufficient energy. The circumference of the cyclotron is 12.56 m. For the energy evaluation, use water conditions and neglect relativistic corrections.
   a. What is the proton energy required to reach the cancer volume?
   b. What is the magnetic field needed to accelerate protons to the energy determined in part a.

E3.2 **Tuning the beam energy:** A tumor volume lies between 10 and 15 cm below the skin. The treatment plan prescribes a PRT. In the treating clinic there is a cyclotron that delivers a 150 MeV proton beam. How do you trim the proton energy to hit the tumor volume and what tools do you use to achieve this goal?

E3.3 **Beam accuracy:** In PRT, a range accuracy of 2 mm at a depth of 10 cm is aimed for. Using a 100 MeV beam, what is the relative and absolute energy fluctuation that can be tolerated?

References

[1]   Wilson RR. Radiological use of fast protons. Radiology. 1946; 47: 487–491.
[2]   Particle Therapy Co-Operative Group, www.ptcog.ch/
[3]   Slopsema R. Proton therapy physics. edited by Paganetti H. Boca Raton, London, New York: CRC
        Handbook; second edition 2019.
[4]   Bryant CM, Henderson RH, Nichols RC, Mendenhall WM, Hoppe BS. et al., Genitourinary
        subcommittee of the particle therapy co-operative group. consensus statement on proton
        therapy for prostate cancer. Int J Part Ther. 2021; 8: 1–16.
[5]   Allen AM, Pawlicki T, Dong L, Fourkal E, Buyyounouski M, Cengel K, Plastaras J, Bucci MK, Yock
        TI, Bonilla L, Price R, Harris RR, Konski AA. An evidence based review of proton beam therapy:
        The report of ASTRO's emerging technology committee. Radiother Oncol. 2012; 103: 8–11.
[6]   Paganetti H. Proton beam therapy. Physics world discovery. IOP Publishing; Book
        Chapter 2017: 1–23.
[7]   Paganetti H, Thomas BT. Proton beam radiotherapy – The state of the art. published in "new
        technologies in radiation oncology". edited by Wolfgang Schlegel W, Thomas Bortfeld T,
        Grosu AL, Berlin, Heidelberg, New York: Springer Verlag; 2006.
[8]   Lilley J. Nuclear physics, principles and applications. New York, London, Sydney, Toronto:
        John Wiley & Sons; 2013.
[9]   Wilson RR. Range, straggling, and multiple scattering of fast protons. Phys Rev. 1947; 71: 385.
[10]   Lomax A. Intensity modulation methods for proton radiotherapy. Phys Med Biol. 1999 Jan; 44:
        185–205.
[11]   Ojerholm E, Kirk ML, Thompson RF, Zhai H, Metz JM, Both S, Ben-Josef E, Plastaras JP. Pencil-
        beam scanning proton therapy for anal cancer: a dosimetric comparison with
        intensitymodulated radiotherapy. Acta Oncologica. 2015; 54: 1209–1217.
[12]   Winterhalter C, Fura E, Tian Y, Aitkenhead A, Bolsi A, Dieterle M, Fredh A, Meier G, Oxley D,
        Siewert D, Weber DC, Lomax A, Safai S. Validating a Monte Carlo approach to absolute dose
        quality assurance for proton pencil beam scanning. Phys Med Biol. 2018; 63: 175001.
[13]   Kanemoto A, Hirayama R, Moritake T, Furusawa Y, Sun L, Sakae T, Kuno A, Terunuma T,
        Yasuoka K, Mori Y, Tsuboi K, Sakurai H. RBE and OER within the spread-out Bragg peak for
        proton beam therapy: In vitro study at the Proton Medical Research Center at the University
        of Tsukuba. J Radiat Res. 2014; 55: 1028–1032.
[14]   Ghithan S, Do Carmo SJC, Ferreira MR, Fraga AF, Simões H, Alvesc F, Crespoa P. On-line
        measurements of proton beam current from a PET cyclotron using a thin aluminum foil.
        J Instrum. 2013; 8: P07010.
[15]   http://geant4.cern.ch/
[16]   Engelsman M, Lu HM, Herrup D, Bussiere M, Kooy HM. Commissioning a passive-scattering
        proton therapy nozzle for accurate SOBP delivery. Med Phys. 2009; 36: 2172–2180.
[17]   Parodi K, Enghardt W. Potential application of PET in quality assurance of proton therapy.
        Phys Med Biol. 2000; 45: N151–N156.
[18]   Knopf AC, Lomax A. In vivo proton range verification: A review. Phys Med Biol. 2013; 58: R131–60.
[19]   Belkic D. Advances in quantum chemistry, theory of heavy ion collisions in hadron therapy,
        ed.. vol. 65; Amsterdam, Boston, Heidelberg, New York: Academic Press – Elsevier: 2013.
[20]   Studenski MT, Xiao Y. Proton therapy dosimetry using positron emission tomography. World J
        Radiol. 2010; 28: 135–142.
[21]   Pshenichnov I, Mishustin I, Greiner W. Distributions of positron-emitting nuclei in proton and
        carbon-ion therapy studied with GEANT4. Phys Med Biol. 2006; 51: 6099–6112.
[22]   Schardt D. Hadrontherapy. In: Basic concepts in nuclear physics: Theory, experiments and
        applications, springer proceedings in physics. García-Ramos JE et al, ed. 2016; 55–86.

[23] Parodi K, Paganetti H, Shih HA, Michaud S, Loeffler JS, Delaney TF, Liebsch NJ, Munzenrider JE, Fischman AJ, Knopf A, Bortfeld T. Patient study of an in vivo verification of beam delivery and range, using positron emission tomography and computed tomography imaging after proton therapy. Int J Radiat Oncol Biol Phys. 2007; 68: 920–934.

[24] Paganetti H, El Fakhri G. Monitoring proton therapy with PET. Br J Radiol. 2015; 88: 20150173.

[25] Verburg JM, Riley K, Bortfeld T, Seco J. Energy- and time-resolved detection of prompt gamma-rays for proton range verification. Phys Med Biol. 2013; 58: L37–L49.

[26] Verburg JM, Seco J. Proton range verification through prompt gamma-ray spectroscopy. Phys Med Biol. 2014; 59: 7089–7106.

[27] Zarifi M, Guatelli S, Bolst D, Hutton B, Rosenfeld A, Qi Y. Characterization of prompt gamma-ray emission with respect to the Bragg peak for proton beam range verification: A Monte Carlo study. Physica Medica. 2017; 33: 197–206.

[28] Hua C, Yao W, Kidani T, Tomida K, Ozawa S, Nishimura T. et al., A robotic C-arm cone beam CT system for image-guided proton therapy: Design and performance. Br J Radiol. 2017; 90: 20170266.

[29] Hoffmann A, Oborn B, Moteabbed M, Yan S, Bortfeld T, Knopf A, Fuchs H, Georg D, Seco J, Spadea MF, Jäkel O, Kurz C, Parodi K. MR-guided proton therapy: A review and a preview. Radiat Oncol. 2020; 29(15): 129.

[30] Corradini S, Alongi F, Andratschke N, Belka C, Boldrini L, Cellini F, Debus J, Guckenberger M, Hörner-Rieber J, Lagerwaard FJ, Mazzola R, Palacios MA, Philippens MEP, Raaijmakers CPJ, Terhaard CHJ, Valentini V, Niyazi M. MR-guidance in clinical reality: Current treatment challenges and future perspectives. Radiat Oncol. 2019; 14: 92.

[31] Greco C, Wolden S. Current status of radiotherapy with proton and light ion beams. CANCER. 2007; 109: 1227–1238.

[32] Wisenbaugh ES, Andrews PE, Ferrigni RG, Schild SE, Keole SR, Wong WW, Vora SA. Proton beam therapy for localized prostate cancer: Basics, controversies, and facts. Rev Urol. 2014; 16: 67–75.

[33] Kamran SC, Light JO, Efstathiou JA. Proton versus photon-based radiation therapy for prostate cancer: Emerging evidence and considerations in the era of value-based cancer care. Prostate Cancer Prostatic Dis. 2019; 22: 509–521.

[34] Damato B, Kacperek A, Errington D, Heimann H. Proton beam radiotherapy of uveal melanoma. Saudi J Ophthalmol. 2013; 27: 151–157.

[35] Groenewald C, Konstantinidis L, Damato B. Effects of radiotherapy on uveal melanomas and adjacent tissue. Eye. 2013; 27: 163–171.

[36] Fleury E, Trnková P, Spruijt K, Herault J, Lebbink F, Heufelder J, Hrbacek J, Horwacik T, Kajdrowicz T, Denker A, Gerard A, Hofverberg P, Mamalui M, Slopsema R, Pignol JP, Hoogeman M. Characterization of the Holland PTC proton therapy beamline dedicated to uveal melanoma treatment and an interinstitutional comparison. Med Phys. 2021; 48: 4506–4522.

[37] Newhauser WD, Zhang R. The physics of proton therapy. Phys Med Biol. 2015; 60: R155–209.

[38] Umezawa M, Ebina F, Fujii Y, Matsuda K, Hiramoto K, Umegaki K, Shirato H. Development of Compact Proton Beam Therapy System for Moving Organs. Hitachi Rev. 2015; 64: 506.

[39] Goto A, Tachikawa T, Jongen Y, Schillo M. Cyclotrons. In: Comprehensive biomedical physics. ed. Brahme A. 2014, Elsevier. Pages 179–195.

[40] Yap J, De Franco A, Sheehy S. Future Developments in Charged Particle Therapy: Improving Beam Delivery for Efficiency and Efficacy. Front Oncol. 2021; 11: 780025.

[41] Zhao T. et al., Commissioning and initial experience with the first clinical gantry-mounted proton therapy system. J Appl Clin Med Phys. 2016; 17: 24–40.

[42] Caporaso GJ, Chen YJ, Sampayan SE The dielectric wall accelerator. Reviews of Accelerator Science and Technology 2009; 1-10-

[43] Hettler C et al., Development of a dielectric wall accelerator proton therapy system, Abstracts IEEE International Conference on Plasma Science (ICOPS), 2013; 1-1-

[44] Hamada N, Imaoka T, Masunaga SI, Ogata T, Okayasu R, Takahashi A, Kato TA, Kobayashi Y, Ohnishi T, Ono K, Shimada Y, Teshima T. Recent advances in the biology of heavy-ion cancer therapy. J Radiat Res. 2010; 51: 365–383.

[45] Jäkel O, Kraft G, Karger CP. The history of ion beam therapy in Germany. Z Med Phys. 2022; 32: 6–22.

[46] Kim J, Park JM, Wu HG. Carbon ion therapy: A review of an advanced technology. Progr Med Phys. 2020; 31: 71–80.

[47] Haettner E, Iwase H, Krämer M, Kraft G, Schardt D. Experimental study of nuclear fragmentation of 200 and 400 MeV/u ^{12}C ions in water for applications in particle therapy. Phys Med Biol. 2013; 58: 8265–8279.

[48] Johnson D, Chen Y, Ahmad S. Dose and linear energy transfer distributions of primary and secondary particles in carbon ion radiation therapy: A Monte Carlo simulation study in water. J Med Phys. 2015; 40: 214–219.

[49] Schlaff CD, Krauze A, Belard A, O'Connell JJ, Camphausen KA. Bringing the heavy: Carbon ion therapy in the radiobiological and clinical. Radiat Oncol. 2014; 9: 88.

[50] Eichkorn T, König L, Held T, Naumann P, Harrabi S, Ellerbrock M, Herfarth K, Haberer T, Debus J. Carbon ion radiation therapy: One Decade of research and clinical experience at heidelberg ion beam therapy center. Int J Radiat Oncol Biol Phys. 2021; 111: 597–609.

[51] Tian X, Liu K, Hou Y, Cheng J, Zhang J. The evolution of proton beam therapy: Current and future status. Mol Clin Oncol. 2018; 8: 15–21.

[52] Marinkovic M, Pors LJ, van den Berg V. et al., Clinical outcomes after international referral of uveal melanoma patients for proton therapy. Cancers (Basel). 2021; 13: 6241.

[53] Main C, Dandapani M, Pritchard M, Dodds R, Stevens SP, Thorp N, Taylor RE, Wheatley K, Pizer B, Morrall M, Phillips R, English M, Kearns PK, Wilne S, Wilson JS. The effectiveness and safety of proton beam radiation therapy in children with malignant central nervous system (CNS) tumours: Protocol for a systematic review. Syst Rev. 2016; 5: 124.

Useful websites

https://meridian.allenpress.com/theijpt/pages/special-issue
https://en.wikipedia.org/wiki/Proton_therapy

Further reading

Paganetti H. Proton therapy physics. 2nd. Boca Raton, London, New York: CRC Press, Taylor and Francis; 2019.

Yajnik S. Proton beam therapy: How protons are revolutionizing cancer treatment. Berlin, Heidelberg, New York: Springer Verlag; 2012.

Goitein M. Radiation oncology: A physicist's-eye view. Berlin, Heidelberg, New York: Springer Verlag; 2008.

Delaney TF, Kooy HF editors. Proton and charged particle radiotherapy. Philadelphia, New York, London: Wolters Kluwer; 2008.

Bethge K, Kraft G, Kreisler P, Walter G. Medical applications of nuclear physics. Berlin, Heidelberg, New York: Springer Verlag; 2004.

Chao AW, Chou W editors. Reviews of accelerator science and technology. Vol. 1. Applications of accelerators. Singapore: World Scientific Publishing; 2008.

Chao AW, Chou W editors. Reviews of accelerator science and technology. Vol. 2. Medical applications of accelerators. Singapore: World Scientific Publishing; 2009.

Amaldi U. Particle accelerators: From big bang physics to hadron therapy. Berlin, Heidelberg, New York: Springer Verlag; 2014.

4 Neutron radiotherapy

Physical parameters of NRT	
Range of 2 MeV neutrons in water	4 cm
Thermal neutrons	<0.1 eV
Epithermal neutrons	1−1 MeV
Fast neutrons	1−20 MeV
Neutron production	fission and reaction
Lifetime of free neutrons	882 ± 3 s
γ_{abs} (^{10}B)	3837 barns
Neutron production	fission, ^{9}Be(p,n)^{9}B, ^{3}H(d,n)^{4}He reactions
Skin sparing effect	yes

4.1 Introduction

Neutrons are hadrons like protons but uncharged. They are more difficult to generate, and because they are neutral, accelerators and magnets are inapplicable to tune their energy or steer the beam. Neutron irradiation requires a neutron source and a radiological shielding environment very different from those used for charged particles. Among the various external beam therapies for cancer, neutron irradiation therapy is not often prescribed at present, and only a few facilities exist worldwide for their application.

Why and when should neutron radiation therapy (NRT) be considered? Neutron radiation treatment is of interest if extended (inoperable) and radiation-resistant tumors are difficult or impossible to destroy with low linear energy transfer (LET) radiation such as photons, electrons, or protons. Fast neutrons, in contrast, have high LET, high relative biological effectiveness (RBE) (see Tab. 7.1 in Volume 2 and Tab. 1.1), and the clinically most crucial oxygen enhancement ratio (OER) is close to 1. While low LET radiation causes single-strand break (SSB) of DNA via generation of toxic radicals, in high LET neutron radiation, the damage is achieved mainly by double-strand breaks (DSB). If a tumor cell is damaged by low LET radiation, it has a certain chance to repair the DNA's SSB and stay active. Due to the very high ionization density along the neutron track, DSB can occur already with a single neutron hit illustrated in Fig. 4.1. The chance of self-repair after a DSB is little.

Since fast neutrons do not depend on oxygen concentration to kill cancer cells, they can control very large tumors. In addition, the RBE of neutrons is not affected by the life cycle of cancer cells or stage of tumor development, as is the case for low LET radiation. Thus NRT has an advantage if the cell division rate is slow and tumorous

https://doi.org/10.1515/9783111168739-004

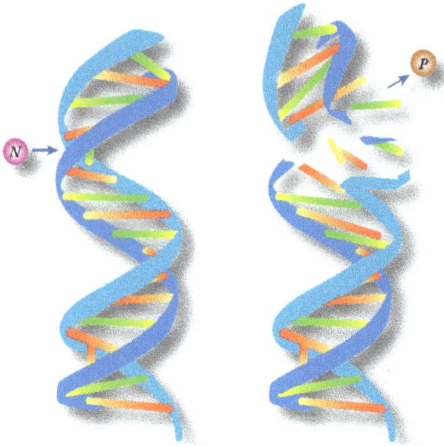

Fig. 4.1: Neutrons kill cancer cells by destroying both strands of DNA. (Reproduced from NIU, Institute for Neutron Therapy, Fermilab, USA).

or malignant cells show signs of oxygen deficiency (hypoxia). Moreover, because of the high RBE, the required tumor dose is about one-third the dose applied with low LET radiation. A full neutron radiation course is delivered in only 5–12 sessions, in contrast to 30–40 sessions needed for low LET radiation.

Two main sources of fast neutrons are of interest for clinical applications:
1. Nuclear fission of the uranium isotope ^{235}U;
2. Nuclear reaction of accelerated protons with light elements.

Neutrons from these sources are either used for external beam radiotherapies (EBRT) or for generation of α-particles after neutron capture. There are only a few places worldwide that offer one or the other type of NRT: in Seattle (USA), Tomsk (RUS), Themba (South Africa), and Munich (Germany). Some countries like the UK have discontinued neutron therapy altogether.

4.2 Neutron energies and lifetime

Some basic properties of neutrons are stated in Chapter 5 of Volume 2. Neutrons, in contrast to protons, cannot be found in nature as a free particle. They are permanently bound in nuclei. However, nuclear reactions can generate free neutrons as a fission byproduct of some actinide isotopes (^{232}Th, ^{233}U, ^{235}U, ^{238}U, ^{239}Pu, ^{241}Pu), by spallation of heavy nuclei upon proton impact or by nuclear reactions of light nuclei with protons. The mass of neutrons $m_n = 1.675 \times 10^{-27}$ kg is slightly larger than the mass of protons ($m_p = 1.673 \times 10^{-27}$ kg). Therefore, free and unbound neutrons decay into the protonic ground state via β⁻ emission:

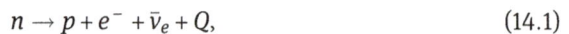

$$n \rightarrow p + e^- + \bar{\nu}_e + Q, \tag{14.1}$$

and simultaneously converting a down quark into an up quark. Here n, p, e^-, $\bar{\nu}_e$, and Q denote neutron, proton, electron, electron-antineutrino, and kinetic energy. The latest result for the $1/e$-lifetime of the neutron decay is 877.75 ± 0.28 s or about 14.6 min [1].

Neutrons are usually very fast when produced via nuclear reaction, fission, or spallation. They can be slowed down by passing them through a moderator of low atomic mass A, where they thermalize via collisions with the nuclei of the moderator. According to classical mechanics, moderation is most efficient via collisions with resting or moving particles of identical mass. Protons and neutrons have about the same mass; therefore, light or heavy water are by far the most efficient moderators. The typical diffusion length of fast neutrons in water to thermal moderation is 4 cm. Further details are discussed in Section 6.4 of Volume 2.

We classify neutrons according to their kinetic energy:
- Cold neutrons: $E_{kin} < 0.020$ eV
- Thermal neutrons: $0.025 < E_{kin} < 0.15$ eV
- Epithermal neutrons: 1 eV $< E_{kin} < 1$ MeV
- Fast neutrons: 0.5 MeV $< E_{kin} < 20$ MeV
- High-energy neutrons: $E_{kin} > 20$ MeV

The boundaries between the different categories are not well defined and more a question of taste. Only the last three categories, epithermal, fast, and high-energy neutrons, are of interest to neutron-radiation therapy. They are either used for direct tissue irradiation or neutron capture by boron injected into tumorous tissue.

Neutrons are difficult to produce and manage. Neutrons offer high LET, high RBE, and low OER radiation treatment, applied to otherwise incurable tumors. !

4.3 Fast neutron production by fission

Nuclear research reactors usually "burn" isotopically enriched ^{235}U. Fission of heavy ^{235}U nuclei takes place by capturing thermal neutrons. The fission products consist of radioactive isotopes, two to three prompt or delayed fast neutrons, fast electrons, and γ-radiation. For maintaining a fission chain reaction, fast neutrons must be slowed down to thermal energies by an adequate moderator. In most cases, water is used because of its excellent moderation and the need to dissipate safely the heat generated by the nuclear fission. After thermalizing, the fraction of neutrons diffusing back to the fuel assembly is used to maintain the controlled ^{235}U fission process. The other thermalized neutrons diffusing in the opposite direction are guided out of the reactor vessel and used for neutron-based research such as structure analysis, tomography, and material irradiation. In contrast to thermal neutrons, fast neutrons cannot be guided to the outside because of the inevitable moderation by the surrounding watery environment. For fast neutron extraction, moderation has to be circumvented.

Figure 4.2 shows a fast neutron extraction system from the research reactor facility FRM II at the Technical University Munich [2, 3]. The cross section indicates the reactor vessel and some neutron beam tubes for the use of thermal and fast neutrons. Beam tube SR10 is specialized for extracting fast neutrons for medical treatment. SR10 faces a sheet of enriched ^{235}U in a converter plate, marked yellow in Fig. 4.2. Thermalized neutrons impinging onto this sheet induce fission reactions releasing fast neutrons with a mean kinetic energy of 1.9 MeV. Part of those fast neutrons exit the reactor via beam tube SR 10. Cooling can be avoided as the heat load at the converter plate is moderate.

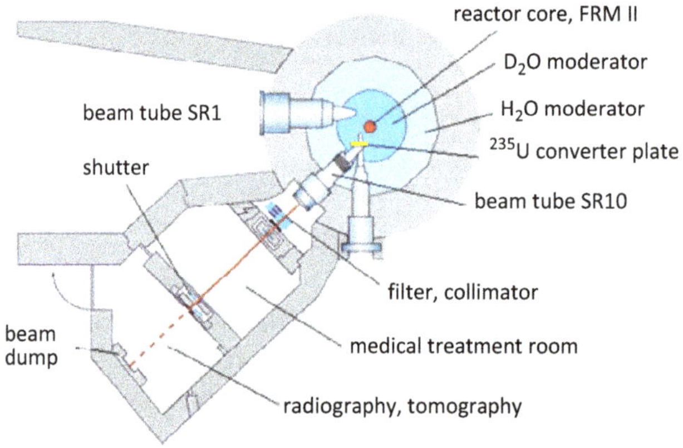

Fig. 4.2: Cross section of the nuclear research reactor FRM II at the Technical University in Munich, Germany. Thermal neutrons are guided out of the moderator (here heavy water) by beam tubes, such as beam tube SR 1. SR10 is used for fast neutron radiation therapy. In front of the SR10 nozzle is a converter plate (yellow) for the production of epithermal neutrons. Beam tubes SR2–SR9 are not shown for clarity. Reproduced with permission from [3].

The beam tube SR10 guides these fast neutrons to the treatment room for medical applications. A combined multisheet filter and collimator assembly stops the thermal neutrons and γ-radiation background from entering the treatment room. Thus a clean fission neutron beam is generated with a mean kinetic energy of 1.9 MeV and a neutron flux of about 10^8–10^9 n/s × cm^2. A multileaf collimator is used for defining the area to be exposed.

4.4 Accelerator-based neutron sources

Fast neutrons can also be generated with the help of charge particle accelerators. First protons or deuterons are accelerated to energies of a few MeV with a linear accelerator (linac) or in a cyclotron. After reaching the specified energy, the projectiles hit isotopically enriched targets to trigger nuclear reactions. The most frequent nuclear reactions used for NRT are shown in Fig. 4.3. The corresponding short notations for these charge exchange and fusion reactions are:

$$^{9}\text{Be}(p, n)^{9}\text{B}, \ Q = -1.85 \,\text{MeV}$$

$$^{9}\text{Be}(d, n)^{10}\text{B}, \ Q = +4.4 \,\text{MeV}$$

$$^{3}\text{H}(d, n)^{4}\text{He}, \ Q = +17.6 \,\text{MeV}$$

In these $X(a, b)Y$ reactions, the Q-value is determined by:

$$Q = (m_{\text{initial}} - m_{\text{final}})c^2 = (m_X + m_a - m_Y - m_b)c^2 \tag{14.2}$$

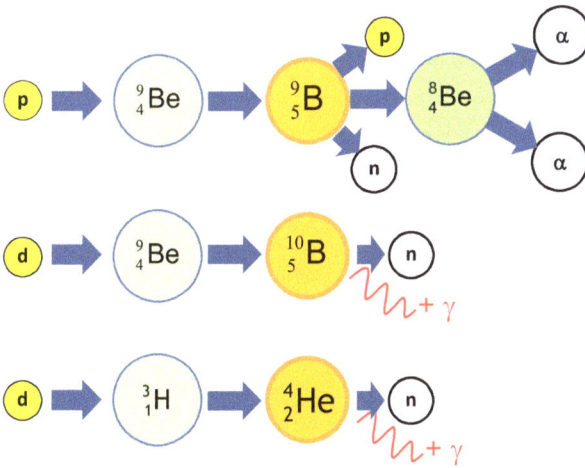

Fig. 4.3: Reaction schemes for the most important productions of fast neutrons by proton and deuterium impact.

4.4.1 ^{9}Be(p,n)^{9}B reaction

The threshold energy for the charge exchange reaction ^{9}Be(p,n)^{9}B is 2.29 MeV proton energy [4, 5]. ^{9}B is not stable but decays immediately into ^{8}Be by emission of a proton; ^{8}Be, in turn, splits up into two α-particles, such that the net reaction is: ^{9}Be (p,n)^{9}B → $2 \times {}^{4}\text{He} + p + n$. With 5 MeV incident protons a wide energy spectrum of fast neutrons can be produced ranging from 0.25 up to 3 MeV. Beam shape assemblies are used to narrow down the neutron energy distribution for specific applications. Proton absorption cross sections for a wide energy range and corresponding neutron yields are plotted in Fig. 14.4 according to Ref [4, 5].

4.4.2 ^{9}Be(d,n)^{10}B reaction

There is no threshold energy for the ^{9}Be(d,n)^{10}B reaction. The neutron yield starts already at low deuteron energies and reaches the highest yield of almost 10^{10} n/s and per μA of beam current at 6 MeV ^{2}H energy. The Q-value starts at neutron energies of 4.4 MeV for low-energy ^{2}H-particles hitting the target and increases continuously with increasing ^{2}H-projectile energy. Therefore this reaction produces faster neutrons than the first reaction. The reaction product, ^{10}B, is a stable isotope, which itself is a strong absorber for thermal neutrons and is used for BNCT [6], as discussed in Section 4.7.

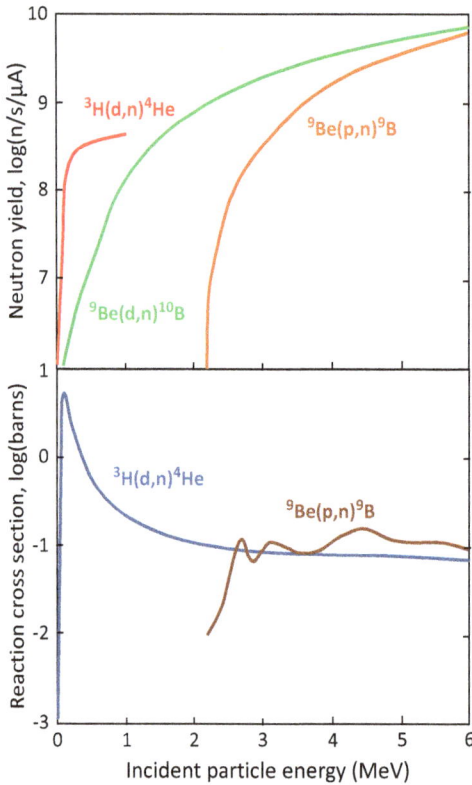

Fig. 4.4: Top panel: neutron yields for ^{9}Be(p,n)^{9}B, ^{9}Be(d,n)^{10}B, and ^{3}H(d,n)^{4}He reactions as function of projectile energy. Bottom: reaction cross sections for ^{9}Be(p,n)^{9}B and ^{3}H(d,n)^{4}He in dependence of projectile energy. Data from Refs. [4, 5].

> **!** Fast neutron beams can be produced by nuclear fission of 235-U or by proton and deuterium bombardment of light nuclei.

4.4.3 ^{3}H(d,n)^{4}He reaction

The ^{3}H(d,n)^{4}He fusion reaction is important for clinical use, as it produces monoenergetic neutrons even for low-energy ^{2}H-projectiles [7]. There is no threshold for the reaction, but a resonance like cross section is located at 100 keV incident particle energy (see Fig. 4.4). The Q-value of the reaction starts at 17.6 MeV, where the kinetic energy is partitioned between neutrons (14.05 MeV) and α-particles (3.5 MeV). The α-particles gain further energy with increasing projectile energy, whereas the neutron energy stays almost constant. The neutron yield saturates already at projectile energies below 1 MeV. This fusion reaction does not feature a threshold, so it is of interest even for low-energy linear accelerators or cyclotron facilities. The main problem is the handling of the radioactive tritium target. This problem has been solved by using metal hydrides, such as Ti^{3}H$_2$ to trap gaseous ^{3}H$_2$ in a "solid target." Furthermore, the target is enclosed in an evacuated glass tube, similar to sealed-off x-ray tubes for x-ray radiography, to hinder tritium from escaping into the environment.

Fast neutrons resulting from these three reactions can be used either for direct neutron irradiation of tumors as an EBRT or as internal radiation therapy after neutron capture in boron nitrate. In either case, a neutron moderator is required to tune down the neutron energy to the appropriate level for the highest dose at the tumor location. The depth-dose relationship for fast neutrons is discussed in the next section.

4.5 Accelerator facility

Charged particles (^{1}H$^+$,^{2}H$^+$) are usually accelerated in radio frequency quadrupole (RFQ) linacs or in *drift tube linacs* (DTL). The linacs run in a pulsed mode with a frequency of 10–20 Hz and the pulses are a few micro- to milliseconds long. A low-pulse repeat frequency (PRF) is needed to provide sufficient travel time of all neutrons from the target to the patient, before the next pulse can be fired. Figure 14.5 shows the schematics of a setup for proton-neutron conversion presently being installed at KEK, Japan [8]. At the end of a RFQ–DTL accelerator, a beryllium target is positioned for proton-neutron conversion. A neutron moderator assembly is used for reducing the neutron energy arriving from the target. Collimators guide neutrons to the tumor, here assumed to be located in the brain. Heavy shielding protects against other sources of radiation, such as γ-radiation and fast ions.

Alternatively, linacs can be replaced by fixed frequency, medium energy cyclotrons delivering a constant energy proton beam. At present, the standard is a 60 MeV, 30 μA proton beam generating a high flux neutron beam for EBRT-NRT or BNCT [9].

Accelerator-based neutron production has the advantage, in comparison to fission neutron sources, that a much smaller facility and less infrastructure is required for the safe operation of a neutron beam. Such facilities could, in principle, be run by those hospitals, which already house an accelerator for isotope production. Indeed, in Japan

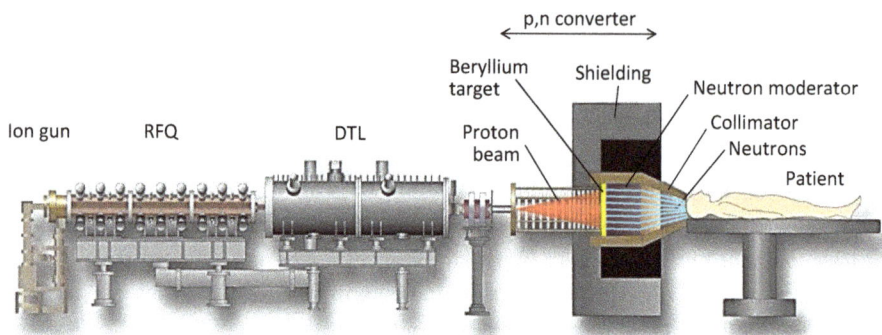

Fig. 4.5: Schematics of a linear proton accelerator irradiating targets to produce epithermal neutrons by a conversion reaction. RFQ = radio-frequency quadrupole linac; DTL = drift-tube-linac (DTL). (Adapted from Ref [6]).

there are many clinics that offer one or the other NRT for a variety of malignant melanoma and intracranial neoplasms [10].

4.6 LET, RBE, and OER of fast neutrons

The LET of charged particles is calculated according to the Bethe–Bloch equation (see eq. (6.32) in Volume 2), which assumes ionization as the main energy loss process. Neutrons are neutral. Therefore, immediate ionization does not work. Nevertheless, neutrons slow down by kicking around atoms in the medium in which they travel and finally stop or rather thermalize. As we have already realized, water is a good moderator. The range of neutrons in water is only about 40 mm before they thermalize. The moderation process of neutrons in water is described in Section 6.4 of Volume 2. Neutrons lose kinetic energy by kinematic energy transfer to the ions in the target. The recoiled ions in the body, mainly hydrogen and oxygen, travel fast and ionize others in the path. These secondary particles are the biologically active and dose relevant species.

The *linear energy transfer* (LET) for high-energy neutrons as function of depth is plotted in Fig. 4.6 [11, 12]. The shape of the LET resembles the depth-dose profile for high-energy photons (Fig. 2.4). In both cases, the LET/dose increases below the skin surface and reaches a maximum, beyond which the LET/dose slowly tails off. Unlike photons, neutrons have a range from start to thermalization. But as neutral particle they do not have a Bragg peak at the end of their travel path, in contrast to protons. However, the recoiled secondary protons and ions have their individual range and Bragg peaks forming a dose cloud around the neutron track. The dose fall-off toward the end of the neutron path is accompanied by an increase in LET and RBE due to these secondary ionization processes. This feature makes dose control of neutron therapy very difficult.

Although similar in shape, the reasons for a buildup of dose below the surface are different for photons and neutrons. The photon energy is first converted to Compton electrons or electron/positron pairs before the relevant dose builds up within the range of fast electrons. In the neutron case, the maximum dose is reached after fast neutrons are slowed down to energies of about 2 MeV, where the LET and RBE are the highest (solid red line). Fission neutrons with an energy of 2 MeV do not show a buildup of dose below the surface. Their dose is the highest already at the skin surface and then drops off continuously, shown by the red-dashed line in Fig. 4.6. Once the neutrons are thermalized, the LET is negligible, indicated by the short length of the dashed line. Also the solid red line ends where the fast neutrons finally stop.

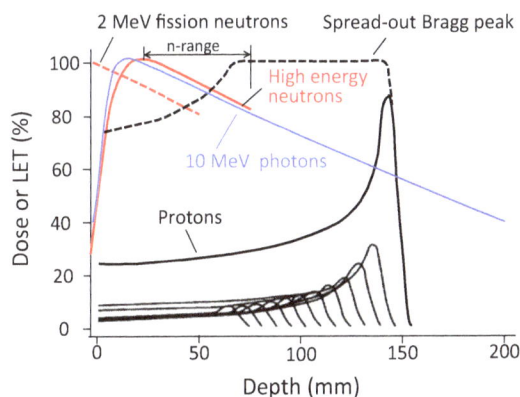

Fig. 4.6: Comparison of the dose-depth profile for protons, photons, and high-energy neutrons (<20 MeV). For fission neutrons of 2 MeV, the maximum dose is at the skin and drops off continuously (red-dashed line). The maximum range of these 2 MeV neutrons is only about 50 mm. Higher energy neutrons show the buildup of dose below the skin. For comparison, a spread-out Bragg peak with a maximum proton energy of 140 MeV is also shown.

Depth-dose measurements of neutrons are carried out in phantoms [11], usually water, which behaves similar to tissue but allows placing a dosimeter (ionization chamber) at various depths to the surface for measuring the energy deposited, as already discussed in Chapter 2. A conversion factor of 1.25 is used to compare water with tissue. For instance, 2 MeV fission neutrons are thermalized within a distance of 40 mm in water, which is assumed to be 50 mm in tissue.

The difference between XRT and NRT is highlighted by the survival curve of mammalian cells upon radiation shown in Fig. 4.7. The XRT survival curve is characterized by a broad initial shoulder described by the linear-quadratic model (see eq. 1.2). In contrast, the survival curve for NRT has almost no shoulder. This difference is typical for low LET versus high LET radiation. For low LET radiation, the total dose to be delivered increases when the dose is fractionated, meaning that the total dose is higher when delivered in a sequence of time intervals. Fractionation is, however, necessary in order to let the healthy cells recover between fractions. In the case of NRT, the shoulder is much less pronounced, and therefore the difference between single exposure and fractionated exposure is negligible. Nevertheless, in NRT, the fractionation is applied for the same reasons as for low LET-XRT.

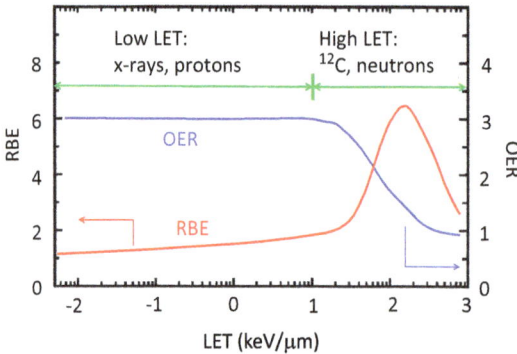

Fig. 4.7: Proportions of surviving mammalian cells to radiation as a function of dose. Dashed lines are for continuous acute radiations, while solid lines are temporally fractionated radiations. Black lines represent neutron radiation and red lines x-ray radiation. Also shown is the RBE of neutrons for single exposure and multiple exposures. (Adapted from http://www.wikiwand.com/en/Cell_survival_curve).

Now we compare the dose accumulated at a survival fraction of 1%. In the example given in Fig. 4.7, the x-ray dose required is 18 Gy in four fractions, whereas it is 7 Gy for neutron irradiation delivered by the same number of fractions. This yields an RBE for neutrons of 2.6. We also notice that the RBE for neutrons increases with the number of fractions. After one fraction, the RBE is only 1.7, but increases to 2.6 after four fractions in this example. The RBE increases even further with more fractions.

Fig. 4.8: Comparison of RBE and OER for low LET and high LET radiation. Adapted from [9, 10].

! Fast neutron have a range from emergence to thermalization. The LET is highest for 2 MeV neutrons.

Figure 4.8 summarizes the interdependencies of LET, RBE, and OER for different radiations. High LET radiation is characterized by an energy loss of more than 50 keV/μm. Then the RBE increases and the sensitivity to the presence of oxygen in the tumor volume decreases. This is an important point and an incentive for using NRT. NRT can damage also those cells, which have a low oxygen supply. According to the explanations provided in Chapter 1, the success of low LET radiation relies on the toxicity of oxygen in the cells. Cancer cells with low oxygen content are difficult to kill with low LET radiation. High LET radiation such as NRT is an option if these cells cannot be oxygenated.

4.7 Boron neutron capture therapy

Boron neutron capture therapy (BNCT) is an effective means of killing cancer cells. Here neutrons only assist the process; fast α-particles cause the cell damage. BNCT proceeds according to the steps illustrated in Fig. 4.9. First, isotopically enriched boron (^{10}B) containing drugs are delivered to the tumor tissue, preferentially by injection into the tumor volume. Then by exposure to a neutron source, thermal neutrons (0.025 eV) are captured by ^{10}B. In response, ^{10}B breaks apart by ejecting energetic a short-range α-particle (E_{kin} = 1.47 MeV) and a ^{7}Li-ion (E_{kin} = 0.84 MeV). The short notation for this reaction is ^{10}B(n,α)^{7}Li. ^{10}B has a particularly large cross section for thermal neutron absorption: σ_{abs} = 3837 barns. The kinetic energy of α-particles and ^{7}Li-ions is deposited mostly within the same cell containing the original ^{10}B isotope. This is because the range of the particles is comparable with the cell size or typically about 10 μm for α-particles and even less for the Li-ions. The α-particles acquire a high-LET of about 150 keV/μm, the ^{7}Li ions reach an LET of about 175 keV/μm [13].

BNCT was performed mainly with neutrons supplied by a nuclear reactor in the past. However, the more modern approach is the use of (p,n) or (d,n) converters by means of a linear proton (deuterium) accelerator as shown in Fig. 4.4. Fast neutrons from such a converter can be slowed down to epithermal energies with an appropriate moderator. The final moderation to thermal energies is achieved within the top layers of the skin [13, 14].

There are few more isotopes with high thermal neutron absorption cross sections. But ^{10}B is the best choice for several reasons [15]:
– ^{10}B is not radioactive and chemically not toxic;
– ^{10}B is readily available (abundance about 20% of naturally occurring boron);
– The emitted particles (^{4}He and ^{7}Li) have high LET;
– The chemistry of boron is well understood and allows it to be readily incorporated into many different chemical structures.

The radiation effect is local and limited to those cells, which have taken up the isotope ^{10}B. The thermal and epithermal neutrons still irradiate other cells, but the RBE of thermal neutrons is not very high and therefore little damage is done to the normal cells.

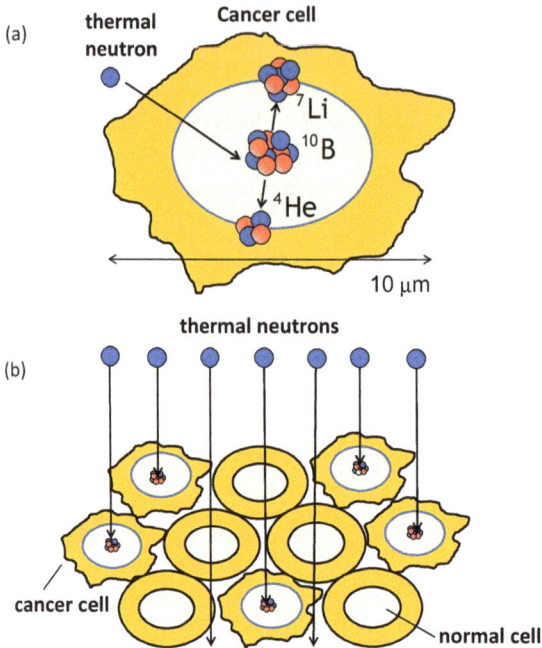

Fig. 4.9: Schematics of radiotherapy of cancer cells via capture of thermal neutrons in boron: (a) shown are the reaction products ^{4}He and ^{7}Li, which have a range in the order of the cell size; (b) ideally, only cancer cells contain ^{10}B isotopes. These cells are then destroyed by thermal neutron capture, while normal cells are unaffected by neutrons.

The standard suppliers of boron are sodium borocaptate (BSH) (chemical formula: $Na_2B_{12}H_{11}SH$) and boron phenylaline (BPA), the respective molecular structures are shown in Fig. 4.10. The challenge is to achieve a boron concentration sufficient to deliver therapeutic doses of radiation to tumor cells with minimal toxicity to surrounding normal tissues. This concentration can be estimated to be at least ~20–35 μg boron ^{10}B per gram of tumor mass. This high concentration should be maintained during irradiation. Diffusion or washout would be counterproductive. Another concern is the purity of sodium borocaptate. Impurities lead to additional and uncontrolled toxicity. In any case, it must be ensured that BSH penetrates readily into most malignant tumor cells and that there is some retention of the drug in the tumor cells for an adequate period of time for radiation treatment, while BSH or BPA is virtually excluded from normal tissues. Under these conditions, BNCT is the only nuclear radiation therapy in which the radiation source is located directly in the cell after neutron capture. It therefore bears some resemblance to brachytherapy, which will be discussed in the next chapter, but even more so to therapeutic nanoparticles, which are discussed in Chapter 7.

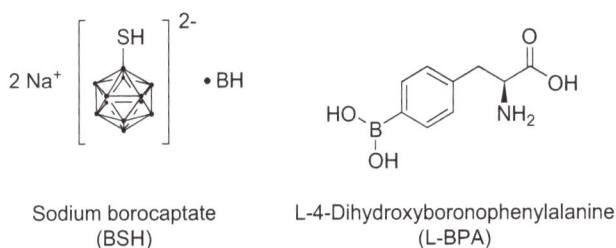

Sodium borocaptate   L-4-Dihydroxyboronophenylalanine
        (BSH)   (L-BPA)

Fig. 4.10: Molecular structure of the boron carrying drugs $Na_2B_{12}H_{11}SH$ (BSH) and boron phenylaline (BPA).

Boron neutron capture therapy is a promising tumor irradiation modality that produces high energy charge particles in tissue cells. Using accelerator based neutron production facilities, BNCT may become more popular in the future.

BNCT appears like a very promising radiotherapy for some specific tumors and melanomas. It is highly effective if tumors are local and close to the surface. BNCT is mainly used for high-grade gliomas, cerebral metastases of melanoma as well as head, neck, and liver cancer. However, in the medical community, the acceptance level is presently not very high. After encouraging experiments in Japan, the USA, and Europe, BNCT treatments have essentially ceased. The decline of interest is mostly due to the insufficient concentration of ^{10}B isotopes at the tumor site and only at the tumor. Here progress is awaited from better-suited boron carriers. The current status of clinical applications can be found in [16].

In the past BNCT could only be performed at nuclear research reactors and the number of operational reactors has declined in recent years. Due to the off-clinical locations of treatment centers, the number of clinical studies has been rather limited, and therefore the advantages of BNCT irradiation modality compared to other radiotherapies are not well documented. Because of this comparably low clinical activity, the development of boron-containing pharmaceuticals is also on a low level. However, with the advent of accelerator-based facilities that can be integrated in a clinical infrastructure, there is hope that the BNCT irradiation modality will become more widely accepted [13, 14]. It may also help make epithermal neutrons more viable in radiotherapy clinics via proton/neutron converters. A review of the present status of BNCT treatment and future perspectives is published in [16] and a critical assessment of future perspectives of NRT is given in [17].

4.8 Summary

S4.1   Neutrons for radiation therapy are produced in nuclear reactors by fission of ^{235}U or by proton-neutron converters using proton accelerators.

S4.2   For accelerator-based fast neutron production, three different nuclear reactions can be used: ^{9}Be(p,n)^{9}B, ^{9}Be(d,n)^{10}B, and ^{3}H(d,n)^{4}He.

S4.3   Tumors with suitable (not to deep) locations can be irradiated directly with fast neutrons, combining high LET, high RBE, and low OER.

S4.4   Fast neutrons have a range from production to thermalization, but they have no Bragg peak. The LET is highest for 2 MeV neutrons.

S4.5   The number of radiation fractions required for a survival chance of 1% is by a factor of 3 less than for low LET radiation.

S4.6   Some tumors can be treated by boron neutron capture therapy (BNCT), which requires thermal neutrons that are absorbed by ^{10}B isotopes with subsequent emission of α- and ^{7}Li particles.

S4.7   The α and ^{7}Li fragments of ^{10}B capture have a very short range in the order of the size of a cell.

S4.8   BNCT is the only nuclear RT which has the source of radiation directly in the tumor cell.

S4.9   The standard boron delivery agents are sodium borocaptate (BSH) and BPA, which are preferentially taken up by tumor cells.

S4.10   NRT is a promising modality not practiced much at present.

? Questions

Q4.1   What are the three main methods of neutron radiation therapy?

Q4.2   How are neutrons distinguished according to their energies?

Q4.3   Why are neutrons used for cancer treatment?

Q4.4   Which types of carcinoma are mainly treated by fast neutrons?

Q4.5   Which are the main accelerator-based nuclear reactions for the production of neutrons?

Q4.6   What is the shape of the depth-dose profile for high-energy neutrons?

Q4.7   What is the reason for dose buildup below the surface in the case of high-energy neutrons?

Q4.8   What is the effect of high LET radiation on the fractionation of dose delivered to the patient?

Q4.9   In BNCT, the radiation source is in the tumor cells. What is the role of neutrons and what type of radiation occurs?

Q4.10   What are the dose-relevant particles in BNCT treatment?

Attained competence checker   + 0 –

I can distinguish between different catagories of neutrons with respect to their energie
I know what energy neutrons have that result from fission of ^{235}U
I can tell the range of fast neutrons in water
I know how to produce fast neutrons
I know what fast neutrons are good for
I am aware of the LET and RBE values of fast neutrons
I know why the OER is an important parameter
I can graph the dose-depth dependence for fast neutrons
I can graphyically show the difference of dose-depth relations for x-rays, protrons, and neutrons
I know how BNCT treatment works
I appreciate the difficulties of BNCT applications

Exercises

E4.1 **Dose**: Thirty micrograms of ^{10}B per gram tumor mass is administered to a patient. Assume that all ^{10}B isotopes capture a neutron over some time. What is the total dose delivered to the tumor?

E4.2 **Neutron exposure**: Taking into account that only those ^{10}B disintegrate which have captured a neutron. Determine the production rate of ^{10}B and the total amount of activated ^{10}B after 1 h neutron exposure. The neutron flux is 10^{12} n/m^{2}s and the neutron capture cross section of ^{10}B is 3835 barn.

References

[1] Gonzalez FM. et al., Improved neutron lifetime measurements with UCNt. Phys Rev Lett. 2021; 127: 162501.
[2] Wagner FM, Kneschaurek P, Kastenmüller A, Loeper-Kabasakal B, Kampfer S, Breitkreutz H, Waschkowski W, Molls M, Petry W. The Munich fission neutron therapy facility MEDAPP at the research reactor FRM II. Strahlenther Onkol. 2008; 184: 643–646.
[3] Genreith C. MEDAPP: Fission neutron beam for science, medicine, and industry. J Large Scale Res Facil. 2015; 1: A18.
[4] Karki A. A study of ^9B spectroscopy via the $^9Be(p,n)$ ^9B reaction using the neutron time-of-flight technique. PhD Thesis. Ohio University; 2013.
[5] Chichester DL. Production and applications of neutrons using particle accelerators. Idaho: Idaho National Laboratory Idaho Falls, vol. 83415: 2009.
[6] Shin JW, Pat TS. New charge exchange model of GEANT4 for 9Be(p,n)9B reaction. Nuclear Instr and Meth B. 2015; 342: 194–199.

[7]   Gagliardi A, Betker AC. The 3H(d, n)4He reaction: A possible source of intermediate energy neutrons. Nucl Instrum Meth Phys Res A. 1989; 284: 356–358.
[8]   Yamamoto T, Tsuboi K, Nakai K, Kumada H, Sakurai H, Matsu A. Boron neutron capture therapy for brain tumors. Transl Cancer Res. 2013; 2: 80–86.
[9]   Mandrillon P. Cyclotrons in radiotherapy. In: Turner S (ed.), CAS-CERN accelerator school: Cyclotrons, linacs and their applications, CERN. Geneva. 1996; 313–327.
[10] Matsumoto Y, Fukumitsu N, Ishikawa H, Nakai K, Sakurai H. A critical review of radiation therapy: From particle beam therapy (Proton, Carbon, and BNCT) to Beyond. J Pers Med. 2021; 11: 825 p. 1–32.
[11] Söderberg J, Carlsson GA. Fast neutron absorbed dose distributions in the energy range 0.5–80MeV – A Monte Carlo study. Phys Med Biol. 2000; 45: 2987–3007.
[12] Söderberg J, Carlsson GA, Ahnesjö A. Monte Carlo evaluation of a photon pencil kernel algorithm applied to fast-neutron therapy treatment planning. Phys Med Biol. 2003; 48: 3327–3344.
[13] Suzuki M. Boron neutron capture therapy (BNCT): A unique role in radiotherapy with a view to entering the accelerator-based BNCT era. Int J Clin Oncol. 2019; 1–8.
[14] Dymova MA, Taskaev SY, Richter VA, Kuligina EV. Boron neutron capture therapy: Current status and future perspectives. Cancer Commun. 2020; 40: 406–421.
[15] Sauerwein W, Wittig A, Moss R, Nakagawa K. Neutron capture therapy: Principles and applications. London: Springer Heidelberg New York Dordrecht; 2012.
[16] Malouff TD, Seneviratne DS, Ebner DK, Stross WC, Waddle MR, Trifiletti DM, Krishnan S. Boron neutron capture therapy: A review of clinical applications. Front Oncol. 2021; 11: 601820.
[17] Jones B. Clinical radiobiology of fast neutron therapy: What was learnt?. Front Oncol. 2020 Sep; 15-: 1537.

Further reading

Barth RF, Vicente MG, Harling OK, Kiger WS, Riley KJ, Binns PJ, Wagner FM, Suzuki M, Aihara T, Kato I, Kawabata S. Current status of boron neutron capture therapy of high grade gliomas and recurrent head and neck cancer. Radiat Oncol. 2012; 7: 146.
Sauerwein WAG, Wittig A, Moss R, Nakagawa Y editors, Neutron capture therapy – Principles and applications. Berlin, Heidelber, New York: Springer Verlag; 2012.

Useful website

https://www-ad.fnal.gov/ntf/what_is/index.html

5 Brachytherapy

Physical parameters and acronyms of brachytherapy

EBRT	external beam radiotherapies
LDR	low dose rate, 0.4–2 Gy/h
HDR	high dose rate >12 Gy/h
^{60}Co emission	1.3 MeV γ-rays
^{60}Co half-life	5.3 years
^{137}Cs emission	0.66 MeV γ-rays
^{137}Cs half-life	30 years
Frequent isotopes for seeds in BT	^{103}Pd, ^{125}I

5.1 Introduction

Brachytherapy (BT) is the application of radioisotopes through direct contact with tumors. The Greek name "brachys" means "near" as opposed to "tele" which means "far." Consequently, all *external beam radiotherapies* (EBRT) discussed so far in Chapters 2–4 can be characterized as *tele-therapies*. Appropriate radioisotopes are required for brachytherapy, which are encapsulated in small tubes (Fig. 5.1) and worked directly into the tumor tissue by a physician. Such therapy is promising if the tumor volume is well-defined and localized and the tumor can be reached either by penetrating through the skin (interstitial), through open cavities (intracavities), or by direct contact with the skin. The main applications of BT are prostate, breast, gynecological, and skin cancers.

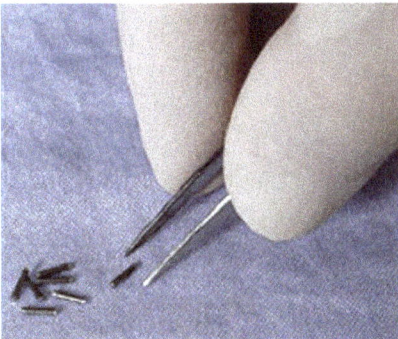

Fig. 5.1: Stainless steel capsules containing radioactive isotopes for brachytherapeutic treatment of tumors. (Reproduced from https://en.wikipedia.org/wiki/Brachytherapy, © creative commons).

Brachytherapy is an important alternative to surgery, chemotherapy, and EBRT via γ-, p-, α-, or neutron irradiation featuring many advantages compared to these latter ones. It guarantees:

https://doi.org/10.1515/9783111168739-005

- an overall comparatively short treatment time;
- local application of radiation;
- conserving healthy tissue;
- lower probability for incontinence and impotence in case of prostate cancer;
- conserving skin and tissue in case of breast cancer;
- single patient visit
- relatively short hospitalization.

However, there are also disadvantages:
- Extended or metastasized tumors cannot be treated by brachytherapy;
- Implantation of radioactive capsules can be a radiation hazard to the clinic staff;
- Only *low dose rate* (LDR) capsules can be implanted by hand;
- *High dose rate* (HDR) capsules are difficult to handle and require special procedures to be described in Section 5.3.

Brachytherapy relies entirely on the availability of radioisotopes emitting γ-radiation with short–to-medium lifetimes and keV to MeV photon energies. These radioisotopes act like miniature x-ray sources with MV power in direct contact with neoplasms. All isotopes of interest result from nuclear fission or thermal neutron capture (NC). They all belong to the class of heavy alkali-metals, transition metals, or rare earth metals. The radioactive isotopes are usually pure elemental metals in the solid-state phase. The use of these isotopes for brachytherapy requires specialized laboratories for the production of isotopes, safe handling, and preparation of appropriate seeds, wires, and pellets.

> **!** Brachytherapy uses gamma-emitting and encapsulated isotopes for irradiation of tumors in direct physical contact.

5.2 Radioisotope selection for brachytherapy

Some important properties for the selection of isotopes suitable for brachytherapy must first be considered. These are summarized as follows:
- Half-life should neither be too short nor too long. Half-life between a few days and a few months are most suitable.
- The emitted radiation should have enough energy to treat the tumor but not too high to make radiation protection a severe problem for patients, clinic staff, and the environment.
- Radioactive material should be in a solid form that can be encapsulated without the danger of spilling. This holds true for parent and daughter radioisotopes and excludes any gaseous or liquid form of isotopes. But pellets, wires, or plaques are appropriate.

- The specific activity, i.e., the activity per mass unit, should be high, so that small amounts of radioactive material already provide sufficient radiation in a short exposure time.

Table 5.1 lists the most common radioisotopes for BT fulfilling these general conditions. All of them are β – γ –emitters, usually from the same parent isotope. Other types of emitters are not used. Because of their short range (see Fig. 6.15 in Volume 2), β-radiation is already stopped within thin stainless steel container walls, while even "soft" γ-radiation of only a few tens of keV photon energy can penetrate through the walls. In practice, most (β, γ)-sources are effectively only γ -sources, unless the β-energy is very high as is the case of ^{106}Ru and ^{90}Sr/^{90}Y. Since β-decay is always associated with a broad energy distribution (see Infobox on β-decay in Chapter 5, Volume 2), only the maximum energy E_{max} is listed in Tab. 5.1. The broad β-energy distribution entails a broad-emission spectrum, in which case only the mean γ -energy is referenced in Tab. 5.1. The reaction mechanisms of the isotopes for brachytherapy are also listed in the third column.

Tab. 5.1: Most frequent radioisotopes for BT, including form of capsules, type of radiation, half-life, and energy of emitted particles.

Radionuclide	Form	Reaction	Type	$T_{1/2}$	Energy MeV
^{137}Cs	Tubes, needles, afterloading	^{235}U(n, γ)^{137}Cs→^{137}Ba fission	γ	30.2 years	0.662
^{131}Cs	Seeds	^{131}Ba(p, γ)^{131}Cs	γ	9.68 days	0.029–0.034
^{60}Co	Tubes, afterloading	^{59}Co (n, γ)^{60}Co→^{60}Ni Neutron capture		5.26 years	1.17 1.33
^{192}Ir	Wires, afterloading	^{191}Ir(n, γ)^{192}Ir→^{192}Pt Neutron capture	γ	73.8 days	0.38 (mean)
^{125}I	Seeds	^{124}Xe(n, e)^{125}I	γ	59.6 days	0.0154 0.0314 0.0355
^{103}Pd	Seeds	^{102}Pd(n, γ)^{103}Pd		17.0 days	0.021 (mean)
^{198}Au	Seeds grains	^{197}Au(n, γ)^{198}Au	γ	2.7 days	0.412
^{106}Ru ^{106}Rh	Plaques	^{235}U(n, β$^-$)^{106}Ru→^{106}Rh, fission	β$^-$	1.02 years 30 s	0.039 3.54
^{90}Sr ^{90}Y		^{235}U(n, γ)^{90}Sr→^{90}Y fission	β$^-$ β$^-$	28.7 years 64 h	0.546 2.15

All isotopes listed in Tab. 5.1 are either fission products of ^{235}U in a nuclear reactor (^{137}Cs, ^{106}Ru, ^{90}Sr), or they are generated by thermal NC in $(n, \beta(\gamma))$-reactions (^{60}Co, ^{192}Ir, ^{125}I). In a few cases, charge particle activation (CPA) via proton or deuterium accelerators is used (^{103}Pd, ^{131}Cs). The standard application of CPA is for the production of light isotopes, which are inaccessible to neutron activation. However, these light isotopes have a half-life time too short to be useful for BT.

Fission products of ^{235}U must be radiochemically separated from all other radioisotopes that are present in irradiated uranium pellets so that the emission spectrum is "clean" when clinically applied and free of spectral lines from contaminated radioisotopes.

Elaborate and risky radiochemistry can be avoided by thermal neutron activation in nuclear reactors (see Section 5.6.6 in Volume 2 for the general background). However, the use of isotopically clean targets to avoid contamination from other isotopes is mandatory. External parameters can then be used to control the activity of the end product: neutron flux, exposure time, density, and intrinsic parameters: thermal NC cross section and lifetime.

Thermal neutron activation usually proceeds by replacing a nuclear fuel rod in a fission reactor with a rod that contains the material to be irradiated. Figure 5.2 shows

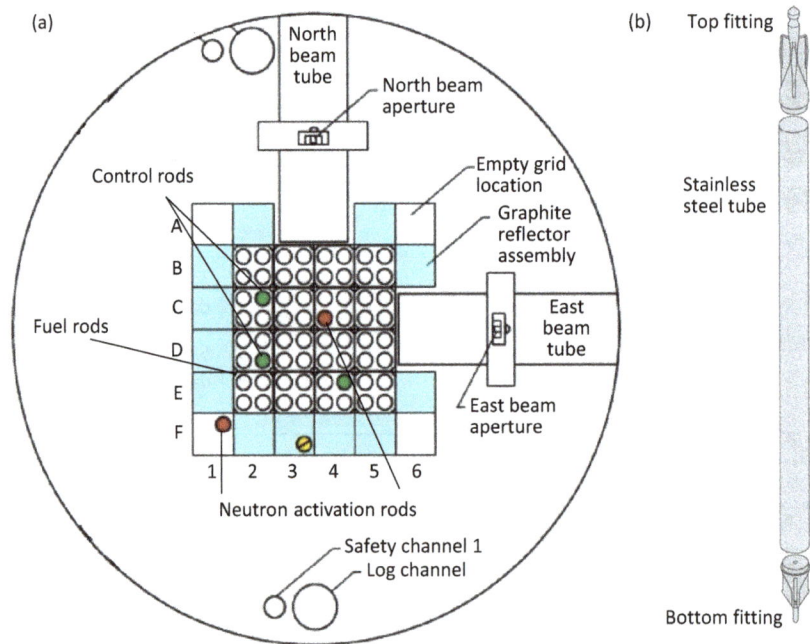

Fig. 5.2: (a) Schematic outline of the fuel rod arrangements in a nuclear safety reactor of the type TRIGA Mark II. The red circles indicate positions where rods for thermal neutron activation can be inserted. (b) Typical design of a fuel rod to be inserted into the reactor tank. Adapted with permission from https://mfc.inl.gov/SitePages/Home.aspx.

a schematic outline of the arrangement of fuel rods in a Mark II TRIGA[1] safety reactor. In the grid, where the fuel rods can be moved in and out, a few places are reserved to be used for neutron activation (red dots). Alternatively, at larger neutron facilities often a pneumatically driven "rabbit" system is used to insert material for irradiation inside the core of the reactor.

Sections 5.2.2–5.2.8 present more details on the nuclear reactions listed in Tab. 5.1.

Infobox: TRIGA nuclear research reactor

Training, research, and isotope production (TRIGA) was the intended use for this nuclear reactor when it was designed in the 50th of the last century by Edward Teller[2] and this team. The other main features of TRIGA reactors are its safety, although not expressed in the acronym. Neverthe-less, because of the inherent safety properties, general atomics has since installed more than 60 TRIGA reactors, mostly at universities, governmental and industrial laboratories, and medical cen-ters in 24 countries. TRIGA reactors operate at an average power of 250 kW up to 1.5 MW, which is much less than large-scale research reactors, operating at powers up to 60–100 MW.

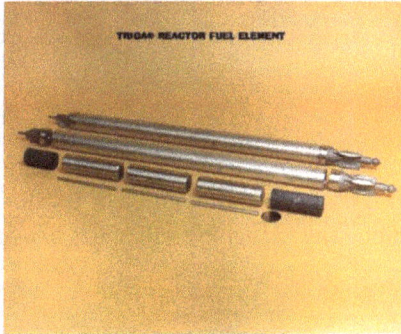

Fig. 5.3: Fuel elements of the TRIGA reactor. Reproduced from http://www.ga.com/triga.

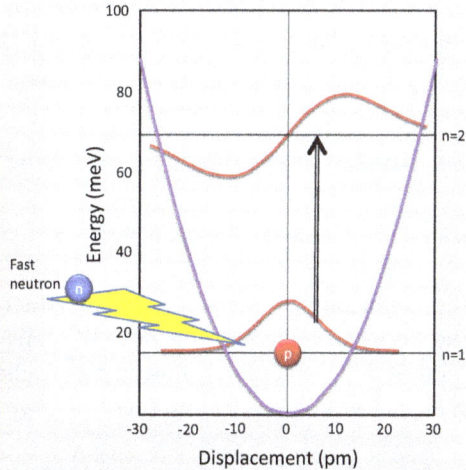

Fig. 5.4: Harmonic potential of protons in ZrH. Fast neutrons are being moderated by exciting protons from the ground level to the upper vibrational level.

As any other nuclear reactor, the TRIGA reactor features nuclear fuel rods for the fission of ^{235}U and production of neutrons. A pool of water moderates the hot neutrons emanating from the fuel rods. Boron carbide (B_4C) control rods strongly absorb thermal neutrons and regulate

---

1 TRIGA stands for Training, Research, and Isotopes, General Atomics.
2 Edward Teller (1908–2003), Hungarian-US American theoretical physicist.

the power output. A graphite reflector surrounding the core keeps the thermalized neutrons close to the fuel rods. However, in contrast to other nuclear reactor types, TRIGA reactors are designed to show a negative temperature coefficient that turns off neutron production whenever the temperature rises. Producing a worst case scenario is therefore impossible in a TRIGA reactor. The key to this inherent safety property is the special design of the fuel elements shown in Fig. 5.3. These rods contain a mixture of fissionable ^{235}U and zirconium hydride (ZrH). In ZrH, the protons reside on interstitial sites in the Zr crystal lattice, acting as local harmonic vibrational oscillators. The discrete energy levels of a quantum mechanical harmonic oscillator are schematically indicated in Fig. 5.4. Fast neutrons released after fission are first moderated by ZrH before entering the water pool. The moderation within the hydride takes place by exciting protons from the ground level to the first excited level.

As soon as the ZrH moderator heats up, the upper level becomes thermally occupied while neutron moderation ceases. Without moderation, the fission process stops. This process is referred to as "prompt negative temperature coefficient of reactivity". It is also the reason for TRIGA reactors being able to be pulsed to high power levels (1000 MW) for just a microsecond. The control rods can be suddenly removed by air pressure, causing a sharp rise in power. However, the sharp rise of hot neutron flux causes simultaneously a saturation of the upper level that inactivates further moderation, shutting off the fission process.

! Most radioisotopes for brachytherapeutical application are produced by neutron capture.

5.3 Specific radioisotopes for brachytherapy

5.3.1 Radioisotopes ^{137}Cs and ^{131}Cs

The radioisotope ^{137}Cs is a fission product of ^{235}U with a half-life of 30.17 years. The decay branches into a pure β^--emission and for the majority of isotopes (94.6%) to β^--emission followed by γ-radiation. The branching scheme is shown in Fig. 5.5 of Volume 2. Formally:

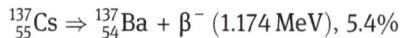

$$^{137}_{55}Cs \Rightarrow {}^{137}_{54}Ba + \beta^- \text{ (1.174 MeV), 5.4%}$$

$$^{137}_{55}Cs \Rightarrow {}^{137}_{54}Ba + \beta^- \text{ (0.512 MeV)} + \gamma \text{ (0.6617 MeV), 94.6%}$$

For BT, only the γ-emission is of interest, as the β^--radiation is stopped in the container walls. The isotope ^{137}Cs is prepared as pellets and encapsulated in stainless steel beads. The beads are then inserted in a hollow stainless steel rod through an inlet; the cross section is shown in Fig. 5.5. The number of beads filled in, controls the total activity. The specific activity remains low, which is ideal for LDR applications. Furthermore, because of the long half-life of 30 years, the source can be used over a long period of time and is well suited for long-term LDR implants.

(a)

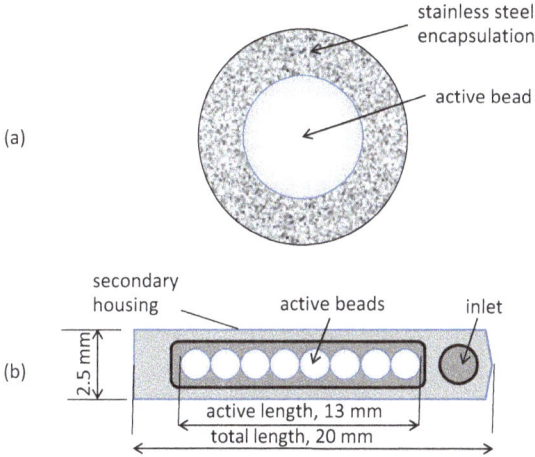

(b)

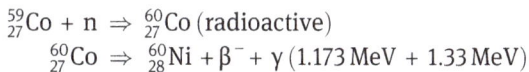

Fig. 5.5: (a) Encapsulation of ^{137}Cs isotopes in stainless steel beads; (b) cross section of a stainless steel rod accepting beads through an inlet.

There is a recent interest in the radioisotope ^{131}Cs as a soft x-ray emitter for radiotherapy. The EC decay of ^{131}Cs has a half-life of 9.68 days, emitting several characteristic x-ray K- and L-lines, the most prominent ones have energies of about 30 keV. Radioisotopes ^{131}Cs are used in seeds for prostate therapy [1, 2].

5.3.2 Radioisotope ^{60}Co

^{60}Co is a radioactive isotope produced by thermal neutron activation. First, the stable ^{59}Co isotope is irradiated by neutrons and forms ^{60}Co, which is unstable with $T_{1/2} = 5.26$ years. The decay via β^- and γ-radiation features two main γ-lines at 1.173 and 1.33 MeV. The ground-state isotope ^{60}Ni is stable. The reaction path is as follows:

$$^{59}_{27}\text{Co} + \text{n} \Rightarrow {}^{60}_{27}\text{Co (radioactive)}$$
$$^{60}_{27}\text{Co} \Rightarrow {}^{60}_{28}\text{Ni} + \beta^- + \gamma \,(1.173\,\text{MeV} + 1.33\,\text{MeV})$$

The decay scheme is shown in Fig. 5.6. ^{60}Co is a pure γ-source as the β radiation is too soft and completely absorbed in the container walls. The main use of ^{60}Co in BT is in HDR afterloading applications. However, in order to get a high specific activity, long neutron irradiation times are required as the neutron absorption cross section is rather low (37 barn). An activity of 130 GBq/mm^3 (14.6 GBq/mg) can be reached. Because Co is corrosive and because Ni is the stable decay end product, Co-containing seeds are usually Ni-plated, which serves simultaneously as a β-filter.

^{59}Co+n → ^{60}Co (radioactive)

β$^-$

1.173 MeV, γ

1.332 MeV, γ

^{60}Ni (stable)

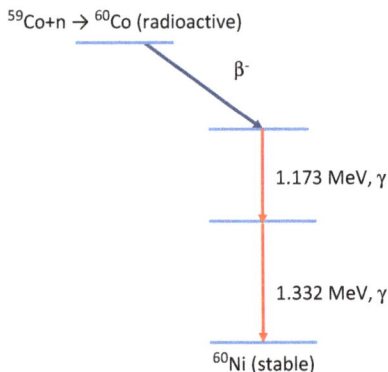

Fig. 5.6: Decay scheme of ^{60}Co after neutron activation.

5.3.3 Radioisotope ^{192}Ir

^{192}Ir is also generated by neutron activation: thermal NC converts ^{191}Ir to radioactive ^{192}Ir, which decays with a half-life of 74 days by β$^-$ (1.459 MeV) and γ-emission to the ground state of ^{192}Pt, formally:

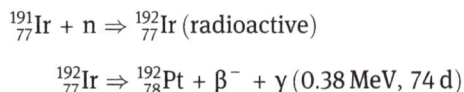

$$^{191}_{77}\text{Ir} + \text{n} \Rightarrow {}^{192}_{77}\text{Ir (radioactive)}$$

$$^{192}_{77}\text{Ir} \Rightarrow {}^{192}_{78}\text{Pt} + \beta^- + \gamma\,(0.38\,\text{MeV},\,74\,\text{d})$$

The γ-emission lines are a complex mixture of transitions, whose average energy is 0.38 MeV. ^{192}Ir radioisotopes are usually fabricated in the form of wires for HDR applications and afterloading insertion. The wire is surrounded by a Pt coating and can be cut into smaller pieces as required for the specific application. The very high specific activity up to TBq/g and the half-life of about 74 days makes this isotope one of the most favorable for HDR applications.

5.3.4 Radioisotope ^{125}I

^{125}I is produced by neutron activation of ^{124}Xe, converting it by electron capture (EC) to radioactive ^{125m}Xe. This intermediate state decays by various γ-emissions to radioactive ^{125}Xe, which, in turn, decays by another EC process with a half-life of 16.9 h to ^{125}I. The isotope ^{125}I is not stable but decays again by EC with a half-life of 59.4 days to ^{125}Te. Formally:

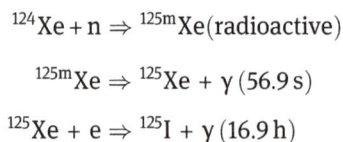

$$^{124}\text{Xe} + \text{n} \Rightarrow {}^{125m}\text{Xe (radioactive)}$$

$$^{125m}\text{Xe} \Rightarrow {}^{125}\text{Xe} + \gamma\,(56.9\,\text{s})$$

$$^{125}\text{Xe} + \text{e} \Rightarrow {}^{125}\text{I} + \gamma\,(16.9\,\text{h})$$

$$^{125}\text{I} + \text{e} \Rightarrow {}^{125}\text{Te} + \gamma \,(35.5\,\text{keV}, 59.4\,\text{d})$$

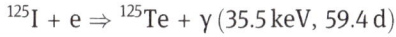

The ^{125}I transition to ^{125}Te features a soft γ-line at 35.5 keV and two characteristic x-ray lines at 27 and 31 keV. These soft γ- and x-ray lines can easily be shielded by thin lead sheets, making radiation protection effective. The radioisotope ^{125}I is mainly used for prostate implants. For this purpose, the radioactive ^{125}I-isotope is incorporated into implantable seeds such as silver rods. Figure 5.7 shows an example of a capsule.

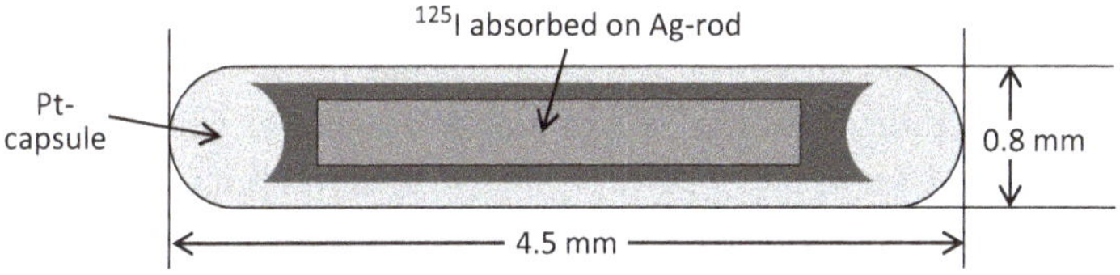

Fig. 5.7: Capsule for encasing ^{125}I radioisotopes. Several other capsule designs are in use.

5.3.5 Radioisotope ^{103}Pd

For prostate implants the radioisotope ^{103}Pd is alternatively used with the decay scheme:

$$^{103}\text{Pd} + \text{e} \Rightarrow {}^{103}\text{Rh} + \gamma \,(21\,\text{keV}, 17\,\text{d})$$

The shorter half-life and lower γ-emission energy are often considered an advantage over ^{125}I treatment. For the preparation of ^{103}Pd there exist several routes using either thermal neutron capture (NC) or charged particle activation (CPA). Some are briefly described:

1. Proton irradiation of rhodium foils via the reaction: ^{103}Rh(p,n)^{103}Pd proceeds by acceleration of protons with a cyclotron. The maximum cross section for this reaction is found at about 10 MeV. The activation is followed by a complex chemical separation process of the radionuclides from the expensive target material [3];
2. Activation can also be achieved by deuterium bombardment of an Rh target via the reaction: ^{103}Rh(d,2n)^{103}Pd;
3. Neutron activation via the capture reaction: ^{102}Pd(n,γ) ^{103}Pd;
4. Photon activation of a Pd-foil via γ,n reaction: ^{104}Pd(γ,n) ^{103}Pd;
5. Proton irradiation of a natural isotopic mixture of Ag-foils with highly energetic protons of about 66 MeV following the reaction: ^{nat}Ag(p, 2p + xn)^{103}Pd [4];
6. Neutron activation via ^{103}Ag(n,γ) ^{103}Pd.

Many more reactions are known and are still being developed. An overview of different charged particle reactions for ^{103}Pd production can be found in [5].

5.3.6 Radioisotopes ^{198}Au and ^{106}Ru

^{198}Au produced by neutron activation of ^{197}Au and decaying with a half-life of 2.7 days via emission of β^- and γ-radiation of 0.4 MeV photons was used in the past in platinum-encapsulated grains for the treatment of tumors in the head and neck region. However, this isotope is only of historical interest, mainly because of the short half-life requiring frequent replacements and because of a low specific activity. It has been replaced by ^{103}Pd and ^{125}I isotopes.

^{106}Ru is a fission product and decays by β^- and γ-emission to ^{106}Rh with a half-life of 1.02 years. The β^- emission energy of this transition has rather low energy of 39.4 keV. However, the daughter nucleus ^{106}Rh is also a β^--emitter with a much higher energy of 3.54 MeV and a half-life of 30 s. The active material is mounted into a silver sheet, forming the plaque's surface for exposure on one side and extra radiation shielding on the opposite side. The shape, form, and size can be adapted to specific applications on the skin. ^{106}Ru/^{106}Rh sources are no longer in use and replaced by ^{90}Sr/^{90}Y – β^- emitter.

5.3.7 Radioisotope ^{90}Sr

^{90}Sr is also a fission product, which decays by β^- and γ-radiation to ^{90}Y with a half-life of 28.8 years. Similar to ^{106}Ru, the β^- energy (0.546 MeV) is too low to be useful. However, the daughter nucleus ^{90}Y also decays by β^- and γ-emission into stable ^{90}Zr, where the β^- radiation has a maximum energy of 2.28 MeV. Formally:

$$^{90}\text{Sr} \Rightarrow {}^{90}\text{Y} + \beta^- \, (\text{E}_{max} = 0.546 \, \text{MeV}, 28.8 \, \text{y})$$

$$^{90}\text{Y} \Rightarrow {}^{90}\text{Zr} + \beta^- \, (\text{E}_{max} = 2.28 \, \text{MeV}, 2.66 \, \text{d})$$

Like for ^{106}Ru, the radioactive ^{90}Sr/^{90}Y material is embedded in silver sheets forming plaques that can be used for cancer treatment via β^--irradiation close to the body surface.

A few more isotopes have been used for BT in the past but are now outphased. Others are still being developed (^{169}Yb, ^{170}Tm). Research is ongoing to find new ways for convenient radioisotope production and isotope separation from targets, which remains an active area in radiochemistry.

5.4 Procedures

5.4.1 Interstitial and contact brachytherapy

There are two main procedures of brachytherapy treatment in terms of placement of the radioactive source: *interstitial* and *contact*.

For interstitial BT, the radioactive source is implanted directly into the carcinoma either temporarily for the time of irradiation or permanently. Typical organs for interstitial placement are breast (temporarily) and prostate (permanently).

For contact brachytherapy, the radioactive source is either placed close to the target tissue through a body cavity such as the vagina, uterus, cervix, or oral cavity. Alternatively, the sources may be applied externally on the skin's surface in case of skin cancer.

Brachytherapy is also distinguished according to the total dose (Gy) and dose rate (Gy/h) administered to the patient:
- LDR: Low dose rate of 0.4 up to 2 Gy/h
- MDR: Medium dose rate between 2 and 12 Gy/h
- HDR: High dose rate above 12 Gy/h
- PDR: Pulsed dose rate, where each pulse can range from LDR to HDR, typically 0.4–1 Gy/h.

An applicator is often used to place radioisotopes interstitially or in contact. An applicator is a hollow tube that is inserted into the body and filled with capsules, seeds, or wires. In afterloading, the applicator is first inserted, and when attached, the radioactive sources are loaded either by hand (manual reloading) or by machine (remote automatic reloading). Hot loading means that the applicator is preloaded and contains radioactive sources when inserted into the patient.

For LDR and PDR applications, the radioactive source typically remains in its implanted location for 24 h before removal. In the case of HDR, the treatment time is a few minutes only but may be repeated several times during a period of 1–2 weeks. PDR is applied for a minute and repeated several times within 24 h. In addition, LDR seeds may be implanted into the carcinoma remaining permanently in the body.

LDR sources can be administered by the clinical personnel using hands or simple tools. However, HDR requires the use of an afterloading machine. For both types of applications, contact or interstitial, first a hollow thin-walled metal or plastic tube (applicator) is inserted into the cancerous region. After surgical placement of the tube or catheter, the hollow tube is connected to an afterloading machine. The afterloading machine contains radioactive wires, mainly ^{192}Ir, which can be slipped into the hollow tube and quickly removed again. The insertion of wires is predefined and computer-controlled using laser scanners. A typical afterloading machine is shown in Fig. 5.8.

Three selected examples (cervical, prostate, and breast cancer) are discussed in the following, which may help to illustrate brachytherapy procedures. Table 5.2 gives an overview on methods and procedures used in various brachytherapeutic applications.

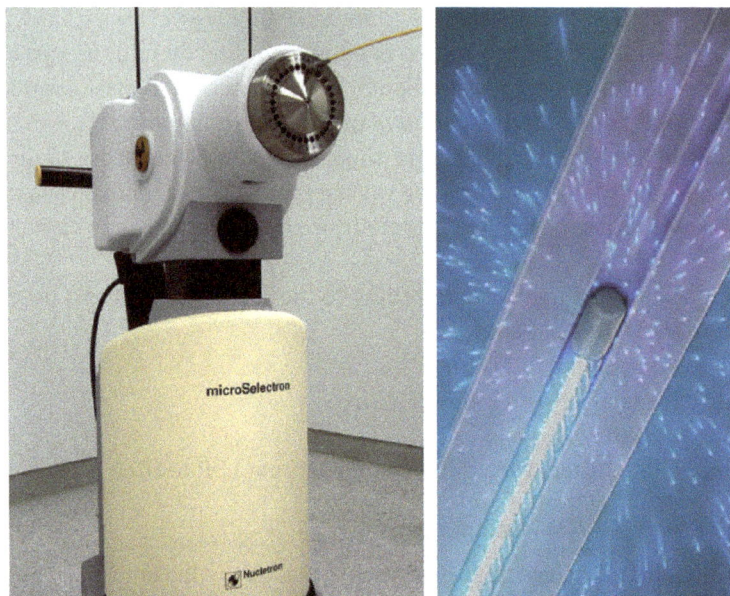

Fig. 5.8: Left: Remote afterloading platform for placement of high dose rate wires into predefined tubes. Right: Radioactive source guided in a tube. (Courtesy Elekta AB, www.elekta.com).

Tab. 5.2: Overview on methods and procedures in brachytherapy.

	LDR	HDR
Methods	Seeds	Applicator, afterloading
Application	Prostate	Cervical, breast
Treatment time	Permanent	Temporary, few minutes
Most frequent isotopes	125I, 103Pd, 131Cs	192Ir

! In brachytherapy we distinguish between interstitial and contact application. The interstitial placement may be temporary or permanent. Furthermore, we distinguish between low dose and high dose application.

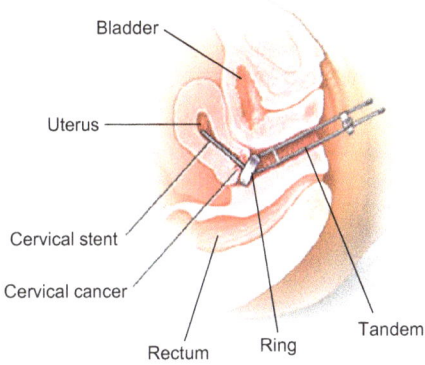

Bladder

Uterus

Cervical stent

Cervical cancer

Rectum   Ring   Tandem

5.4.2 Cervical cancer

Cervical cancer is a typical example where BT has been successfully applied. Either LDR or HDR BT may be performed. But HDR brachytherapy is preferred since it does not require hospitalization and limits radiation exposure to hospital staff.

For cervical cancer, HDR brachytherapy is delivered using intracavitary applicators in direct contact with the cancerous tissue. The applicator consists of a tandem, ring, and a stent (see Fig. 5.9). First a cervical stent is placed into the cervical portion of the uterus for fixation. Following cervical stent placement, the ring and tandem are inserted. The tandem is placed into the uterus where the cervical stent is located. Then, the ring is placed next to the cervix where the tumor is located. Following placement, the applicator is connected to the afterloading machine where a radioactive source (usually ^{192}Ir) is located. When the treatment is initiated, the radioactive wire from an afterloading machine will quickly travel into the tandem and the ring where it will deliver the prescribed dose of radiation to the tumor. Once the treatment is completed, the applicators can be removed quickly. This treatment can be combined with other oncological therapies, such as chemotherapy and EBRT.

5.4.3 Prostate cancer

Brachytherapy is widely accepted as a therapy for early-stage *prostate cancer* as long as the cancerous tissue is confined to the prostate gland. The permanent placement of short-lived radionuclide sources emitting low-energy photons is often used as the primary treatment. However, attempts have also been made to use fractionated or single session HDR brachytherapy in combination with EBRT.

Radioactive seeds are usually implanted permanently into the prostate to provide a local and homogeneous radiation distribution whenever prostate cancer is treated via BT. The LDR seeds decay over weeks to months and remain in place even after complete decay. Seeds are implanted according to a well-defined treatment plan and placement is continuously monitored by ultrasound imaging through the rectum or by CT scans. A template helps to find the correct seed position to be inserted interstitially in the area between the testicles and the anus as shown in Fig. 5.10. The method is known as transperineal interstitial permanent prostate brachytherapy. It is considered one of the most effective and efficient methods for managing prostate cancer [6].

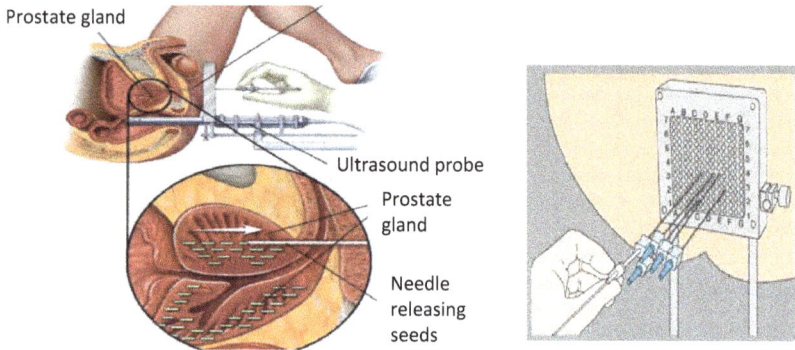

Fig. 5.10: Left: placement of seeds in prostate gland controlled by ultrasound imaging. Right: matrix for locating seed positions. (From https://en.wikipedia.org/wiki/Prostate_brachytherapy, © creative commons).

5.4.4 Breast cancer

The last example shows *breast cancer* treatment using interstitial flexible plastic tubes across the extension of the breast, Fig. 5.11 [7]. After placement of the tubes, ^{192}Ir wires can be inserted and halted at predefined points for a specified time before moving to the next position. Repeating this procedure several times guarantees a homogenous distribution of radiation across the breast according to a predetermined radiation treatment plan. The main benefit of breast brachytherapy compared to EBRT is a high dose of radiation that can be precisely applied to the tumor while sparing radiation to healthy breast tissues and underlying structures such as the ribs and lungs.

Other less common applications of BT are the irradiation of intraocular melanoma on the rear side of the eyeball using an eye plaque loaded with ^{125}I seeds as an alternative to PRT (see Section 13.3.6).

! Cervical, prostate, and breast cancer are the most frequent application of brachytherapy.

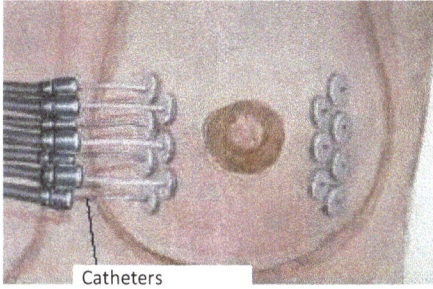

Catheters

Fig. 5.11: Treatment of breast cancer with BT by first inserting flexible plastic tubes guiding radioactive wires from an afterloading machine. (Reproduced from Reference [7], © creative commons).

5.4.5 New developments

Several research groups are presently developing intensity-modulated BT (IMBT) methods for delivering asymmetric dose distributions to actual tumor shapes. Two types of IMBT can be distinguished: static and dynamic. For LDR implants, static shielding of organs at risk (OAR) is used, such as in the case of ocular melanoma. For dynamic IMBT, rotatable shieldings limit the dose and direct the radiation in predetermined directions [8].

Appropriate imaging techniques always guide brachytherapeutic treatments. For prostate radiotherapy, ultrasound imaging is often used; in other applications, C-arm fluoroscopes. In recent years, 2D imaging methods have been replaced by 3D imaging via CT and MRI for tumor control and definition of the clinical target volume. Improved tumor volume delineation helps protect the OAR. Furthermore, MRI has increasingly been used for image-guided radiotherapy (IGRT) because of the better soft-tissue contrast compared to CT. Hence, CT- and MRI-based brachytherapy refine the local control and reduces serious side effects such as toxicity to the OAR compared with earlier 2D image-based BT. The advanced IGRT and IMBT methods promise to maximize the target coverage and minimize the doses to OARs. An overview of recent developments is given in [8].

5.5 Dosimetry

The application of LDR or HDR requires a detailed treatment plan to determine the total dose to be administered and the dose rate for each fraction. For this purpose, treatment planning systems (TPS) are commercially available and can be refined by the radiation therapist for special conditions.

We have already discussed dosimetry aspects in Chapters 7 of Volume 2. However, BT requires special consideration because of the direct contact of radioactive sources with carcinoma. This is a dosimetry topic with internationally agreed recommendations by the American Association of Physicists in Medicine and the European Society for Radiotherapy and Oncology. The last updated report was published in [9] and references to earlier reports can be found in this publication. The purpose is to provide dose-estimation methods and dosimetric parameters for specific photon-emitting sources used for BT with photon energies exceeding 50 keV (mainly ^{137}Cs, ^{60}Co, ^{192}Ir). While dosimetry of LDR sources is well established, new developments of HDR sources, PDR applications, afterloading techniques, and newly developed isotopes require frequent updates of the recommended procedures.

In the simplest case of dose simulation, the tissue is replaced by an infinitely extended spherical phantom with the density of water. Radioactive sources are homogeneously distributed inside the phantom and treated as point sources (Fig. 5.12(a)). Absorption of radiation is only considered by the phantom, not by the other sources. However, the scattering of radiation by the other sources is taken into account. The dose calculation is performed according to the superposition principle. Each source radiates with a dose that drops off according to the $1/r^2$ law. The dose rate $\dot{D}(r,\theta)$ at the point $P(r,\theta)$ at distance r and angle θ of a single-elongated source of length L in comparison to the dose at the standard position $P(r_0,\theta_0)$ and at a right angle ($\theta_0 = 90°$) to the elongated source is given, according to the recommendations of the AAPP-RTTG,[3] by the expression (Fig. 5.12(b)) [10]:

$$\dot{D}(r,\theta) = S_K \Lambda \, \frac{G(r,\theta)}{G(r_0,\theta_0)} g(r) F(r,\theta) \tag{5.1}$$

This lengthy expression is a factorization of several contributions to the dose rate: S_K is the air kerma strength of the source, Λ is the dose rate constant in water, $G(r,\theta)$ is a geometry function considering the actual shape of the source as compared to a

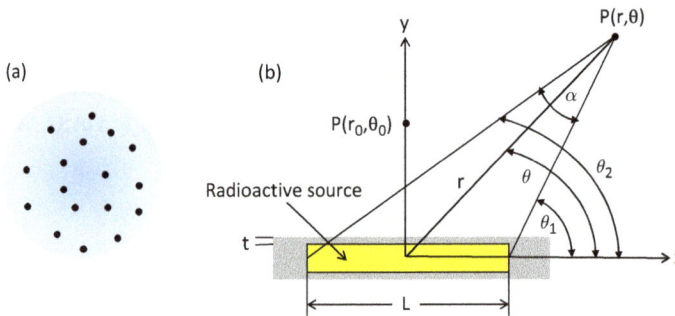

Fig. 5.12: (a) Phantom for calculation of dose of implanted radioactive sources; (b) geometry for calculating dose and dose rate at particular points in the phantom.

---

3 American Association of Physicists in Medicine – Radiation Therapy Committee Task Group.

point source, $g(r)$ is the radial dose function, and $F(r, \theta)$ is the anisotropy function. All these factors are explained below in more detail.

The air kerma strength S_K is derived from the reference kerma rate in a vacuum $\dot{K}_{vac}$. As explained in Section 7.3 of Volume II, kerma stands for *the kinetic energy release in matter* and is defined as the energy fluence of a photon beam (x-ray or γ-radiation) that is converted into kinetic energy of charged particles via the photo effect and/or Compton effect. Accordingly, $\dot{K}_{vac}$ is defined as the dose rate in vacuum attributed to all photons of energy larger than a cutoff energy E_{min} at a distance d (usually 1 m) from the source center. The distance d can be any distance that is sufficiently large in comparison to the source size, such that the source can be treated as a point source. The term "in vacuum" refers to the fact that an actual measurement in the air should be corrected for photon attenuation and scattering in air as well as from encapsulation material and for photon scattering from nearby objects like walls, furniture, and ceiling. The cutoff energy E_{min} (typically 5 keV) excludes all low-energy photons that increase the overall background noise without contributing to the dose in tissue at distances exceeding 1 mm in tissue. The kerma rate has the units: $[\dot{K}_{vac}]$ = Gy s^{-1} at 1 m. More convenient units in use are mGy h^{-1} and mGy s^{-1} always at a reference distance of 1 m in vacuum.

The kerma strength S_K is defined as the kerma rate in vacuum at a reference distance d_{ref} times the reference distance squared [11]:

$$S_K = (\dot{K}_{vac}) \times d_{ref}^2 \qquad (5.2)$$

The SI unit of S_K is $[S_K]$ = Gy m^2 h^{-1}.

The dose rate constant Λ is defined as the dose rate in water at a distance of 1 cm from the source on the transverse plane ($\theta_0 = 90°$) per unit kerma strength:

$$\Lambda = \frac{\dot{D}(r_0, \theta_0)}{S_K} \qquad (5.3)$$

Hence, the dose rate constant Λ is a conversion factor for converting dose rates in vacuum (air) to dose rates in water. The SI unit of the dose rate constant Λ is $[\Lambda]$ = Gyh^{-1}/(Gyh^{-1}m^2). The dose rate constant, Λ, depends on the type of radioisotope, its encapsulation, and overall construction. Dose rate constant values for seeds from different manufacturer are listed in [11]. They vary between 0.7 and 1.04 in units of cGy h^{-1} U^{-1}. The units of U are μGy-m^2/h.

The geometry function $G(r, \theta)$ takes into account the variation of the relative dose from a spatially extended source as compared to a point source. For a point source:

$$G(r, \theta) = 1/r^2 \qquad (5.4)$$

> ! Because of the three dimensional arrangement of radioactive sources (capsules), standard do-
> simetry is not applicable and must be replaced by a dosimetry of point sources that determines
> the local dose by the superposition principle of all individual sources in the tumor volume.

for a cylindrically shaped line source, $G(r, \theta)$ approximates to:

$$G(r, \theta) = \frac{\alpha}{Lr \sin \theta}, \text{ for } \theta \neq 0 \tag{5.5}$$

L and α are defined in Fig. 5.12(b).

The radial dose function $g(r)$ accounts for effects of attenuation (absorption and scattering) in water on the transverse plane of the source ($\theta_0 = 90°$), excluding the normal $1/r^2$ falloff, which is already included in the geometry function $G(r, \theta)$.

The anisotropy function $F(r, \theta)$ collects all anisotropic dose distributions outside of the source, including absorption in the source encapsulation of thickness t. It is defined as unity on the transverse plane but decreases outside the transverse plane.

In the simplest form neglecting anisotropy terms the dose rate is:

$$\dot{D}(r, \theta) = S_K \Lambda \frac{g(r)}{r^2} \tag{5.6}$$

The manufacturer of seeds provide data sheets on the dose rate and isodose distribution. The first task of the local radiologist is to measure and confirm the data sheet. The confirmed and updated values can then be used for calculating dose rate distributions in phantoms and eventually in the body. For this task, commercial software is available. A simple converter from γ-activity to dose rate and vice versa is available for point sources with and without shielding at [12]. The estimate uses the expression for the kerma in eq. (7.19) of volume 2

$$K = \Psi_{ph} \left(\frac{\mu_{tr}}{\varrho} \right) \tag{5.7}$$

where μ_{tr}/ϱ is the mass-energy transfer coefficient and $\Psi_{ph} = \Phi_{ph} \times E_{ph}$ is the photon energy fluence. For the photon energy range 0.02–2 MeV, which is the range used in brachytherapy, the mass-energy transfer coefficient in air is about $\mu_{tr}/\varrho \cong 25 \text{ cm}^2/\text{kg}$. Then we can estimate the dose at distance r from the source and during the accumulated time t as:

$$D(E_{ph}, r) = \frac{A E_{ph} t}{4\pi r^2} \left(\frac{\mu_{tr}}{\varrho} \right), \tag{5.8}$$

where A is the source activity. An example is calculated in exercise E5.3.

Model calculations based on the dose rate expression in eq. (5.1) suffer from several shortcomings. Variations of composition, finiteness of tissues and organs, neglect of radiation shielding usually applied in clinical practice, etc., are not considered. Nevertheless, it is a good starting point, which can be refined for more realistic circumstances in advanced TPS. A number of specific TPS computer programs for brachytherapy applications are commercially available to help predict desired dose rates and total prescribed dose to avoid common treatment mistakes.

5.6 Summary

S5.1 Brachytherapy is an irradiation modality of cancerous tissue which is administered in direct contact of radioisotopes to the tumor.

S5.2 The advantage of BT over external beam radiation therapy is the comparatively short treatment time and a local application that conserves healthy tissue.

S5.3 In BT, mostly only the γ-radiation from radioisotopes is effective; the concomitant β-radiation is absorbed in surrounding container walls.

S5.4 Brachytherapy can be interstitial into the tumor or in outside contact with the tumor.

S5.5 The most common radioisotopes for BT are ^{137}Cs, ^{192}Ir, and ^{103}Pd.

S5.6 Most radioisotopes for BT are generated by neutron capture, but ^{103}Pd is produced by a number of other procedures, including proton and γ-irradiation.

S5.7 The most common applications of BT are in treating cervical cancer, prostate cancer, and breast cancer.

S5.8 High dose rates require the use of applicators in combination with afterloading machines.

Questions

Q5.1   What kind of radiation therapy is BT?
Q5.2   What is the antonym of BT?
Q5.3   What are the main applications of BT?
Q5.4   How is BT applied?
Q5.5   What are the main advantages of BT compared to external beam therapies?
Q5.6   Which type of radiation is mainly active in BT?
Q5.7   Which are the main isotopes used for BT?
Q5.8   How are the isotopes produced?
Q5.9   How are the isotopes applied?
Q5.10 What are the two main procedures by brachytherapy treatment?
Q5.11 What is typical low dose rate (LDR) application, and what is a high dose rate (HDR) application in terms of Gray per hour?
Q5.12 The dosimetry for BT requires special consideration. What is different compared to standard dosimetry?

Attained competence checker

	+	0	–
I can distinguish between teletherapies and brachytherapies			
I know what LDR and HDR means			
I know which isotopes are mainly used for brachytherapy			
I know under what circumstances applicators are used			
I can name four main applications of brachytherapy for tumor treatment			
I am aware of the most frequently used isotopes for brachytherapy			
I know how the isotopes used for brachytherapy are produced			
I know which type of radiation is dose relevant in brachytherapy			
I recognize that the radiology of brachytherapies can be rather complex			

Exercises

E5.1 **Co-activation**: How long does it take to irradiate ^{59}Co with neutrons according to the reaction ^{59}Co + n $\rightarrow$ ^{60}Co in order to reach 75% of the saturation value. The half-life of ^{60}Co is $T_{1/2}$ = 5.26 years.

E5.2 **Co activity:**
  a. What is the activity of 1 μg ^{60}Co? The half-life of ^{60}Co is $T_{1/2}$ = 5.26 years.
  b. What is the specific activity of ^{60}Co?

E5.3 **Dose estimation**: Determine the dose of a ^{137}Cs source, assuming an activity of 100 MBq after exposure time of 1 h at a distance of 100 cm.

References

[1]   Murphy MK, Piper RK, Greenwood LR, Mitch MG, Lamperti PJ, Seltzer SM, Bales MJ, Phillips MH. Evaluation of the new cesium-131 seed for use in low-energy x-ray brachytherapy. Med Phys. 2004; 31: 1529–1538.

[2]   Hubley E, Trager M, Bar-Ad V, Luginbuhl A, Doyle L. A nomogram to determine required seed air kerma strength in planar 131Cesium permanent seed implant brachytherapy. J Contemp Brachytherapy. 2019; 11: 91–98.

[3]   Sonck M, Fenyvesi A, Daraban L. Study on production of ^{103}Pd and characterisation of possible contaminants in the proton irradiation of ^{103}Rh up to 28 Mev. Nucl Instrum Methods Phys Res B. 2000; 170: 281–292.

[4]   Mausner LF, Kolsky KL, Awasthi V, Srivastava SC. Production of 103Pd by proton irradiation of silver. J Labelled Comp Radiopharm. 2001; 44: S775–S777.

[5]   Tarkanyi F, Hermanne A, Kiraly B, Takacs S, Ditroi F, Csikai J, Fenyvesi A, Uddin MS, Hagiwara M, Baba M, Ido T, Shubin YN, Ignatyuk AV. New cross-sections for production of ^{103}Pd; review of charged particle production routes. Appl Radiat Isot. 2009; 67: 1574–1581.

[6]   Skowronek J. Low-dose-rate or high-dose-rate brachytherapy in treatment of prostate cancer – Between options. J Contemp Brachytherapy. 2013; 5: 33–41.

[7] Njeh CF, Saunders MW, Langton CM. Accelerated Partial Breast Irradiation (APBI): A review of available techniques. Radiat Oncol. 2015; 5: 90.

[8] Lim YK, Kim D. Brachytherapy: A comprehensive review. Prog Med Phys. 2021; 32: 25–39.

[9] Perez-Calatayud J, Ballester F, Das RK, DeWerd LA, Ibbott GS, Meigooni AS, Ouhib Z, Rivard MJ, Sloboda RS, Williamson JF. Dose calculation for photon-emitting brachytherapy sources with average energy higher than 50 keV: Report of the AAPM and ESTRO. Med Phys. 2012; 39: 2904–2929.

[10] Nath R, Anderson LL, Luxton G, Weaver KA, Williamson JF, Meigooni AS. Dosimetry of interstitial brachytherapy sources: Recommendations of the AAPM Radiation Therapy Committee Task Group No. 43. American Association of Physicists in Medicine. Med Phys. 1995; 22: 209–234.

[11] Rivard MJ, Coursey BM, DeWerd LA, Hanson WF, Saiful Huq M, Ibbot GS, Mitch MG, Nath R, Williamson JF. Update of AAPM Task Group No. 43 Report: A revised AAPM protocol for brachytherapy dose calculations. Med Phys. 2004; 31: 633–674.

[12] http://www.radprocalculator.com/Gamma.aspx

Further reading

Hoskin P, Coyle C, eds. Radiotherapy in practice – Brachytherapy. Oxford, New York, Athens: Oxford University Press; 2011.

Podgorsak EB. Radiation physics for medical physicists. 2nd edition. Berlin, Heidelberg, New York: Springer Verlag; 2014.

Khan FM, Gibbons JP. The physics of radiation therapy. 5th edition. Philadelphia, Baltimore, New York, London: Wolters Kluwer; 2014.

Chao AW, Chou W, eds. Reviews of accelerator science and technology. Vol. 2. Medical applications of accelerators. Singapore: World Scientific Publishing; 2009. p. 133–156.

Baltas D, Sakelliou L, Zamboglou N. The physics of modern Brachytherapy for oncology. Boca Raton, London, New York: CRC Press, Taylor & Francis; 2007.

Mayles P, Nahum A, Rosenwald JC, eds. Handbook of radiotherapy physics. Theory and practise. Taylor & Francis. Boca Raton, London, New York: CRC Press; 2007.

Useful websites

Neutron capture activation calculator: https://www.ncnr.nist.gov/resources/activation/
Overview over all isotopes: www.periodictable.com/Isotopes/039.90/index.p.full.dm.html
Podgorsak EB, technical editor. Radiation oncology physics: A handbook for teachers and students. Available from: http://www-pub.iaea.org/mtcd/publications/pdf/pub1196_web.pdf

Part E: **Diagnostics and therapeutics beyond radiology**

6 Laser applications in medicine

Laser physical parameters

Nd:YAG crystal laser	$\lambda = 1.064$ μm, $L_{water} = 6$ mm
CO_2 gas laser	$\lambda = 10$ μm, $L_{water} = 0.1$ mm
Q-switch	electro-optic, acousto-optic
Femtosecond pulse generation	mode coupling
fs Nd:YAG laser peak power density	10^{12} W/cm^2
Laser interaction with tissue	photothermal, photochemical, ablation, explosive evaporation

6.1 Introduction

Lasers are so ubiquitous in our daily life that it is easy to forget the sophisticated technology behind their implementation. *LASER* is an acronym for *light amplification by stimulated emission of radiation*. Groundbreaking work that led to the invention of lasers was done in the 1950s of the last century using microwaves called MASER. Maser and laser are based on the principle of population inversion, an idea that Albert Einstein[1] put forward in one of his publications [1]. Townes,[2] Basov,[3] and Prokhorov[4] shared the Nobel Prize in 1964 in Physics for their fundamental work, which led to the technical realization of laser devices.

When we compare light emitted from an old-fashioned light bulb with a tungsten wire inside and light from a laser pointer, we notice three fundamental differences, schematically illustrated in Fig. 6.1. The light bulb is a black-body radiator, emitting polychromatic, incoherent, and multidirectional electromagnetic radiation. Our photoreceptors in the retina detect part of the emitted light spectrum. Laser light, in contrast, has a high monochromaticity, and the light propagation is unidirectional. Furthermore, laser light features phase coherence. Light emitted from black body radiators has no coherence at all. Light bulbs have been abolished in recent years, but as a demonstrator for black body radiation, they still serve a useful purpose.

---

1 Albert Einstein (1879–1955), German-American theoretical physicist, Nobel laureate in physics 1921.
2 Charles Hard Townes (1915 –2015), US American physicist and Nobel laureate in physics 1964.
3 Nikolay G. Basov (1922–2001), Russian physicist and Nobel laureate in physics 1964.
4 Alexander M. Prokhorov (1916 –2002), Russian physicist and Nobel laureate in physics 1964.

https://doi.org/10.1515/9783111168739-006

Light bulb   Green laser

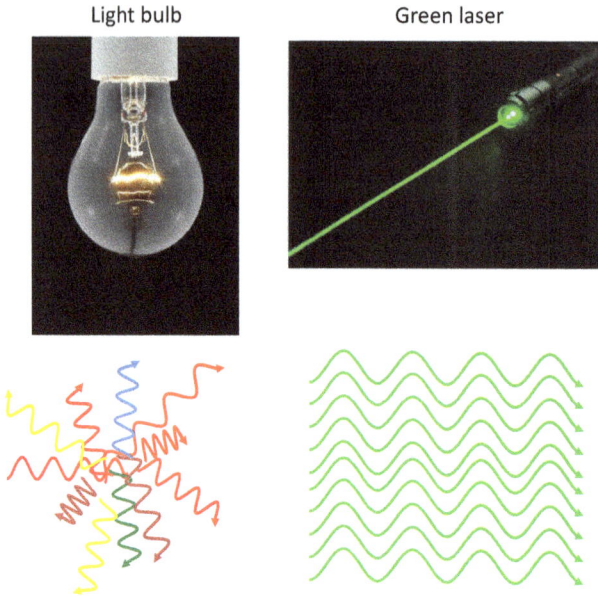

Fig. 6.1: Light from a light bulb and from a laser pointer. Light from the light bulb is polychromatic, propagates in all directions, and lacks phase coherence. In contrast, light from a laser is monochromatic, unidirectional, and features phase coherence. (Source for light bulb: Frankfurter Allgemeine, 18.03.2009).

All laser sources consist of three main components: (1) an active medium; (2) an optic pump; and (3) a resonator. The optical pump populates an unoccupied upper atomic energy level in an active medium with electrons. The resonator warrants that stimulated transitions of electrons from the excited state to the ground state dominate over spontaneous transitions. The stimulated emission of radiation amplifies the intensity in the resonator, which is then released from the resonator cavity. Lasers are available in a huge variety of specifications concerning wavelength, power, tunability, and pulsation. Because of this wide variety, we have to confine ourselves to some basic features, relevant to medical applications. Five main areas of laser application in medicine can be identified:

1. Dermatology
2. Plastic surgery
3. Oncology
4. Ophthalmology
5. Dentistry

Between some of these areas, there is a certain amount of overlap. For instance, dermatology overlaps with plastic surgery and oncology.

Laser light operating from the far infrared to the ultraviolet spectral range is characterized as nonionizing radiation. Nevertheless, potential health hazards are considerable. This is because the intensity of laser light is focused on a tiny spot. An example may illustrate the potential danger. A 100 W light bulb may have a quantum efficiency of at most 10%, i.e., only 10% of the electric energy is converted in light intensity. The emitted light is distributed over a solid angle of 4π. At 1 m distance from the light bulb, the intensity is: $I = P/A = 10\ \text{W}/4\pi r^2 = 10\ \text{W}/4\pi\ 1\text{m}^2 = 0.8\ \text{W}/\text{m}^2$. In case of a laser with the same power but focused on an area of 1 mm radius, the intensity is $10\ \text{W}/\pi \times 10^{-6}\ \text{m}^2 = 3 \times 10^6\ \text{W}/\text{m}^2$, which is a factor ~$10^6$ higher than for the ordinary light bulb. At a 1 m distance, the intensity is still the same since laser light is highly collimated. Even for a 1 mW Laser (pointer) the intensity is still 380 W/m², which is 400 times higher than for the quoted light bulb.

Laser in the visible spectral range is categorized in classes according to their power output and potential danger:

- Class 1 is inherently safe when enclosed in light–tight equipment like CD players or spectrometers;
- Class 1M is safe, except when passed through magnifying optics;
- Class 2 lasers have power up to 1 mW. If the eye is hit, the natural blink reflex (closing of eye lids within 0.25 s) prevents damage.
- Class 2M is also safe, if not viewed through optical instruments;
- Class 3R (formerly 3a) includes lasers with power up to 5 mW, carrying a small risk of eye damage within blink time reflex. Staring into such a beam can likely damage a spot on the retina because of the focusing properties of cornea and lens;
- Class 3B is likely to cause immediate damage of the retina upon exposure;
- Class 4 laser has the power to burn skin and even scattered light may cause skin and eye damage. Medical, industrial, and scientific lasers are in this class.

Lasers operating in the nonvisible infrared and ultraviolet region are particularly dangerous as the eye lid reflex is not operable. On the other hand, laser light in this spectral range is highly absorbing and does not reach the retina, but can nevertheless cause severe damage to cornea and skin.

The remainder of this chapter is organized as follows. First we introduce the basic physical concept of a laser and give an overview on their specifications. Then we discuss the interaction of laser light with tissue, which results in different effects depending on intensity, wavelength, and pulse duration. These different effects lead directly to various areas of medical applications discussed in the last part. Understanding the laser–tissue interaction is hence particularly important.

Lasers are sources of strongly collimated, monochromatic, and phase coherent electromagnetic radiation.

6.2 Laser basics

6.2.1 Two-level system

We start with a two-level energy system as depicted in Fig. 6.2(a) characterized by an energy difference $\Delta E = E_2 - E_1 = hf$. Furthermore, we assume that the lower energy level E_1 is occupied by one electron, whereas the upper level E_2 is unoccupied. If a photon with the energy $E_{photon} = hf$ hits this two-level system, with a certain probability the photon will transfer all its energy to the electron in the ground level and lifts it up to the excited upper state. This transition may be called resonant absorption or stimulated absorption. It occurs with a certain quantum efficiency, which depends on the nature of the two-level system it is embedded in. We have already seen such a case in Chapter 11 of Vol. 1 for the cis–trans isomerization of the retinal in our eyes. A transition with a quantum efficiency of about 20% generates such a transition, which initiates vision. If dipole allowed, the lifetime of the electron in the upper state is likely to be very short in the order of a few femto- to picoseconds. Then the electron recombines with the hole in the ground state by emitting a photon of the frequency corresponding to the energy difference. The photon is emitted is any arbitrary direction within a solid angle of 4π, as indicated in Fig. 6.2(b). This transition is not what we aim for. We aim for a stimulated emission, which amplifies the photon field in the incident photon direction to yield an amplified and directional photon output, as sketched in panel (c) of Fig. 6.2. How can this be achieved? As this simple graph suggests, it requires a condition where the electron is already in the excited state before the incident photon arrives. This condition is referred to as *population inversion*. Then the incident photon and the photon from stimulated emission can overlap coherently, intensifying the photon field.

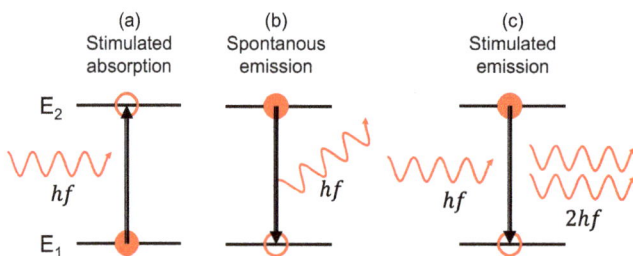

Fig. 6.2: Resonant photon absorption (a) and spontaneous emission (b) in a two-level system. Stimulated emission requires that the electron is in the excited state before the photon arrives, as indicated in panel (c).

Now we take a statistical approach, considering a total number of N identical atoms, each one featuring a two-level energy system in a photon field of density ρ. There may be N_1 atoms in the ground state and N_2 atoms in the excited state. To determine the population of the levels and, in particular, the population difference for any given temperature, we need to determine the rate of transitions from 1 to 2 and vice versa. Level 1 is populated by spontaneous and stimulated emission from 2 to 1 minus stimulated absorption from 1 to 2. Vice versa, level 2 is populated by stimulated absorption from 1 to 2, minus spontaneous and stimulated emission from 2 to 1. The respective rate equations are therefore:

$$\frac{dN_1}{dt} = (B_{21}N_2\rho + AN_2) - B_{12}N_1\rho \tag{6.1}$$

$$\frac{dN_2}{dt} = (B_{12}N_1\rho) - (B_{21}N_2\rho + AN_2) \tag{6.2}$$

Here B_{12}, B_{21}, and A are the Einstein coefficients for stimulated emission, stimulated absorption, and spontaneous emission, respectively. Assuming that

$$B_{21} = B_{12} = B, \tag{6.3}$$

we notice that in equilibrium:

$$\frac{dN_1}{dt} = -\frac{dN_2}{dt}. \tag{6.4}$$

In fact, we are interested in the population difference $\Delta N = N_1 - N_2$ since population inversion requires $\Delta N < 0$. ΔN can be evaluated from the rate equations:

$$\frac{d(\Delta N)}{dt} = 0 = AN - A(\Delta N) - 2B\rho(\Delta N) \tag{6.5}$$

with $N = N_1 + N_2$, yielding for the population difference:

$$\Delta N = \frac{N}{1 + 2\frac{B}{A}\rho} = \frac{N}{1 + 2\frac{\rho}{C}} \tag{6.6}$$

where $C = A/B$. Even if the spontaneous emission could be completely suppressed, the difference ΔN remains positive, meaning that population inversion in a two-level system is intrinsically not possible. In fact, the limit $C \to 0$ implies $\Delta N = 0$, i.e., equal population of the upper and lower level.

6.2.2 Three-level system

In a three-level system, shown in Fig. 6.3, we assume that electrons in the ground energy level E_1 can be excited to an upper level E_3 either by electron collision in a plasma (gas laser) or by optical pumping (solid state laser). The excited state has a short lifetime and decays fast, but not to the ground state. Instead, the transition takes place to an intermediate energy level E_2. The lifetime of E_2 is assumed to be much longer than the lifetime of E_3 since the transition $E_2 \to E_1$ is often dipole-forbidden. The rate equations for the three-level system can then be simplified because stimulated absorption from 1 to level 2 and spontaneous emission from level 2 to 1 via the photon field of the optical pump are missing. Optical pumping goes immediately and resonantly into level 3. Thus we have the rate equations:

$$\frac{dN_1}{dt} = AN_2 - BN_1\rho \tag{6.7}$$

$$\frac{dN_2}{dt} = BN_1\rho - AN_2 \tag{6.8}$$

Again in equilibrium:

$$\frac{dN_1}{dt} = -\frac{dN_2}{dt} \quad \text{and} \quad \frac{d\Delta N}{dt} = 0. \tag{6.9}$$

(a)
Optical
pumping

(b)
Population
inversion

(c)
Stimulated
emission

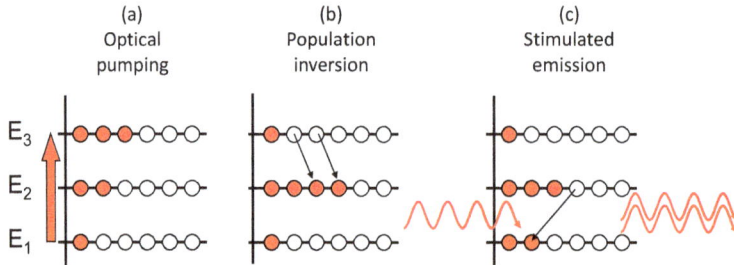

Fig. 6.3: (a) Energy-level E_3 is populated by optical pumping. (b) E_3 is depleted fast compared to the lifetime of energy-level E_2. (c) With this population inversion, stimulated emission from energy level 2 to 1 with corresponding amplification of the photon field becomes possible.

For the occupational difference we then obtain:

$$\Delta N = N \frac{1 - \frac{\rho}{C}}{1 + \frac{\rho}{C}} \tag{6.10}$$

The last equation can be negative for $\rho > C = A/B$, which is the case for a very high photon density in the resonator space. Hence population inversion becomes indeed feasible. It should be pointed out that these rate equations are a simplified version of real laser activity since coupling to lattice vibrations (phonons) and cooperative effects is neglected. Nevertheless, the population inversion is properly reproduced.

6.2.3 Basic laser components

When the transition $E_2 \rightarrow E_1$ takes place, the emitted photon with energy $hf_{21} = E_2 - E_1$ is trapped in an optical cavity with mirrors on both ends such that the photon will bounce back and forth, forming a standing wave with the boundary condition

$L = n\lambda/2$, where L is the cavity length and λ is the wavelength of the laser light. This condition holds for flat parallel mirrors. The emitted photon will then stimulate further $E_2 \rightarrow E_1$ transitions and thus will amplify the photon density in the cavity. As the photons in the cavity form standing waves, the stimulated emission is in phase with the photons that are already present, resulting in phase coherence between the trapped photons. Furthermore, the optical cavity has an optical axis. Photons traveling parallel to the optical axis are in phase with the standing wave. Photons inclined against the optical axis either escape the cavity or vanish by destructive interference. We may imagine the optical cavity as an acoustic cavity like an open organ pipe. The boundary conditions are the same and the standing wave would generate a single-frequency tone that becomes louder with time because of stimulated amplification.

Figure 6.4 depicts a basic outline of a laser consisting of an optically active medium, flash lamps for optical pumping, mirrors on both ends of an optical cavity, one of them semitransparent to let the laser beam out of the cavity. Parallel to the optical axis, the photon field is amplified, off-axis photons are lost and do not contribute to the amplification process. The optic medium can be a gas, a crystalline solid, or an organic material.

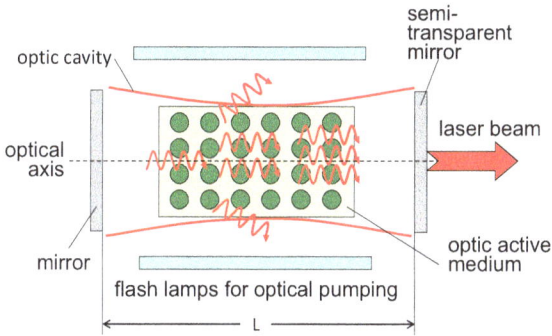

Fig. 6.4: Basic components of a laser, consisting of an optic medium, flash lamps for optical excitation, and mirrors for trapping the photons that form standing waves.

6.2.4 YAG laser

Figure 6.5 shows the energy scheme of a widely distributed Nd-YAG laser. The optic active medium is an yttrium–aluminum garnet (YAG) crystal with the chemical composition $Y_3Al_5O_{12}$ which is doped with neodymium ions. YAG is a transparent colorless crystal. It can be doped with many different substitutional ions on the yttrium and aluminum crystal lattice sites. With doping, the crystal becomes colored. The color depends on the absorption bands in the visible regime. The most prominent example is doping with neodymium Nd^{3+}–ions substituting for yttrium. The relevant electronic transitions for the laser process are depicted in Fig. 6.5. The Nd^{3+}-ions are optically pumped from the Nd^{3+}-ground state with the spectroscopy notation $^4I_{9/2}$ to a higher lying pump band, mainly $^4F_{5/2}$. In the past, pumping was achieved with a broadband discharge lamp. Discharge lamps are now replaced by more efficient narrow-band laser diodes. The pump band becomes quickly depopulated by transitions to the $^4F_{3/2}$ atomic state. The lasing transition takes places from the $^4F_{3/2}$ excited state to $^4I_{11/2}$. The final transition from $^4I_{11/2}$ to the ground state $^4I_{9/2}$ is radiation less, meaning that the energy difference is carried to the lattice in a nonradiative process such as in form of lattice vibrations. Hence, the Nd-YAG laser is characterized by a four-level system, as schematically shown in Fig. 6.5. A more detailed discussion of the relevant optical transition is given in the Infobox on Nd^{3+}-spectroscopy.

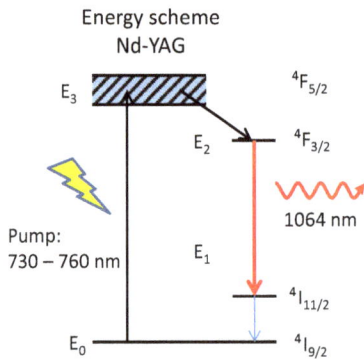

Fig. 6.5: Energy scheme and optical transitions of Nd^{3+} in YAG.

6.2.5 Laser types, wavelengths, and units

Depending on the application, there is a large variety of lasers to choose from. A short list is given here as an overview:
1. Gas-laser: He–Ne, Ar, CO_2
2. Dye-laser: dyes dissolved in water or in methanol, ethylene, glycol, etc.
3. Excimer laser: Nobel gas – halogenates
4. Solid-state laser: Nd-YAG, Ruby, Ti:sapphire
5. Heterostructure semiconductor laser: GaAs, GaAlAs, and GaN
6. Free electron laser

The wavelength regions where the various laser sources have their fundamental emission lines are indicated in Fig. 6.7. Lasers can also be excited to higher harmonics, filling in the wavelength gaps seen in this chart. Free electron lasers with extremely high brightness, emitting light from the extreme ultraviolet to the hard x-ray regime are presently only of interest for science and technology and therefore will not be discussed any further.

Infobox I: Nd³⁺-spectroscopy

Here we take a closer look at the electronic states and transitions of Nd^{3+} as a representative example for other laser transitions. Nd is one of the lanthanide atoms, which are characterized by sequentially filling the inner 4 f electronic shell from zero (La, [Xe] 4 f^0 $5d^1$ $6s^2$) to a maximum of 14 electrons (Lu, [Xe] 4 f^{14} $5d^1$ $6s^2$). The outer three electrons $5d^1$ $6s^2$ are responsible for chemical bonding, while the inner 4 f shell remains chemically protected and exhibits almost atomic like spectroscopic levels. The electronic configuration of lanthanide atoms in the 3^+ ionic state is then denoted as: [Xe] 4 f^n. Consequentially the electronic configuration of Nd^{3+} is [Xe] 4 f^3. These three electrons in the 4f shell occupy the atomic levels according to the first Hund's[5] rule (Fig. 6.6(b)): all spins are parallel for less than a half-filled shell and the lowest three m_L levels 3, 2, and 1 are occupied. This yields an orbital momentum of $L = 6$ (spectroscopic notation: I) and a total spin of $S = 3/2$. Combining L and S to the total angular momentum J again according to Hund's rule for half-filled shells, we find for the lowest level $J = L - S = 6 - 3/2 = 9/2$, the next one is $J = L - S + 1 = 11/2$, and up to $J = L + S = 15/2$. These four ^4I levels are energetically split due to spin–orbit coupling, as shown in Fig. 6.6(a). They are further energetically fanned out into sublevels due to crystal field effects, which are neglected here.

The next higher term in the level scheme is the ^4F state, which is also split by spin-orbit coupling. The lowest of the sublevels is the $^4F_{3/2}$ state, which is triply degenerated with the electron configurations 4f(+3, +2, −2), (+3, +1, −1), (0, +2, +1)). Only one of the configurations (+3, +2, −2) is shown in Fig. 6.6(b), and only "spin-up" configurations are considered.

The lasing transition from $^4F_{3/2}$ to $^4I_{11/2}$, marked in dark red in Fig. 6.6(a), requires a change of the angular momentum by $\Delta L = 3$. This corresponds to an octupolar transition with a long $1/e$ lifetime of about 250 μs. The long lifetime has two advantages. Stimulated emission becomes possible and the transition with a wavelength of $\lambda = 1064$ nm is very sharp according to the Heisenberg uncertainty principle $\Delta E \Delta t \leq h$.

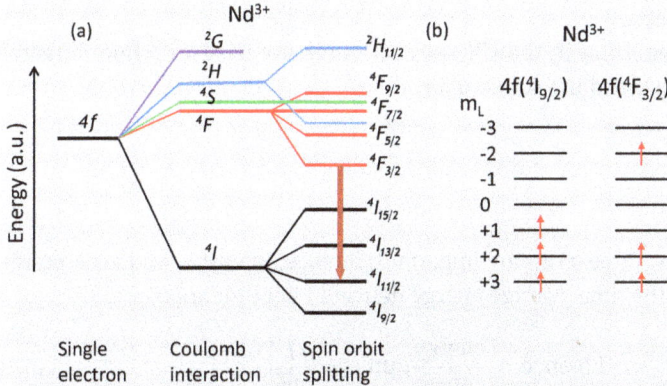

Fig. 6.6: (a) Energy levels of 4f Nd^{3+}. The dark red arrow corresponds to the lasing transition; (b) electronic configuration of the $^4I_{9/2}$ (ground state) and $^4F_{3/2}$ (excited state) levels.

There are several other transitions that can be used for laser activity, such as the transition from $^4F_{3/2}$ to $^4I_{15/2}$. Furthermore, Nd-ions can be replaced by Yb, Er, Ce, or Cr for a selection of additional lasing transitions with different wavelengths.

---

5 Friedrich Hermann Hund (1896–1997), German theoretical physicist.

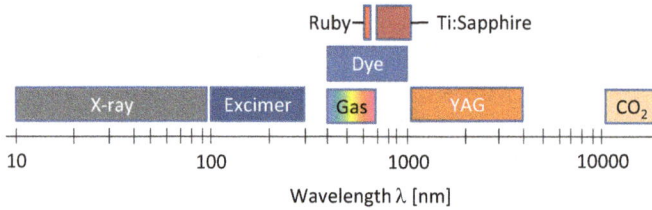

Fig. 6.7: Laser types and their wavelength regions.

Laser sources may be used in a continuous mode (continuous wave (CW) laser) or in a pulse mode with pulse width ranging from milliseconds (ms) to femtoseconds (fs) and with variable pulse repeat frequency (PRF). Pulsing is achieved by three different methods. For low frequencies it is sufficient to pulse the pumping flash lamp. For intermediate frequencies the quality factor of the resonator is switched, which is referred to as *Q-switching*. Very short fs-pulses are achieved by *mode coupling*. These three methods will be explained further below.

In medical practice three types of lasers are most frequently used:
1. Nd-YAG solid-state laser, emitting in the infrared spectral region at 1064 nm, is used for treating hemangioma, glaucoma, cataract, removing tattoos, caries, and for fragmenting stones in the urinary bladder.
2. CO_2 gas laser emits in the far-infrared at 10 μm and is applied in the areas of neurosurgery, dermatology, plastic surgery, ophthalmology, and general diagnostics and therapy.
3. Excimer lasers emitting in the UV spectral range are used for photorefractive keratectomy, discussed in Section 6.5.1.

6.2.6 Laser specifications

For laser applications, three terms are important: fluence, intensity, and peak power. Laser fluence is the time average energy delivered per unit area:

$$\text{Fluence} = \frac{\text{energy}}{\text{area}}; \text{ units} = \left[\frac{J}{m^2}\right] \tag{6.11}$$

Note that the laser fluence defined here corresponds to the energy fluence defined in Section 7.3 of Volume 2 for x-rays. In the case of a Gaussian beam profile, the spot size of the laser beam is taken at the I_0/e intensity level, where I_0 is the intensity at the center.

Intensity is defined as fluence per time, which yields power per area (also called power density):

$$\text{Intensity} = \frac{\text{energy}}{\text{area} \times \text{time}} = \frac{\text{power}}{\text{area}}; \text{unit} = \left[\frac{W}{m^2}\right] \qquad (6.12)$$

The peak power of a pulsed laser is defined:

$$\text{Peak power} = \frac{\text{laser pulse energy}}{\text{pulse length}}; \text{unit} = \left[\frac{J}{s}\right] \qquad (6.13)$$

The peak power density is defined accordingly:

$$\text{Peak power density} = \frac{\text{laser pulse energy}}{\text{area} \times \text{pulse length}}; \text{unit} = \left[\frac{J}{m^2 s}\right] \qquad (6.14)$$

Often fluence and intensity are quoted in units of joule per square centimeter and watts per square centimeter, respectively. It is paramount to differentiate between time average and single pulse specifications. The average laser power may be harmless, while the peak power causes ablation.

In electrodynamics, the poynting vector $\vec{S}$ relates the electric field $\vec{E}$ to the intensity of electromagnetic waves. In a vacuum, the time-averaged energy flow, i.e., energy per time that goes through a cross-sectional area A, is expressed by [2]:

$$\langle S \rangle = \varepsilon_0 c \langle \vec{E}^2 \rangle = \frac{1}{2} \varepsilon_0 c \vec{E}_0^2, \qquad (6.15)$$

where $\varepsilon_0 \left(= 8.85 \times 10^{-12} \ C^2/Nm^2 \right)$ is the dielectric constant of the vacuum, $c \ (\cong 3 \times 10^8 \ m/s)$ is the speed of light, and $E_0 \ (N/C = V/m)$ is the electric field amplitude. The SI unit of S is $[S] = W/m^2$. Therefore, the pointing vector S and beam intensity I have the same units.

The peak intensity is accordingly:

$$I_p = S_p = \varepsilon_0 c \vec{E}_0^2, \qquad (6.16)$$

Examples for determining fluence, intensity, peak power, and electric field are given in the exercise section.

6.3 Laser pulsation

The time-averaged interaction of laser intensity with tissue is roughly speaking a local increase of temperature. However, in a short pulse, much higher power can be concentrated, leading to very different reactions of the tissue, such as photocoagulation, photoablation, and evaporation. Therefore the ability to pulse a laser beam opens many more treatment possibilities than a CW operation. We distinguish three methods for pulsing a laser: mechanical, Q-switch, and mode coupling, described below.

6.3.1 Mechanical switching

At low frequencies of up to about 10 kHz, pulsing and pulse shaping of a laser beam can be achieved mechanically by switching the pumping flash lamp on and off. Alternatively, the exit laser beam can be chopped with a rotating shutter (Fig. 6.8) or mirror outside of the resonator cavity. In this case, each pulse has the same intensity as if the laser were operated in CW mode. In other words, by mechanical switching, there is no amplification effect of the laser intensity, as is the case for the Q-switched and mode-locked laser.

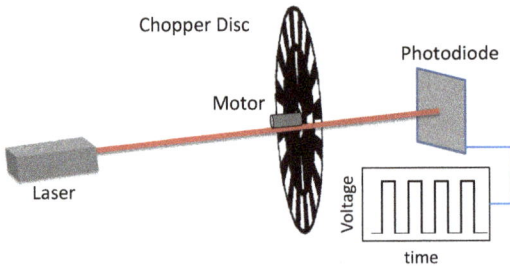

Fig. 6.8: Pulsation of a laser by chopping a continuous beam with a chopper disc.

6.3.2 Q-switch

Q-switching is an alternative switching scheme for frequencies up to a few kilohertz and for gaining much higher pulse intensities in each nanosecond long sharp pulse. Q-switches are fast shutters placed inside of a laser resonator. The name "Q-switch" refers to a change in the quality factor of the laser resonator. In the low Q state, the Q-switch is closed and the cavity loss is high, preventing lasing. Losses occur at the mirrors and are due to absorption and scattering in the cavity. Contrary, in the high Q state, the Q-switch is open, losses are negligible, and lasing takes place.

A laser resonator containing a Q-switch is schematically shown in the upper panel of Fig. 6.9. The sequence of events during one operational cycle is plotted in the lower panels. The resonator consists of a mirror pair, guiding the continuously pumping light source through the laser crystal. Reflector and semitransparent output coupler complete the basic design. The Q-switch is positioned inside the cavity in front of the output coupler.

At the start of a cycle, the Q-switch is closed, preventing lasing. While the Q-switch is closed, the pumping light source is continuously on and increases the stored energy (gain) in the laser crystal, i.e., electrons are being pumped into the upper state, but stimulated emission is hindered by interrupting the cavity resonance. When the Q-switch opens by an external stimulus, the cavity loss quickly drops to zero. All the stored energy (gain) is now emitted in a short light pulse. The pulse duration is determined by the time it takes to empty the inverted electron state. The buildup of stored energy during the low Q state and short rise and fall times of the Q-switch govern high pulse peak power and short pulse lengths. A laser may emit 100 mW in cw mode. The same laser in Q-switched mode with a PRF of 1 kHz, a pulse length of 1 ns and 100 µJ pulse energy has a peak power of about 10 kW. The lower the repetition rate, the higher the output power. High-power YAG lasers typically have a PRF of only 10 Hz and pulse repetition time of 100 ms.

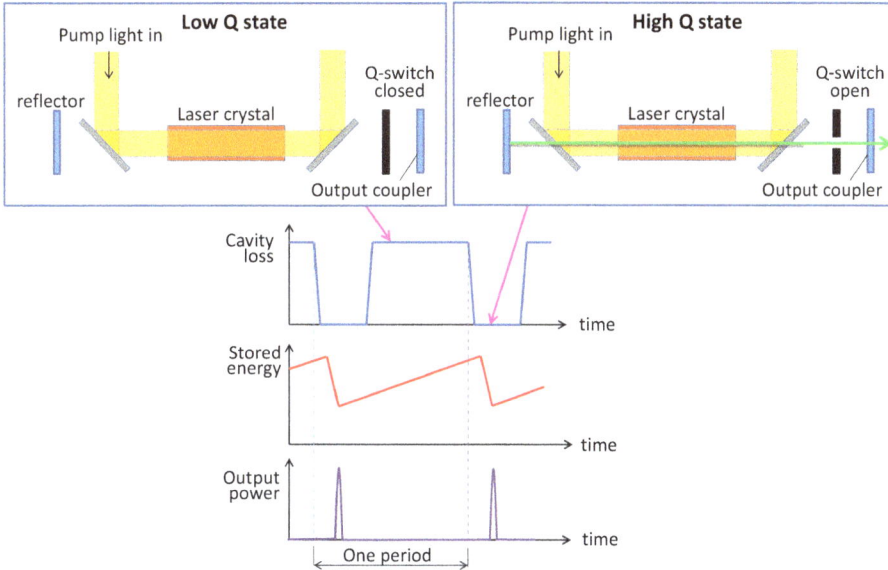

Fig. 6.9: Upper panels: Laser cavity in the low Q state with closed Q-switch (left) and high Q state with open Q-switch (right). Bottom panels: Plotted are the cavity loss, stored energy, and output power as a function of time in a Q-switched laser cavity; the dashed vertical lines denote one cycle of the operation, which results in a single laser pulse. One cycle typically lasts anywhere from about 10 µs to 1 ms and is controlled by the Q-switch (Adapted from Reference [3] with Copyright 2017 by PennWell).

Q-switches are distinguished with respect to their active or passive switching. The most common Q-switches are active devices powered externally: electro-optic (EO) and acousto-optic (AO) modulators. The Q-switch in EO modulators is a Pockels cell [3]. The Pockels[6] cell rotates the polarization of light passing through when a voltage is applied. By inserting this crystal between two crossed polarizers, P and P', as shown in Fig. 6.10(a), light can only pass through this assembly when the applied

---

6 Friedrich Carl Alwin Pockels (1865–1913), German physicist.

voltage activates the Pockels cell and rotates the polarization of the light from the *y*-
to the *x*-direction. EO modulators are typically used at low pulse-repetition rates (up
to a few kilohertz) and high pulse energies, commonly in the range of multiple mJ
per pulse.

For higher pulse-repetition rates (tens of kilohertz or more) and lower pulse energies, AO modulators are usually preferred. According to Fig. 6.10(b), in an AO modulator, a piezocrystal is excited by radiofrequency to set up a strain-wave modulation
that acts as a grating. Light passing through the modulator is diffracted at a fixed
angle when the modulator is activated. If the light beam is deflected, the Q-switch is
in the off-state, preventing lasing; otherwise, the beam is transmitted in the on-state,
allowing lasing.

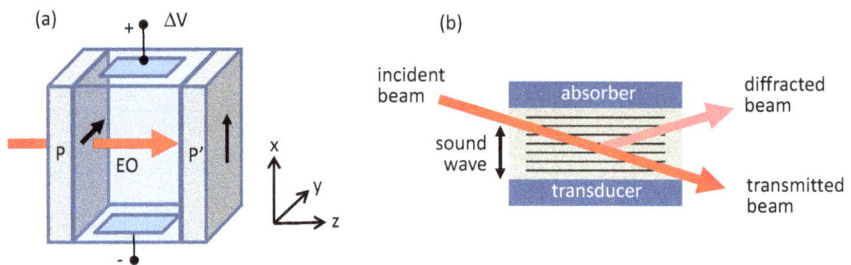

Fig. 6.10: (a) Elecro-optic (EO) modulator rotates the polarization of the beam when activated;
(b) acousto-optic (AO) modulator deflects the light when the transducer is activated.

6.3.3 Mode locking

Mode locking is an advanced laser operational mode that leads to fs pulsing. fs
pulsing finds more and more applications in technology and medicine [4, 5]. In simple terms, the operational principle can be described as follows. In laser cavities,
the standing waves fulfill the boundary condition:

$$L = n\lambda/2, \quad (n = 1, 2, \ldots) \tag{6.17}$$

where L is the cavity length and λ is the wavelength of the laser light (Fig. 6.4). If
the optical transition has a narrow bandwidth, this condition can be fulfilled for
only a few wavelengths close to the center wavelength. Conversely, if the optical
transition has a broad band, the boundary condition can be fulfilled by thousands
of different wavelengths as $L \gg \lambda$. All allowed wavelengths within a Gaussian width
$\delta\lambda$ will overlap and interfere, where the interference of waves with different wavelengths and random phase will lead to a beating effect (Fig. 6.11(a)). The beating
effect results in a chaotic sequence of pulses with different intensities and pulse
lengths. If it can be managed that different wavelengths couple in a fixed-phase

relationship, the interference of these coupled modes yields a periodically reoccur-ring wave packet with a period of:

$$T = 2L/c \tag{6.18}$$

and a pulse length of:

$$t_p = \frac{2L}{cN} = \frac{T}{N} \tag{6.19}$$

where N is the number of modes in the cavity fulfilling the boundary condition (Fig. 6.11(b)) and c is the light speed. The number of modes that can participate in phase-locked interference is limited by the bandwidth $\delta\lambda$ that still allows stimulated emission:

$$N = \frac{4L\,(\delta\lambda)}{\lambda_0^2}, \tag{6.20}$$

where $\lambda_0 \gg \delta\lambda$ is the center wavelength. Now we can rephrase the pulse length as follows:

$$t_p = \frac{2L}{cN} = \frac{\lambda_0^2}{2c(\delta\lambda)}. \tag{6.21}$$

Thus the pulse becomes shorter, the broader the bandwidth $\delta\lambda$ is. The reasons are similar to the ones for diffraction gratings in optics. The higher the number of slits in a grating, the sharper is the central diffraction maximum, and the lower is the intensity of side maxima. This very short pulse travels around the cavity in a time $T = 2L/c$. Every time it reaches the output mirror, the laser emits part of the pulse intensity. Therefore, the PRF is given by the reciprocal round trip time of the laser cavity:

$$PRF = \frac{1}{T} = \frac{c}{2L} \tag{6.22}$$

The PRF corresponds to the frequency difference of two neighboring modes that ful-fill the boundary condition:

$$n\lambda_n = (n+1)\lambda_{n+1} = 2L. \tag{6.23}$$

Rephrasing in terms of frequency differences, we find:

$$\Delta f = f_{n+1} - f_n = \frac{c}{2L}. \tag{6.24}$$

As an example, we assume a central wavelength of 500 nm and a bandwidth of 5 nm. Then the pulse width t_p is 83 fs, and for a cavity length of 2 m the PRF is 75 MHz. For a bandwidth of 100 nm; the pulse width is reduced to 4 fs while the PRF remains the same. The sequence of pulses and pulse width are schematically shown in Fig. 6.12.

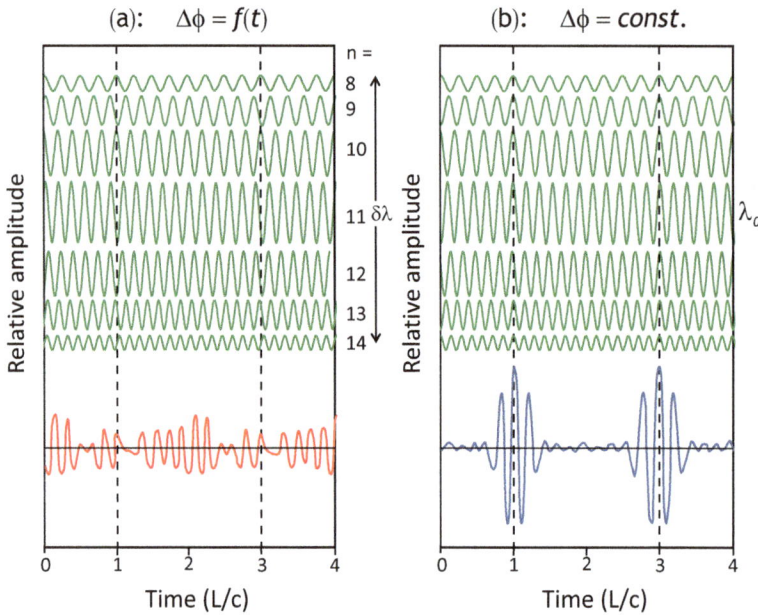

Fig. 6.11: Interference of waves with random and time-dependent phase (a) and with a fixed-phase relationship (b) are compared. The dashed lines are guides to the eye for identifying the phase relation of waves with different wavelengths. The time scale is in units of L/c. n is the number of wavelengths fitting between the vertical dashed lines, and $\delta\lambda$ is the wavelength spread of participating modes with center wavelength λ_0. (Adapted from References [4, 5]).

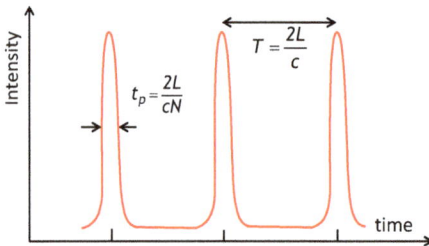

Fig. 6.12: Pulse sequence of a mode-coupled laser. t_p is the pulse length and T is the pulse period. Peak intensity times pulse length yields the fluence of the laser.

A continuous laser can be turned into a mode-locking pulsed operation if the gain is higher in mode-locking than in CW mode. Mode-locking can be achieved by using active or passive switches, similar to the Q-switched lasers. Active mode-locking requires gating of the cavity at the intrinsic PRF, determined by the round trip time of the pulses using, for instance, an AO modulator. On the other hand, passive mode-locking can be performed by using the so-called Kerr[7] lens effect in the gain material, illustrated in Fig. 6.13. The Kerr lens effect occurs through non-linear optical properties of the laser crystal at very high light intensities. The high peak intensity of the mode-locked pulse changes the refractive index of the laser crystal locally. According to the beam profile, the refractive index becomes higher at the center of the crystal than at the edges. This refractive index gradient acts as a lens and leads to self-focusing of the beam, enabling it to pass through an internal aperture. In contrast, in CW operation, the beam hits the aperture, and the beam intensity suffers from clipping.

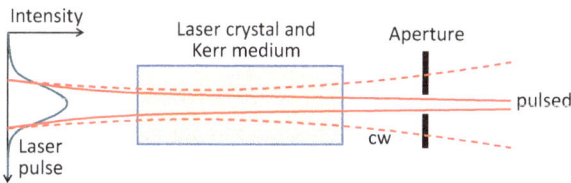

Fig. 6.13: A mode-locked laser with beam focusing due to nonlinear optical properties of the laser crystal at very high intensities.

6.4 Laser interaction with tissue

6.4.1 Laser beam penetration depth

When a laser beam hits the skin surface, three main effects occur: the laser beam is partially transmitted, reflected, and scattered. Since the skin surface is rough on the length scale of the wavelength used, the (specular) reflection can be neglected and diffuse scattering predominates (Fig. 6.14). The diffuse scattering of photons of rough surfaces is the fundamental reason why we "see" objects around us. With specular reflection, our visual perception could easily be fooled (fata morgana).

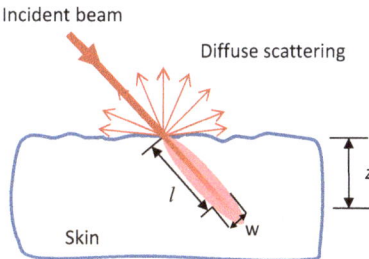

Fig. 6.14: Incident laser light causes diffuse scattering outside and penetration inside. z is the penetration depth and l is the penetration length. The width w is due to the scattering of light in the tissue.

---

7 John Kerr (1824 – 1907), Scottish physicist and pioneer of electro and magneto-optics.

Scattering of light also occurs below the surface. This is the main reason for the finite penetration depth of laser light in the skin and tissue. However, the details depend on the light's wavelength, the intensity, the area of exposure, and the time of exposure. For instance, if the laser beam causes photocoagulation, the penetration depth changes over time from deep to shallow. When UV light strikes the skin, it produces melanin known as tanning of the skin to protect it from genetic damage. Tanning reduces the penetration depth of light.

In spite of the huge parameter space for light interaction with the skin/tissue, the penetration depth of laser light is often described by the simple Lambert–Beer law, according to which the transmitted intensity decays exponentially at depth l described by:

$$I(\lambda, l) = I_0 \exp(-\alpha(\lambda)cl) = I_0(\lambda) \exp\left(-\frac{l}{L_{op}}\right) \tag{6.25}$$

Here L_{op} is the optical penetration length, α is the attenuation coefficient, and c is the molar concentration of a solute in a solvent. The molar concentration c applied to tissues is an ensemble average over different molecules (proteins, membranes, cartilage, etc.). As the penetration depth and the attenuation coefficient cannot be determined in vivo, the best alternative is a water solution containing the main chromophores of the tissue: melanin and hemoglobin. Melanin is responsible for the color of the skin and the tan developing under sunshine. The attenuation coefficient and the penetration depth of water, HbO_2, and melanin are shown as a function of wavelength in Fig. 6.15. Water has a transparency window in the visible range of the optical spectrum. In this window, the attenuation coefficient of HbO_2 and of melanin is particularly high. Oxygenated HbO_2 and deoxygenated Hb (not shown) can clearly be distinguished at the wavelength 560 nm and between 600 and 700 nm. The absorbance properties of Hb and HbO_2 are discussed in Section 8.6.4 of Volume 1. Also indicated in Fig. 6.15 are the main laser wavelengths for various applications discussed further below. According to Fig. 6.15, the primary absorption of laser light in the visible regime is due to blood and melanin. Obviously, the blood vessels can be better seen under white skin than under dark skin. Typical penetration depths for different lasers in tissue are listed in Tab. 6.1.

Tab. 6.1: Emitted wavelength and penetration depth in tissue of typical lasers used in medicine [6].

Laser type	Emitted wavelength (nm)	Penetration depth (mm)
CO_2	10 6000	0.10
Nd:YAG	1 064	6.00
Ar+	488, 514	2.00
Excimer	126, 351	0.01

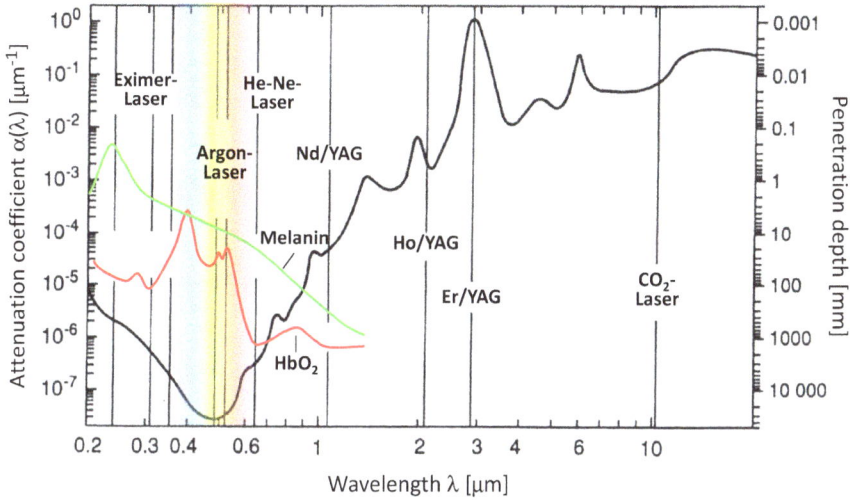

Fig. 6.15: Attenuation coefficient of water (black line), melanin (green line), and oxygenated hemoglobin (red line) plotted as a function of wavelength. Note the double logarithmic scale of the plot. The vertical lines indicate the wavelength emitted by some standard laser sources.

6.4.2 Laser-tissue interaction

In general, laser-tissue interaction depends on the absorbance properties of the tissue exposed at a particular wavelength. In addition, the laser-tissue interaction is also a question of laser intensity and exposure time. A single short pulse may just cause a heat wave, whereas a longer exposure time may cause coagulation or evaporation. We distinguish five main types of laser–tissue interactions distinguished by power density and pulse length [6, 7]:
1. Photothermal interaction
2. Photoablation
3. Plasma-induced ablation
4. Photomechanical interaction
5. Photochemical interaction

These main interaction types are related to the characteristic laser properties, in particular to their average power density and pulse length. Power density versus pulse length is plotted on a double-logarithmic scale in Fig. 6.16 [6, 7]. The dashed blue lines indicate the energy fluence within a corridor of 1 J/cm^2 (left-dashed line) and 10^3 J/cm^2 (right-dashed line). In the left top area, the laser flux of single pulses is very high due to very short pulses. This is the area of fs-laser sources. As we continue along the diagonal dashed lines, the power density decreases and the pulse length increases. In the bottom right corner, we find lower power CW lasers. The circles indicate areas where dominantly one or the other process takes place as listed above and described in more detail in the following subsections. We start on the lower right end and successively increase the laser flux up to the top left corner. The photochemical interaction is a special type of laser-tissue interaction mediated by uptake of drugs, a topic that will be discussed at the end of this section.

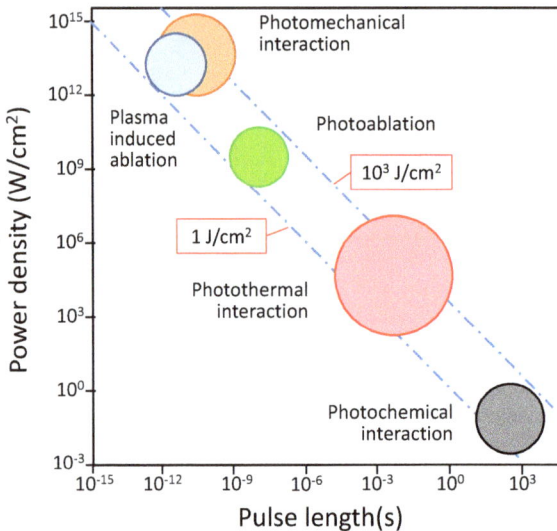

Fig. 6.16: Main types of laser–tissue interaction in relation to laser characteristics. Note the double-logarithmic plot of single pulse laser power density versus pulse length. (Adapted from Refs. [6, 7] by permission of Springer Verlag).

> ❗ Pulsation of laser photons allows achieving very high pulse intensities for specific ablation and cutting processes.

6.4.3 Photothermal interaction

For not too high intensities, the laser photons couple to the vibrational modes of molecules and increase the temperature locally. From taking a sun bath, we know that photons are converted to heat in the skin. Photons are scattered in the tissue over some diffusion length and finally become absorbed by molecular vibrations, which increases their kinetic energy that we recognized as heat. These processes also occur by the absorption of laser light on the skin but covering a much smaller area with a much higher intensity. The result can be a denaturation or destruction of the tissue [8]. Figure 6.17 shows an example of a laser beam hitting the surface of the skin at normal incidence. While the laser beam intensity is attenuated, it spreads out to the sides by scattering and heat conduction, yielding the indicated isointensity contour lines [7]. The modeling of heat conduction in tissues is described in Infobox II. The rule of thumb is that the heat diffusion length is about 1 μm in 1 μs.

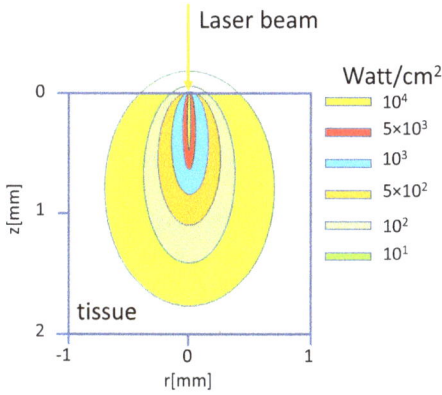

Fig. 6.17: Light scattering and heat conduction enlarge the volume of the laser interaction. Adapted from [7].

Infobox II: Heat conduction in tissue

Heat conduction is a well-known experimental and theoretical problem, and the results depend strongly on the boundary conditions imposed. Here we treat the case of temperature gradient superposed by local heat sources representing local laser exposure.

The absorption of laser light at a depth z below the skin surface over an area with radius r constitutes a heat source $S(r,z,t)$ during the exposure time t. The heat source is proportional to the laser intensity $I(r,z,t)$:

$$S(r,z,t) = \alpha I(r,z,t). \tag{6.26}$$

α is the attenuation coefficient. The local heat source leads to an increase of the local temperature:

$$dT = \frac{1}{mC_m} dQ, \tag{6.27}$$

where m is the tissue mass and C_m its specific heat.

The local heat produced by a laser spot for some time t is transported to other parts of the tissue by heat conduction, heat convection, and heat radiation. Heat radiation can be neglected, and heat convection is not considered. Heat conduction is the primary mechanism.

Now we combine the equation for heat conduction (eq. (4.19)/Vol. 1):

$$j_Q = -\lambda \operatorname{grad}(T), \tag{6.28}$$

and the continuity equation for heat flow (eq. (8.16)/Vol. 1):

$$\mathrm{div} j_Q = -\frac{dq}{dt}.$$ (6.29)

Here q is the heat density, i.e., the heat content per unit volume $q = \Delta Q/\Delta V$, and j_Q is the heat flow: $j_Q = qv$, where v is the flow velocity through a cross-sectional area A. The continuity equation states that the temporal change in the heat content per unit volume (dq/dt) is determined by the divergence of the heat flow j_Q through a cross-sectional area A.

Together with eq. (6.28) we derive the differential equation for the temperature diffusion (see exercise E6.8):

$$\frac{dT}{dt} = \varepsilon \nabla^2 T,$$ (6.30)

where $\varepsilon = \lambda/\rho C_m$. Now adding an internal heat source like the laser spot, we obtain:

$$\frac{dT}{dt} = \varepsilon \nabla^2 T + \frac{1}{\rho C_m} S$$ (6.31)

According to [6], this equation has the general solution:

$$T(r,z,t) = T_0 + \frac{\tilde{T}}{(4\pi \varepsilon t)^{3/2}} \exp\left(-\frac{r^2 + z^2}{4\varepsilon t}\right),$$ (6.32)

where T_0 is the temperature at $t = 0$ and $\tilde{T}$ is an integration constant. According to eq. (6.32), the spread of the temperature has the shape of a Gauss curve that evolves in time. The time-dependent thermal penetration depth describes the spatial extent of the heat transport in the z-direction, where:

$$z_{\mathrm{therm}} = \sqrt{4\varepsilon t}$$ (6.33)

is the distance over which the temperature has decreased to $1/e$ of its peak value, and

$$\tau_{\mathrm{therm}} = z_{\mathrm{therm}}^2/4\varepsilon$$ (6.34)

is the time to reach this distance, also called the thermal relaxation time. If we compare the thermal penetration depth z_{therm} with the optical penetration depth L_{op} in eq. (6.25), and set them equal:

$$L_{op} = \sqrt{4\varepsilon \tau_{\mathrm{therm}}},$$ (6.35)

then we can conclude that for laser exposure times $t < \tau_{\mathrm{therm}}$, heat has not diffused to the optical diffusion depth and thus can not cause any damage. Vice versa, at times $t > \tau_{\mathrm{therm}}$, there is sufficient time for thermal damage by the laser beam. A characteristic time for water is 1 µs for a thermal penetration depth of 0.7 µm.

Depending on the intensity and duration of the laser beam exposure, we can distinguish the following three main effects [9, 10]:

1. *Hyperthermia* implies local and moderate heating of tissues to several degrees above body temperature for several tens of minutes. This local hyperthermia causes cell death due to changes in enzymatic processes. Hyperthermia from mere laser beam exposure is difficult to control but has been used effectively,

mediated by plasmonic [11] and magnetic nanoparticles [12] (see discussions in Chapter 7 on magnetic nanoparticles).

2. *Coagulation* occurs when the local temperature reaches 50 to 100 °C within a few seconds leading to complete and irreversible tissue destruction through denaturation of proteins and fibers. Boiling an egg is a vivid example of the denaturation process that occurs during clotting.

3. *Volatilization* implies the loss of tissue material through evaporation. When this loss is local, there is a temperature gradient between the hot spot and adjacent tissue. Within the edges of the hotspot is a zone of coagulative necrosis, i.e., cell death by coagulation. Tumors can be destroyed by volatilization and coagulation, preferentially in the skin due to the limited penetration of the laser light.

Two examples may illustrate how coagulation and volatilization can be used for specific treatments. Cutaneous angiomas, also known as port vine stain, is a reddish skin color due to abnormal blood vessels in the upper dermis, sometimes occurring upon birth. These stains can be removed by a laser beam absorbed preferentially by hemoglobin. Then hemoglobin, instead of the dermis, becomes coagulated. According to the plot in Fig. 6.15, an Ar-laser would be suitable for such a treatment. The second example concerns the treatment of virus-infected warts, which may appear anywhere on the skin, from hand, feet, to genitals. They can be treated by volatilization with a CO_2 laser. Upon volatilization, the lesion disappears into smoke, and the coagulation necrosis seals the blood vessels in the skin without bleeding.

6.4.4 Photoablation

Laser-induced processes that cause vaporization of tissue material without thermal lesions at the margins are referred to as photoablation. This effect requires a very energetic laser beam emitting at short wavelengths in the UV. Two properties are combined to make photoablation effective: the high electric field at short wavelengths and the high flux ($\sim 10^9$ W/cm^2) of the laser field breaking chemical bonds without generating heat at the edges. The resultant molecular fragments then expand in a plasma plume carrying off the thermal energy without destroying the surrounding tissue. For this reason, excimer lasers are used for photoablation emitting at 193 nm (ArF), 248 nm (KrF), or 308 nm (XeCl). The temporal evolution of the ablation process is illustrated in Fig. 6.18, according to Monte Carlo simulations [13]. It consists of four main processes: (1) absorption of high-energy UV photons; (2) promotion to repulsive excited states of the molecules *n* the tissue, causing chemical bond breaking; (3) ejection of material fragments (no necrosis); (4) crater formation in the tissue and ablation. Photoablation is mainly used in ophthalmology and, in

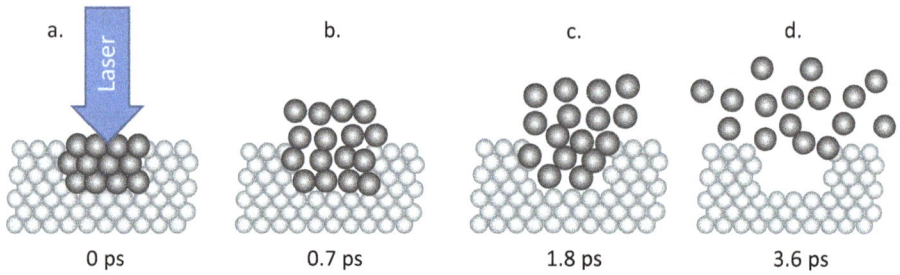

Fig. 6.18: Photoablation using a high flux, short-wavelength pulsed laser. Monte Carlo simulations show the temporal evolution of material ablation after laser impact: (a) Absorption of high-energy UV photons; (b) promotion to repulsive excited states, leading to dissociation of chemical bonds; (c) ejection of material fragments (no necrosis); (d) crater formation in the tissue and ablation. Adapted from Ref [13].

particular, for photorefractive keratectomy (LASEK and LASIK). Both methods are discussed further below.

6.4.5 Plasma-induced ablation

Increasing further the laser flux, plasma-induced ablation occurs for pulses shorter than nanoseconds, illustrated in Fig. 6.19. The plasma-induced ablation is solely based on tissue ionization in contrast to plasma creation followed by shockwave propagation, discussed in the next section. Nd:YAG and Ti:sapphire laser sources are used for plasma-induced ablation with picosecond to femtosecond pulse length and flux on the order of 10^{12} W/cm^2. Such a high power density leads to electrical fields of 10^7 V/cm, comparable with electric fields of electrons orbiting in atoms. Therefore, Q-switched or mode-locked laser sources can ionize molecules in biological tissue. Due to avalanche amplification, this process creates a very high density of free electrons for a very short time with values reaching 10^{18} electrons/cm^3. Hot free electrons from ionization are being accelerated to high energies and collide

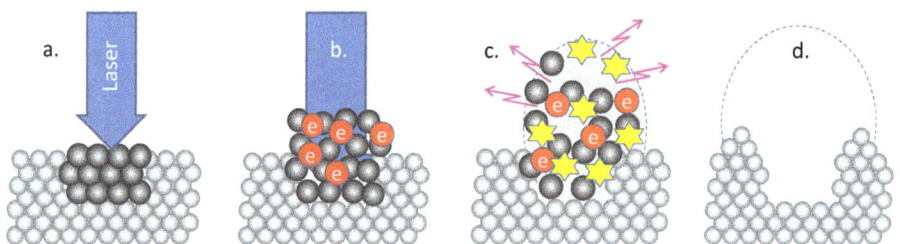

Fig. 6.19: (a) Pulsed laser exposure; (b) tissue ionization; (c) ionized plasma formation; (d) crater formation. Adapted from [14].

with other molecules leading to further excitation and ionization. Light electrons and heavy ions move with different velocities, typical for a highly ionized plasma. The plasma allows a well-defined removal of tissue by "optical breakdown" (plasma explosion) without evidence of thermal or mechanical damage to surrounding tissue. The applications are similar to those of photoablation.

6.4.6 Photomechanical interaction

Laser sources may cause photomechanical effects either by plasma creation, explosive vaporization, or cavitation. In either of these three cases, the photomechanical effect is associated with the generation of high pressure shock waves, as indicated in Fig. 6.20.

Plasma creation: Using a nano or picosecond-pulsed Nd:YAG laser source, very high flux can be concentrated over a very small area reaching flux values between 10^{11} and 10^{14} W/cm^2. This high flux creates a plasma by ionizing atoms (panel b). At the border of the ionized area, a steep pressure gradient occurs that causes the shock wave to propagate outwards (panel c). The expansion of the shock wave causes mechanically destructive effects. This effect is, for instance, used for removing a thin gray coating on an artificial lens that may develop after cataract treatment. By carefully focusing the laser, the tissue membrane on the backside of the lens can be vaporized without harming the retina.

Explosive evaporation: If the exposure time of the laser is shorter than the characteristic time for heat diffusion in the tissue, the laser accumulates heat in a local spot that finally causes an explosive vaporization of the target tissue. As thermal diffusion is rather slow, the laser pulse width can be comparatively long for this effect to occur. A Q-switched Nd:YAG laser with a pulse length of 100 μs is typically employed for removal of tattoos via explosive vaporization. The large ink particles in the skin explode and redistribute into smaller particles, which are reabsorbed by the skin and eventually removed. In any case, it is important that the laser beam is absorbed by the ink pigments and not by the skin. Colored tattoos require several laser lines for removal.

Cavitation: If a laser pulse creates a hot spot that is not only spatially confined but also mechanically hindered to explode, then the high-pressure gas bubble will implode as soon as the laser pulse is turned off. The subsequent implosion is known as cavitation (panel d). Using this phenomenon, urinary stones can be fragmented by a pulsed laser beam with microsecond pulse length. For this purpose, an optical fiber is placed under endoscopic control through the urethra into the bladder. The stone is broken into pieces small enough to naturally clear out the fragments through the urethra.

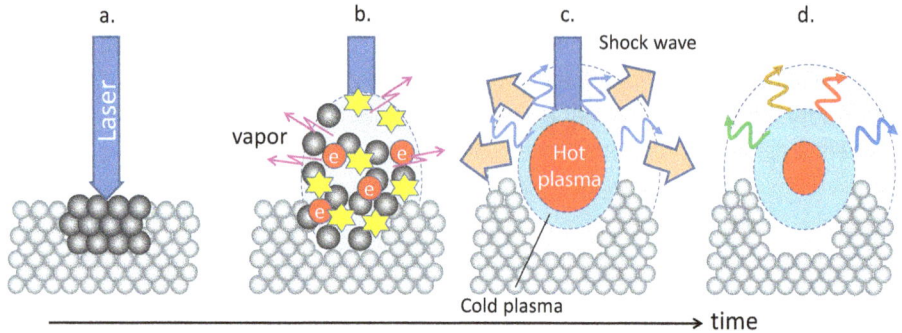

Fig. 6.20: Schematic of the laser-induced breakdown process: (a) plasma ignition; (b) thermal evaporation; (c) plasma expansion emitting a shock wave; (d) element specific emission, de-excitation, and cavity formation. Adapted from [15].

6.4.7 Photochemical interaction

Photochemical interaction of lasers with tissue is used for tumor therapy and requires first the uptake of photosensitive drugs [16, 17]. The treatment steps are explained in Fig. 6.21. They have similarities to the boron–neutron capture therapy presented in Chapter 4. First, the tumor area must be identified before a photosensitizer is injected into this area. The drug should have the property to clear out from normal tissue but remain and eventually accumulate in the tumor cells. The clearance may take a few hours up to a few days. Then the area is irradiated by a low-power laser beam. The photosensitizers become excited by the laser beam from the ground state to an excited level $S \rightarrow S^*$. The laser light has to be chosen to cover the absorption maximum of the photosensitizer S. The excitation energy is then transferred to free and ambient oxygen molecules in the tissue. Free oxygen in the ground state is in a spin triplet state, i.e., the highest occupied molecular level is occupied by two unpaired electrons, one from each oxygen atom, see Fig. 6.22. Because of this spin pairing, oxygen gas and liquid are paramagnetic. Upon energy transfer from the photosensitizer, O_2-molecules are switched from the triplet to the excited singlet state. In the singlet state, O_2^* is extremely reactive, as we have learned about already in Chapter 1 on cancer treatment. The procedure is called *photodynamic treatment* (PDT). It is confined to the area of the administered drug, contrasting treatments that involve dissipation of heat. Unlike cancer therapy by ionizing radiation aiming at the destruction of DNA in the tumor cell, PDT relies on the attack of cellular components by phototoxic singlet oxygen molecules. If sufficient oxidative damage has occured, the target cells will react by necrosis within the illuminated area and eventually total tumor recession. The necrosis sets in within a few days after laser exposure. PDT is a noninvasive cancer treatment for tumors just under the skin or at the surface of inner organs that can be reached by a light source, for instance, with the help of an endoscope.

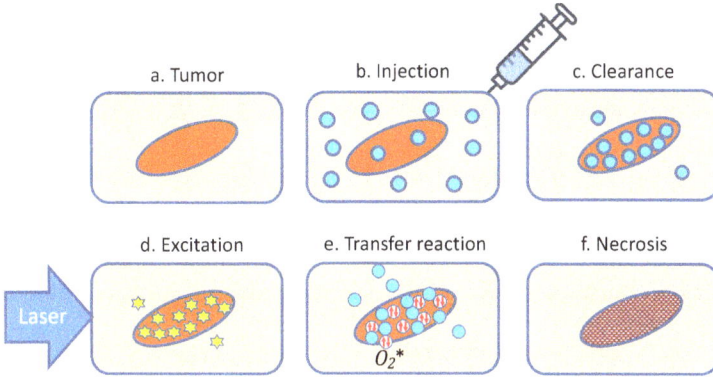

Fig. 6.21: Photodynamic cancer treatment with the help of a photosensitizer that transfers the excitation energy from a laser to ambient oxygen. Triplet oxygen transforms to phototoxic singlet oxygen, killing cancer cells. (a–f) are the different steps taking place during the treatment process.

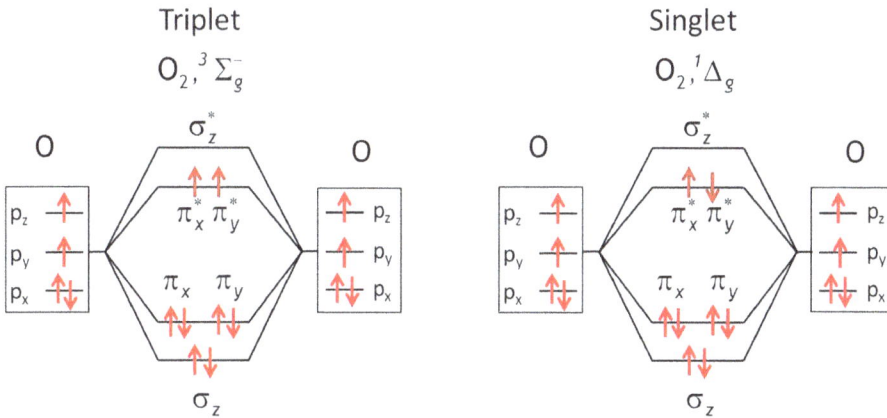

Fig. 6.22: Atomic and molecular orbitals of oxygen in the triplet and singlet state. Singlet oxygen is not stable and decays back to triplet oxygen. Only the 2p orbitals are sketched, the completely filled 2s orbitals are not shown.

This treatment excludes cancers that have grown deep into an organ or those already spread. A number of photosensitizers exist for PDT, distinguished as porphyrins, chlorophylls, and dyes.

In contrast to apoptosis, necrosis is a sudden and unprogramed cell death due to trauma, photocoagulation, or, in the present case, due to contact with phototoxic singlet oxygen molecules. Necrosis does not follow an apoptotic signal transduction pathway. Instead, various receptors are activated, resulting in the loss of cell membrane integrity and an uncontrolled release of cell content into the extracellular space. Hence, the cell fragments are not properly disposed of by phagocytosis. Therefore it often appears necessary to remove necrotic tissue by surgical means, a procedure known as debridement. However, recent studies indicate that the mechanisms of apoptotic and necrotic cell removal are not as different as assumed in the past [18]. This will likely change debridement procedures in the future. A recent updated review is provided in [19].

Table 6.2 gives an overview on the different laser tissue interactions discussed in this section, including their main applications in medicine.

Tab. 6.2: Laser-tissue interaction for different lasers and laser flux.

Interaction	Laser type	Flux or intensity (W/cm^2)	Pulse length (s)	Tissue reaction	Treatment
Photothermal	CO_2	10^6	$10^0–10^{-3}$	Coagulation	Hyperthermia, stains, warts
Photoablation	Excimer	$10^7–10^8$	$10^{-7}–10^{-8}$	Material removal	LASIK, LASEK
Plasma-induced ablation	Nd:YAG Ti:sapphire	10^{12}	$10^{-12}–10^{-15}$	Removal of tissue by optical breakdown	Refractive corneal surgery
Photomechanical	Nd:YAG Ti:sapphire	$10^{11}–10^{14}$	$10^{-7}–10^{-13}$	Shock wave, Plasma cavitation	Lens and stone fragmentation, tattoo removal
Photochemical		1	CW	Photodynamic therapy	Skin tumors

6.5 Laser applications in ophthalmology

Several laser applications have already been mentioned, many more are used in medicine. A concise and uptodate overview is given in [20]. Aside from tumor treatment, the main applications are in ophthalmology and brief accounts on these methods are presented here.

6.5.1 Photorefractive keratectomy

Myopia and hyperopia (see Chapter 11 of Volume 1) can be corrected by laser application. In both cases, the curvature of the cornea is altered to bring the focus onto the retina: in myopia, the curvature is too large and must be reduced; in hyperopia, the curvature is too small and must be increased.

Before we go into the various corrective laser procedures used for these corrections, we first take a closer look at the cornea structure shown in the cross section in

Fig. 6.23. The cornea has an average thickness of 555 µm and is thinner in the middle than at the edge. The central and most important layer is the stroma having an average thickness of 460 µm. The stroma is protected externally and internally by lamina layers (lamina limitans anterior and posterior) and by epithelial and endothelial layers. Photo-refractive keratectomy implies a change in the stroma. For accessing the stroma, an incision through the upper layers is required. Two different methods are commonly applied: laser-assisted interstitial keratomileusis (LASIK) and laser-assisted epithelial keratomileusis (LASEK). Both methods use ArF excimer lasers with a wavelength of 193 nm. The advantages of this type of laser are a high absorption coefficient of the UV light, the possibility of sharply focusing the laser beam for precise tissue ablation, no heat development in the treated tissue, and no transparency of the laser light in the different layers and parts of the eye. Therefore, the laser light cannot reach or damage the retina. Both methods, LASIK and LASEK, and their differences are explained below [21].

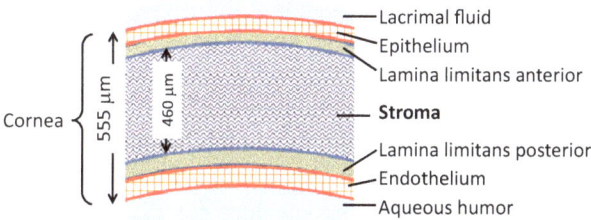

Fig. 6.23: Structure of the cornea composed of several layers that protect the central stroma. The schematic outline is not to scale, but the respective thicknesses are indicated.

LASIK is a method for correcting the refractive properties of the cornea, which contributes to 75% of the total refractive power of the eye. In the first step, a thin and sharp 100–140 µm deep cut is performed by an oscillating metal blade called microkeratome to open the cornea cover and to get access to the stroma. This is shown schematically in panels 2 and 3 of Fig. 6.24. The cut contains components of the epithelium and the stroma. The layered cut, called flap, is not lifted off. Instead, a hinge is left on the side (panel 4) so that the "flap" can be closed again after surgery. Then the stroma is ablated with an excimer laser according to the predefined refractive properties aimed for. Plano-LASIK implies a flattening of the stroma within the area of the laser beam using one broad beam shown in panel 5 of Fig. 6.24. In contrast to a broad beam, with a focused excimer fine laser beam, the exact required and predefined curvature can be ablated, as indicated in frame 6 of the same figure. After finishing ablation, the flap is closed and sealed (panels 7 and 8). The opening of the cover can also be done with a fs laser, which provides a sharper cut and avoids complications that may occur with the use of metal blades. Furthermore, better adhesion between flap and stroma is achieved upon closing the flap.

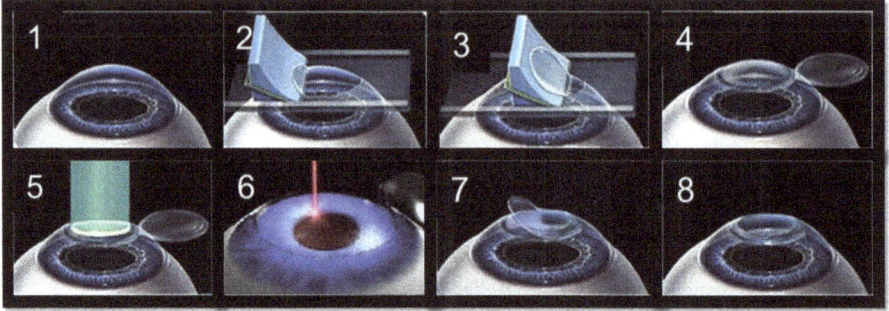

Fig. 6.24: Surgical procedures in LASIK. (Adapted from http://eyeclinicpc.com/lasik.html). The eight steps are explained in the text.

LASEK: In contrast to LASIK, in LASEK the 50 μm epithelial layer is dissolved in alcohol, moved to the side, and disposed of. The disposed part does not contain stromal components. Then the treatment of the stroma follows in the same manner as for LASIK. Since there is no flap for closing, the eye needs to be protected for a few days by a contact lens until the epithelium has regenerated. LASIK and LASEK are only distinct in the last step of the treatment: LASIK recycles the epithelial layer, whereas LASEK disposes it and protects the eye by a contact lens. The epithelium has the capability to regenerate, whereas the stroma does not. Changes of the stroma are permanent.

Correcting myopia requires a flattening of the cornea achieved by material removal. In contrast, hyperopia is corrected by an increase of the refraction power of the eye and entails material addition. However, this cannot be realized by ablation methods. Nevertheless, hyperopia can be corrected with LASIK methods using a "trick." Instead of ablating the center of the stroma like for correcting myopia, a ring around the center is ablated, resulting in an increased curvature at the center by elastic relaxation. The ring depth determines the curvature at the center and thereby the refraction power that can be achieved. Both correction methods for myopia and hyperopia are compared in Fig. 6.25.

In all cases, the success of the treatment depends on a careful and thorough characterization of all optical properties of the eye, which includes the refraction power, astigmatism, thickness, and thickness gradients of the cornea (pachymetry performed by ultrasound), and the entire topography of the eye surface. During surgery, markers are placed on the eye that allows tracking of eye movement and corresponding correction of the laser delivery.

Although LASIK/LASEK surgery has a large success rate, one problem that may occur after surgery should be mentioned. The epithelium layer consists of basal cells, which have the capability to divide and therefore to regenerate. In contrast, the stroma consists of fibrous collagen material that holds a large amount of water, controlling the transparency of the cornea. Stroma and the bordering membranes (lamina limitans) do

Myopia correction

Optical correction zone

Hyperopia correction

Optical correction zone

Transition zones

Transition zones

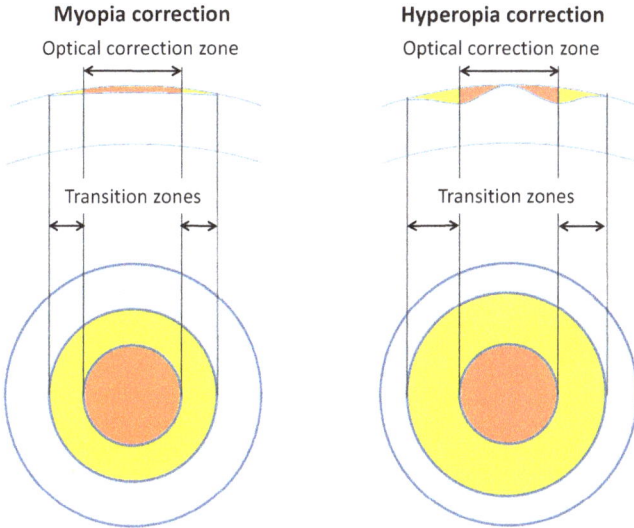

Fig. 6.25: Myopia and hyperopia correction with LASIK are compared. Myopia requires a flattening of the cornea curvature at the center within the optical zone. Hyperopia entails a steepening of the curvature. Myopia correction is achieved by removing material at the center; hyperopia correction is performed by removing a thin ring around the center.

not contain regenerative cells. This is an important property for a lasting LASIK/LASEK refractive correction. However both epithelium and stroma are infiltrated by nerve fibers (innervation). In fact, the cornea holds the densest sensory innervation of the body, more than the skin. These nerve fibers have many tasks, such as controlling lacrimal fluid and the cornea reflex that closes the eye lids in case of lacrimal fluid surplus or any other irritation of the epithelium. Upon LASIK or LASEK surgery, a portion of nerve fibers are damaged and patients may experience persistent eye pain. Some of the nerve fibers regenerate only within a year after surgery. Further information on LASIK and LASEK can be found in the review of [21].

6.5.2 Diabetic retinopathy

As a result of diabetes, two conditions may develop on the retina, summarized as diabetic retinopathy:

a. diabetic macular edema;
b. proliferative retinopathy.

Diabetic retinopathy is the most frequent cause of blindness for ages from 25 to 65 years in Europe and North America. In both cases, the application of lasers promises curative treatment.

Diabetic macular edema: Macula edema refers to some leakage in the macula that potentially causes a decrease of vision. The normal retinal blood vessels develop tiny little blebs along the blood vessel walls, see Fig. 6.26. These blebs leak blood that can cause fluid to accumulate in the macula, degrading vision and color sensitivity. The leakage points can be sealed off with a focal laser. A small lens is placed on the cornea, and an argon laser is focused on the blebs *near* the macula, avoiding the macula, which would cause blindness. Argon lasers are used for this treatment because of the high absorption of the laser beam at about 500 nm by hemoglobin. During treatment, the eye must be kept as still as possible for a few minutes to ensure accurate application of the laser beam and to avoid damage of the macula. The focal laser treatment has temporary success but may be repeated as new blebs show up. As treatment is underway, the patient experiences bright flashes of light but no pain.

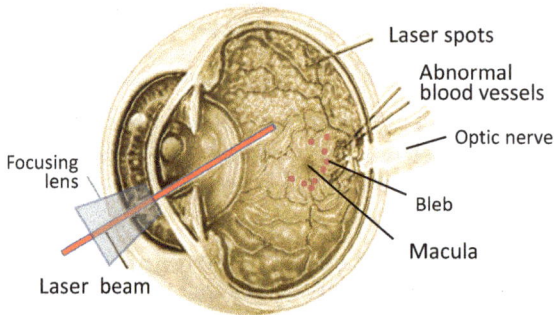

Fig. 6.26: Laser treatment of diabetic retinopathy removing blebs close to the macular and creating laser spot patterns in peripheral part of the retina for reducing oxygen demand. (Adapted from http://www.ferchoeyeclinic.com/).

Proliferative diabetic retinopathy (PDR): The appearance of diabetic retinopathy is associated with the proliferation of a protein called *vascular endothelial growth factor* (VEGF) in the retina. VEGF stimulates the production of new blood vessels in the retina to bring more oxygen to the tissue to compensate for insufficient retinal blood circulation because of diabetes. Blindness may develop from massive proliferation of these abnormal and leaking blood vessels causing either retinal detachment or a rare type of painful glaucoma that is not to be mistaken for the regular type of glaucoma. Proliferation can be stopped by suppressing the production of VEGF. Production of VEGF is suppressed indirectly by decreasing the oxygen demand. The oxygen demand, in turn, is reduced by destroying small portions of the peripheral retina, which are less vital for vision. A pattern of laser spots is imprinted with a focused argon laser, causing local photocoagulation. The procedure is known as *pan-retinal photocoagulation* (PRP). The treatment consists of a greater number of larger, more intense burns, more rapidly applied than in macular laser therapy, which may cause pain. PRP is used not only to stop the proliferation of abnormal blood vessels but also to weld back together the retina and choroidal layers in the case of detachment. New developments aim at minimizing retinal damage by delivering microsecond laser pulses. These micropulses appear to cause fewer retinal injuries [22]. It has been speculated that the success of PRP may likely be due to a significant reduction of the choroidal thickness under the fovea, reducing blood flow [23].

To be successful in the long run, both laser methods – focal and PRP – must be accompanied by a general treatment of diabetes that is responsible for the development of diabetic retinopathy.

6.5.3 Cataract and glaucoma

Femtosecond laser-assisted cataract surgery (FLACS) is the latest addition to fs laser applications in ophthalmology [24]. A near-infrared Nd:YAG with a wavelength of 1053 nm is used for replacing the knife to access the capsular bag for replacing the lens. The laser cut is monitored by optical coherence tomography (OCT). The laser is mainly used as a knife, but also helps in fragmenting the old lens for removal. The main advantage is the all-contactless procedure, more gentle treatment of the eye, and earlier recovery. Several years after cataract surgery, a thin membrane forms on the back side of the new lens, which permanently impairs vision. In this case, again, an Nd:YAG laser is used to explosively vaporize and remove the membrane and its debris.

Glaucoma is an eye condition expressed by abnormal intraocular pressure (IOP). If not treated in time, it will lead to blindness. The IOP is controlled by a balance of production rate and outflow of aqueous humor. The IOP will rise if the production rate is higher than the drainage. The problem can be cured by reducing the production rate or by increasing the depletion rate. The production rate can be controlled to some extent by pharmaceuticals. If this is not sufficient, there are three laser-assisted methods to support glaucoma treatment:

1. Laser trabeculoplasty (LTP) (Fig. 6.27) : The laser beam is pointed onto the trabecular meshwork to increase the drainage through the canal of Schlemm. For this treatment a double-frequency Nd:YAG laser is used, emitting at a wavelength of 532 nm in the green visible region.
2. Controlled cyclophotocoagulation (Coco) (Fig. 6.28): Instead of increasing the drainage, the IOP can be lowered by suppressing the production rate of aqueous humor. To this end, a laser beam is directed to the ciliary body, where the aqueous production takes place. Two methods are applied: either the laser beam is directly focused on the ciliary body causing photocoagulation, or a narrow endoscope is used to deliver the laser beam through the cornea. The endoscope couples a light source, video imaging, and diode laser to visualize the ciliary processes directly during controlled laser application [25, 26]. This results in an efficient and secure procedure for reducing IOP.

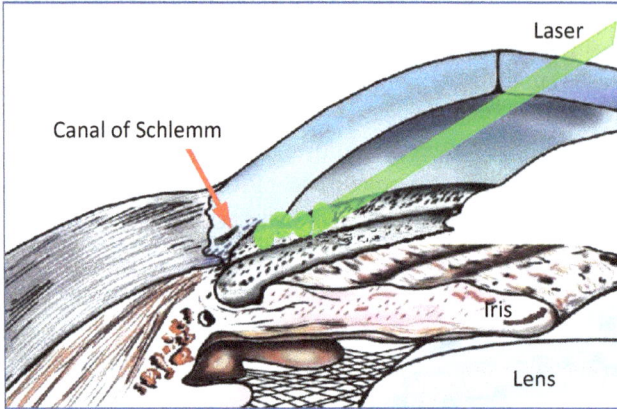

Fig. 6.27: Laser trabecular plastic surgery for increasing the drainage of aqueous humor. (Courtesy A.R.C. Laser Technology).

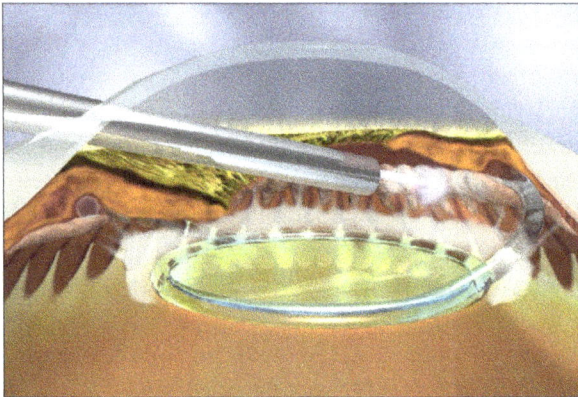

Fig. 6.28: Endoscope-controlled photocoagulation of the ciliary body to suppress the production rate of aqueous humor. (Reproduced from [22] by permission of John Wiley & Sons, 2017).

6.6 Summary

S6.1   Laser light is monochromatic, coherent, and directional electromagnetic radiation.

S6.2   All laser sources consist of three main components: (1) active medium; (2) optic pump; and (3) resonator.

S6.3   Laser safety is an important issue, in particular for laser light that is not visible.

S6.4   Laser emission can be described by a three or four-level system.

S6.5   Laser sources are distinguished according to their active medium as gas, dye, excimer, solid-state, and semiconductor laser.

S6.6   Laser sources can be operated in a continuous mode or in pulsed mode.

S6.7   Pulsing of a laser is achieved by mechanical switching, Q-switching, or mode-locking.

S6.8   Q-switched lasers require a fast optical switch in the resonator, which can be an electro-optic or an acousto-optic modulator.

S6.9   Laser penetration into the skin and eye depends on the wavelength.

S6.10   In the visible wavelength regime, hemoglobin and melanin mainly contribute to the absorption of laser light.

S6.11   Laser-tissue interaction can be categorized according to average power density, peak pulse power density, and pulse length.

S6.12   The main types of interaction are photothermal, photoablative, photomechanical, and photochemical.

S6.13   Photochemical interaction requires administering a photosensitizer.

S6.14   Photochemical reactions produce highly oxidizing singlet oxygen molecules.

S6.15   Photorefractive keratectomy can be used for correcting myopia and hyperopia.

S6.16   Two methods are applied: LASIK and LASEK. They are distinguished by cutting through the epithelium and stroma and by the recycling of the epithelium.

S6.17   Diabetic retinopathy either removes blebs on the retina or reduces the oxygen supply in the retina.

S6.18   Enhanced intraocular pressures can be reduced by increasing the drainage of aqueous humor through the canal of Schlemm or by reducing the production rate in the ciliary body.

Questions

Q6.1   What does the acronym LASER stand for?

Q6.2   What are the basic characteristics of laser light?

Q6.3   What are lasers mainly used for in medicine?

Q6.4   Lasers can be operated in continuous or in pulsed mode. What is the reason for laser pulsation?

Q6.5   What are the three methods for the pulsation of a laser?

Q6.6   Which types of Q-switches are you aware of?

Q6.7   How is mode-locking achieved?

Q6.8   What types of laser source are mainly used in medicine?

Q6.9   What laser flux is required to cause plasma-induced ablation and what is it used for?

Q6.10   How does photodynamic treatment proceed?

Q6.11   What is the difference between LASEK and LASIK?

Q6.12   Which types of laser treatments are common in ophthalmology?

| **Attained competence checker** | + | 0 | − |

I know what distinguishes a laser from a light bulb

I appreciate the many uses of laser sources

I know how a laser source can be pulsed

I know that lasers can be harmful, particularly those in the invisible wavelength range

I know in which frequency range water is most transparent to laser light

I know which molecules mostly contribute to the attenuation of laser light

I know how to determine the fluence of a laser

I can distinguish between thermal effects of laser sources, ablation, and plasma formation

In ophthalmology, laser sources are often used. I can name the three most important applications.

I can distinguish between LASIK and LASEK

Exercises

E6.1 **Laser peak energy and peak power**: A laser is specified as having a peak energy of $E_p = 100$ µJ, a pulse length of $\Delta t = 10$ fs, and a spot size of $A = 1$ cm^2. What is the peak fluence and the peak power density of this laser?

E6.2 **LASIK eye surgery**: LASIK uses a pulsed laser to reshape the cornea. A typical LASIK laser has a footprint of 1×1 mm^2, a wavelength of 193 nm, and a pulse length of 15 ns. Each pulse contains a photon energy of 1.0 mJ. What is the power and the power density (=intensity) of each laser pulse in watts?

E6.3 **Average power of a pulsed laser**: A pulsed laser has a peak power of 5 MW, a pulse repetition rate of 50 Hz, a pulse width of 1 ns, and a footprint of 1 mm^2.
What is (a) the peak fluence and (b) time average fluence of the laser beam?

E6.4 **Electric field**: A pulsed laser has a peak intensity of 100 MW/m^2. Determine the electric field amplitude of the laser's electromagnetic wave.

E6.5 **Flux**
Verify the statement in Section 6.4.3: " . . . flux on the order of 10^{12} W/cm^2 . . . leads to electrical fields of 10^7 V/cm . . .."

E6.6 **Spectroscopic terms**: Determine L, S, and J for the spectroscopic terms:
$^4I_{9/2}, {}^4F_{3/2}, {}^4F_{5/2}, {}^4I_{11/2}$

E6.7 **Degenerate states**: Show all three degenerate 4f electron configurations of the $^4F_{3/2}$ state in Nd^{3+}. What is common in these three spin configurations?

E6.8 **Temperature spread**
Derive the eq. (6.30): $\frac{dT}{dt} = \varepsilon \nabla^2 T$ using the continuity equation and the heat conduction equation.

References

[1]   Einstein A. Zur Quantentheorie der Strahlung. Physikalische Zeitschrift. 1917; 18: 121–128.
[2]   Jackson JD. Introduction to electrodynamics. 3rd Edition. New York, London, Sydney, Toronto: Wiley, 1998.
[3]   Arrigoni M, Bengtsson MS, Schulze M Solid state lasers: Laser pulsing: The nuts and bolts of Q-switching and modelocking. Laser Focus World. Issue 06/ 01/2012.
[4]   Abramczyk H. Introduction to laser spectroscopy. Amsterdam, Boston, Heidelberg, London: Elsevier; 2005.
[5]   Lakowicz JR editor, Topics in fluorescence spectroscopy. 1991; vol. 1: Berlin, Heidelberg, New York: Springer Press.
[6]   Niemz MH. Laser-tissue interactions: Fundamentals and applications. 2nd. Berlin, Heidelberg, New York: Springer- Verlag; 2007.
[7]   Boulnois JL. Photophysical processes in recent medical laser developments: A review. Lasers Med Sci. 1986; 1: 47–66.
[8]   Jacques S, Patterson M. Light–tissue interactions. In: eds. Chunlei Guo C, Singh SC. CRC handbook of laser technology and applications. 2nd. Boca Raton, London, New York: CRC Press; 2021.
[9]   Ma J, Yang X, Sun Y. et al., Thermal damage in three-dimensional vivo bio-tissues induced by moving heat sources in laser therapy. Sci Rep. 2019; 9: 10987.
[10] Ghanbari M, Rezazadeh G. Thermo-vibrational analyses of skin tissue subjected to laser heating source in thermal therapy. Sci Rep. 2021; 11: 22633.
[11] Carroll L, Humphreys TR. Laser-tissue interactions. Clin Dermatol. 2006; 24: 2–7.
[12] Hirsch LR, Stafford RJ, Bankson JA, Sershen SR, Rivera B, Price RE, Hazle JD, Halas NJ, West JL. Nanoshell-mediated near-infrared thermal therapy of tumors under magnetic resonance guidance. Pnas. 2003; 100: 13549–13554.
[13] Garrison BJ, Srinivasan R. Laser ablation of organic polymers: Microscopic models for photochemical and thermal processes. J Appl Phys. 1985; 57: 2909–2914.
[14] Pasquini C, Cortez J, Silva LMC, Gonzaga FB. Laser induced breakdown spectroscopy. J Braz Chem Soc. 2007; 18: 463–512.
[15] Guo LB, Zhang D, Sun LX, Yao SC, Zhang L, Wang ZZ, Wang QQ, Ding HB, Lu Y, Hou ZY, Wang Z. Development in the application of laser-induced breakdown spectroscopy in recent years: Topical review. Front Phys. 2021; 16: 22500.
[16] Dolmans DE, Fukumura D, Jain RK. Photodynamic therapy for cancer. Nat Rev Cancer. 2003; 3: 380–387.
[17] Huang Z. A review of progress in clinical photodynamic therapy. Technol Cancer Res Treat. 2005; 4: 283–293.
[18] Poon IKH, Hulett MD, Parish CR. Molecular mechanisms of late apoptotic/necrotic cell clearance. Cell Death Differ. 2010; 17: 381–397.
[19] Correia JH, Rodrigues JA, Pimenta S, Dong T, Yang Z. Photodynamic therapy review: Principles, photosensitizers, applications, and future directions. Pharm. 2021; 13: 1332.
[20] Lin JT. Progress of medical lasers: Fundamentals and applications. Med Devices Diagn Eng. 2016; 2: 36–41.
[21] Ambrosio R, Wilson S. LASIK vs LASEK vs PRK: Advantages and indications. Semin Ophthalmol. 2003; 18: 2–10.
[22] Gawęcki M. Micropulse Laser Treatment of Retinal Diseases. J Clin Med. 2019; 13: 8: 242, p. 1–18.

[23] Okamoto M, Matsuura T, Ogata N. Effects of panretinal photocoagulation on choroidal thickness and choroidal blood flow in patients with severe nonproliferative diabetic retinopathy. Retina. 2016; 36: 805–811.

[24] Levitz LM, Dick HB, Scott W, Hodge C, Reich JA. The latest evidence with regards to femtosecond laser-assisted cataract surgery and its use post 2020. Clin Ophthalmol. 2021; 15: 1357–1363.

[25] Siegel MJ, Boling WS, Faridi OS, Gupta CK, Kim C, Boling RC, Citron ME, Siegel MJ, Siegel LI. Combined endoscopic cyclophotocoagulation and phacoemulsification versus phacoemulsification alone in the treatment of mild to moderate glaucoma. Clin Exp Ophthalmol. 2015; 43: 531–539.

[26] Richter GM, Coleman AL. Minimally invasive glaucoma surgery: Current status and future prospects. Clin Ophthalmol. 2016; 10: 189–206.

Further reading

Silfvast WT. Laser fundamentals. 2nd. Cambridge, New York, Melbourne: Cambridge University Press; 2004.

Thyagarajan K, Ghatak A. Lasers: Fundamentals and applications. Berlin, Heidelberg, New York: Springer-Verlag; 2011.

Niemz MH. Laser-tissue interactions: Fundamentals and applications. 2nd. Berlin, Heidelberg, New York: Springer- Verlag; 2007.

Welch AJ, van Gemert MJC editors, Optical-thermal response of laser-irradiated tissue. Berlin, Heidelberg, New York: Springer-Verlag; 2011.

Kaschke M, Donnerhacke KH, Rill MS. Optical devices in ophthalmology and optometry: Technology, design principles and clinical applications. Weinheim, Germany: WILEY-VCH Verlag GmbH & Co. KGaA; 2014.

Bhattacharya B. Laser in ophthalmology. Step by step. New Delhi, Ahmedabad, Bengaluru, Chennai, Hyderabad: JAYPEE BROTHERS MEDICAL PUBLISHERS (P) LTD; 2009.

Jelinkova H editor, Lasers for medical applications, diagnostics, therapy and surgery. Amsterdam, Boston, Heidelberg, London, New York, Oxford: Elsevier; 2013.

Bäuerle DW. Laser processing and chemistry. 4th. Berlin, Heidelberg, New York: Springer-Verlag; 2011.

7 Nanoparticles for nanomedical applications

Acronyms and physical parameters	
FID	free induction decay
HD	hydrodynamic diameter
MNP	magnetic nanoparticle
PEG	polyethylene glycol
PES	photoelectron emission spectroscopy
PTT	photothermal therapy
RES	reticuloendothelial system
SERS	surface enhanced Raman scattering
SLP	specific loss power
SPIO	superparamagnetic iron oxide
SPR	surface plasma resonance
USPIO	ultrasmall superparamagnetic iron oxide
Size of nanoparticles	5–200 nm
Coating	PEG
Magnetic moment of Fe_3O_4	$4\,\mu_B$
Magnetic moment of Gd^{3+}	$8\,\mu_B$
Positive contrast enhancement	Gd^{3+}
Negative contrast agent	Fe_3O_4

7.1 Introduction

Nanomedicine is a multidisciplinary field of science and technology that has emerged over the last 10 to 20 years. It includes medical physics, materials science, biochemistry, biomedicine, pharmacy, polymer science, virology, and possibly more fields. Nanomedicine is broadly defined as "the medical application of nanotechnology for the diagnosis, treatment and general management of human health" [1]. As such, nanomedicine promises higher sensitive diagnostics and more precise treatment of certain diseases, especially cancer.

This short chapter will provide a brief overview on nanomedical procedures with main focus on physical aspects and methods. References at the end of this chapter will guide to review articles and books covering further aspects of this rapidly developing field. From a physics and materials science point of view, nanomedicine uses *nanoparticles* (NP) for:

- contrast enhancement of various imaging modalities;
- early diagnostics of cancer cells other than imaging;
- hyperthermia of tumor cells;
- delivery of targeted pharmaceuticals.

https://doi.org/10.1515/9783111168739-007

These four main applications of nanoparticles can be divided into two categories: diagnostics and therapeutics.

What does the term "nano" mean? In this context, the prefix "nano" refers to the length scale from (sub-)nanometers (10^{-9} m) to about 200 nm. Atoms, molecules, and proteins fall within this range. However, the science of atoms, molecules, and proteins is not nanoscience per se. In addition to their size, nanomaterials are mostly produced artificially. By scaling materials from the macro-size to the nano-size, they exhibit new physical properties, including mechanical, electrical, optical, and magnetic properties. Often these properties can be fine-tuned by the size and shape of the NPs. Furthermore, with increasing the surface/volume ratio, catalytic activities at surfaces play a prominent role. This distinguishes nanomaterials from atoms, molecules, and proteins, which have their intrinsic size, and no possibility for scaling.

Nanoparticles used for nanomedical applications range in diameter from about 5 to 200 nm. The surface is protected by a polymer or silica coating and a functionalized shell is used to target-specific receptors. A generic structure is sketched in Fig. 7.1, which can vary depending on the specific task.

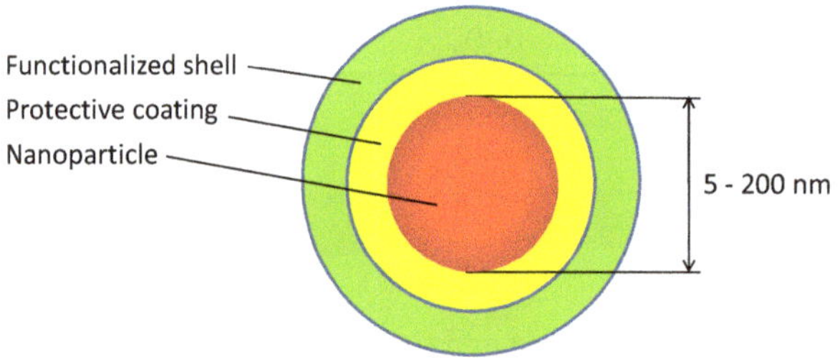

Fig. 7.1: Schematics of a nanoparticle used for nanomedical applications composed of an inorganic core, a polymer coating, and functionalized shell containing affinity ligands for specific targets.

The ideal nanoparticle finds its own way to the target tissue, its position is monitored by multimodal imaging methods, and after accumulation in the target area, it either delivers personalized drugs, or it can be stimulated from the outside to hyperthermic or chemical activity. In short, the "dream" nanoparticle is expected to "find, tell, and heal." A new term for nanoparticles exhibiting this multimodality has been coined: *theranostic nanoparticles* [2]. The expectations raised by nanomedicine are enormous, especially for those methods that use theranostic nanoparticles. The interconnectivity of medical nanoparticles is illustrated in Fig. 7.2. Precise targeting is the prerequisite for all other modalities. Most nanoparticles combine at least two modalities, such as imaging and therapy.

Before theranostic nanoparticles can be successfully applied in nanomedical treatment protocols, a number of issues must first be considered and eventually resolved. Precise knowledge of their path through the body determines their ultimate

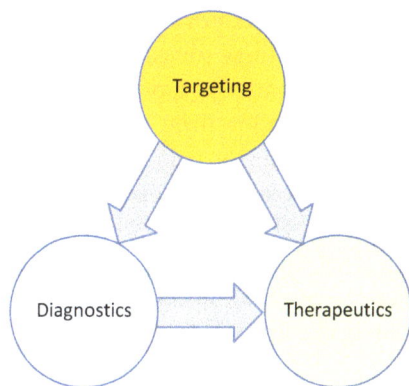

Fig. 7.2: Nanomedical triangle, consisting of the main goals: diagnostics of diseases, targeting diseased tissues, and providing an exact therapy. Ideally, multifunctional nanoparticles can address all three modalities.

effectiveness [3]. The pathway includes possible biodegradation, bioassimilation, elimination in liver, spleen, and kidneys, or permanent deposition in tissues. Toxicity to healthy tissues, immune response, and long-term health effects require attention. The parameter space is vast and includes particle size, shape, core-shell structure, material choice, electric charge, and functional coatings. Once the main issues are resolved, the benefits of nanoparticles for diagnostics and therapeutics have to be verified in vivo by clinical studies. Obviously, there are many tasks for many scientists and practitioners in this new field to make nanomedicine a success story. In this chapter, we do not emphasize the synthetic aspects of nanoparticles, but instead focus on their physical properties.

The concept of nanoparticles moving around and transporting some payloads to their destination is used throughout biology. Cells and, on a smaller scale, mitochondria, blood cells, liposomes, ferritin, extracellular vesicles (exosomes), and viruses, such as SARS COVID I and II, are some examples of natural nanoparticles. These naturally occurring nanoparticles are not the subject of the present chapter.

In Section 7.2, we will present some general biological aspects that apply to all kinds of artificial NPs; in Section 7.3; we discuss the magnetic properties of NPs and their applications in nanomedicine, followed by a treatment of nonmagnetic metallic NPs in Section 7.4. Nanoparticles for microscopy and spectroscopy are discussed in 7.5. The chapter is concluded with a brief note on theranostic nanoparticles in Section 7.6.

7.2 Pathway of nanoparticles through the body

7.2.1 Reticuloendothelial system (RES)

After intravenous injection, NPs invariably encounter the body's *reticuloendothelial system* (RES) [4]. RES is part of the immune system and consists of phagocytic or scavenger cells. More specifically, these cells are made up of monocytes in the bloodstream and macrophages in the tissues. In the liver, macrophages are known as Kupffer[1] cells. Once administered, NPs are quickly recognized as aliens by these surveillance phagocytes. Upon recognition, the NPs are enveloped by plasma proteins, a process known as *opsonization* that increases the monocytes' appetite to swallow them like packmen. After incorporation, the NPs are transported to the liver or spleen and disposed of there, where macrophages and Kupffer cells degrade the coating and excrete the remainder in the feces (see Fig. 7.3). The opsonization and elimination process can be rapid, varying from 1 to 30 h. Hydrophobic particles, negatively charged particles, and particles larger than about 50 nm in size are eagerly scavenged by monocytes. However, long circulation times and a low elimination rate are desirable to keep the NPs in the bloodstream until they reach organs and targeted tissues for diagnostics and therapeutics. Therefore, it is necessary to increase the circulation time, for example, by bypassing the RES system. Hence, an important strategy of NP design is to make them sufficiently small, hydrophilic, and charge-neutral. Alternatively, the NPs can be camouflaged (stealth NP), to avoid being detected as aliens by the RES. Another strategy is to use the NPs as Trojan horses that deliver their drugs to the target sites.

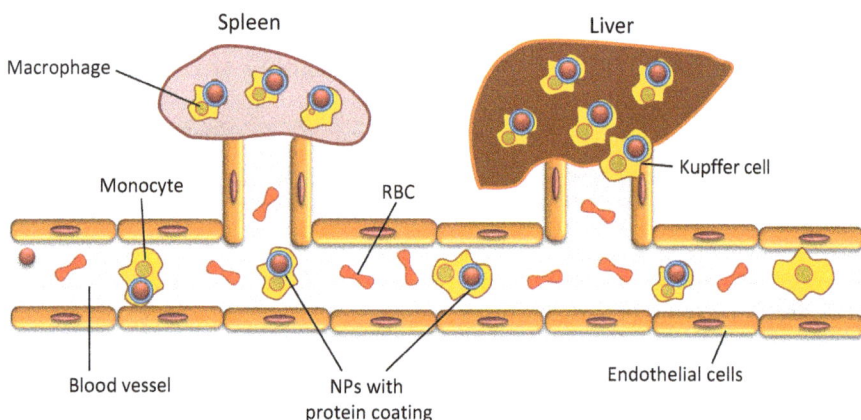

Fig. 7.3: Nanoparticles are rapidly recognized by the reticuloendothelial immune system (RES) that coats them first with proteins before monocytes carry them off to spleen and liver, where they are decomposed and finally excreted. RBC = red blood cell.

7.2.2 Clearance

Contrast agents such as iodinated iopromide (Fig. 8.20 of Volume 2) for x-ray imaging or Gd chelates for MRI (Fig. 3.38 of Volume 2) have a short half-life in the systemic circulation due to their rapid renal clearance [5]. They are too small for

---

1 Karl Wilhelm Kupffer (1829–1902), Baltic-German physiologist.

the RES to be detected and not large enough to be retained in the bloodstream. Any molecules or particles smaller than 10 nm are removed by the glomerular filter in the kidneys (see Section 10.6 of Volume 1). To keep nanoparticles in the bloodstream, they should have a total hydrodynamic diameter (HD) between 10 and 50 nm. On the other hand, rapid clearance is beneficial for perfusion studies of liquids through the kidneys using radioisotopes, such as for the MAG3 scan (see Section 9.3 of Volume 2).

7.2.3 Enhanced permeation and retention effect (EPR)

After intravenous administration, NPs circulate throughout the body and passively accumulate in tumor tissues through the *enhanced permeation and retention effect* (EPR) [6]. The EPR effect is caused by leaking and ruptured blood vessels in the tumor volume combined with porous tissue, shown schematically in Fig. 7.4. The leaky vasculature allows the NPs to penetrate the endothelial wall, from where they seep through the porous tumor tissue. In addition, the lymphatic drainage system is underdeveloped in cancerous tissues, which increases the retention time of NPs in these areas. Therefore, NPs have ample time to either discharge their anticancer drug load or undergo hypothermic treatment. In any case, the passive accumulation of NPs already helps to identify the location and extent of cancerous volumes.

7.2.4 Coatings

Whatever the core of NPs contains: metals, oxides, or semiconductors, the NPs should be water soluble and biocompatible. Silica coatings can well meet these conditions. Silica-coated NPs have excellent water-solubility properties, their shell thickness can be easily adjusted, and their biocompatibility is excellent. However, the dielectric silica shell accumulates electric charges and is sensitive to the pH value of the embedding medium, which can lead to precipitation and gelation. Therefore, silica coatings need to be protected with another organic shell, usually *polyethylene glycol* (PEG), to improve hydrophilicity and colloidal stability. PEG with the general formula $C_{2n}H_{4n+2}O_{n+1}$ is a biocompatible and water-soluble polymer of variable chain length, also used in pharmacy for drug delivery and cosmetics. The varying length allows adaptation of the NP's hydrodynamic diameter. In addition, PEG shells impede protein adsorption, suppress opsonization, and thus bypassing RES. PEG coatings can be further modified to target-specific organs or tumors, as outlined in the next section. A generic nanoparticle is sketched in Fig. 7.5 according to the design principles outlined so far. The review article [7] discusses many additional design concepts.

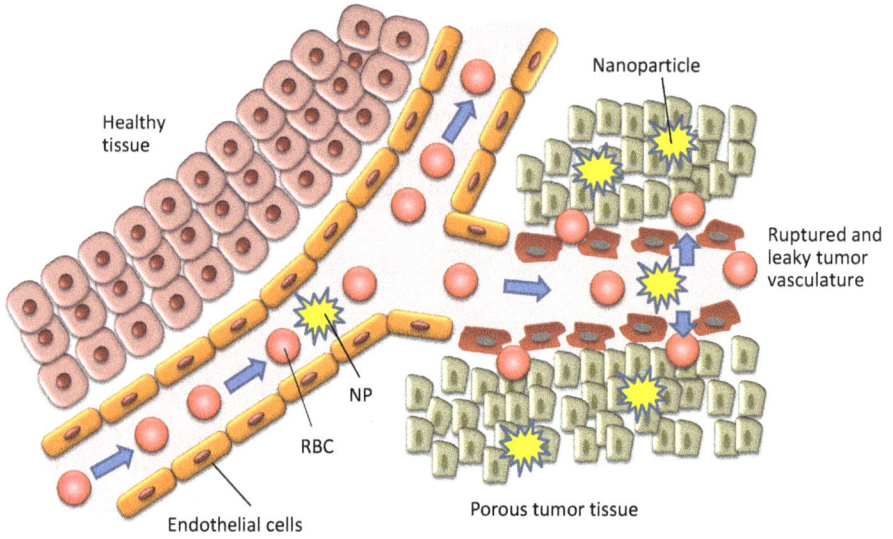

Fig. 7.4: Vasculature in healthy and tumorous tissue. In tumorous tissue, the vasculature is ruptured and leaky, allowing nanoparticles to penetrate through and percolate into the carcinoma even without specific targeting. Hence, tumorous tissues provide enhanced permeation and retention of nanoparticles (adapted with permission from [6]).

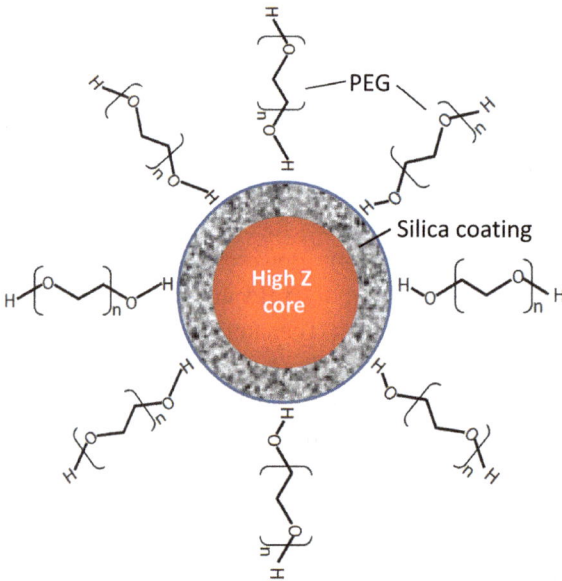

Fig. 7.5: Generic nanoparticle for x-ray contrast enhancement, containing a high Z material in the core. The NP is coated with silica and covered with a polyethylene glycol (PEG) shell, consisting of a chain with varying length.

7.2.5 Antigen-Antibody

Viruses and bacteria have antigens on their surface that are different from the antigens produced by the immune system of our body. Antigens consist of proteins or carbohydrate chains. If foreign antigens try to penetrate our body, they are recognized as foreign by antibodies. Antibodies, in turn, are proteins that are able to recognize certain foreign particles based on the shape of their antigens. Antibodies are produced by the body's immune system when it detects harmful substances, such as antigens. Antibodies are Y-shaped molecules. They fit with antigens like a lock and key. The basic unit of an antibody is a tetramer of four polypeptides: two identical light chains and two identical heavy chains, each consisting of a constant and a variable region, see Fig. 7.6(a). The chains are held together by disulfide bonds.

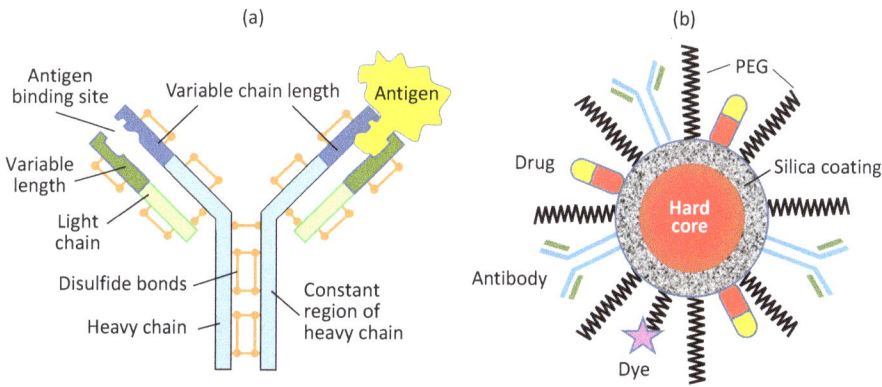

Fig. 7.6: (a) Schematic structure of an antibody. The Y-shaped molecule consists of two heavy chains and two light chains. Both have a fixed part and a part of variable-length. The chains are held together by disulfide bonds. Each arm of the antibody contains one antigen binding site, which acts according to the lock and key principle. (b) Nanoparticle coated with silica and conjugated with PEG, antibodies, drugs, and eventually further contrast agents or targeting ligands.

Our body does not produce antibodies that bind to its own antigens. Therefore, any particles bound to antibodies are foreign. The concept of antigen-antibody has been used to fight cancer cells. Extensive research in recent years has shown that antigen structures on tumor cells differ from antigens on normal cells [8]. The next task is then to find antibodies that clearly target the tumor-specific antigens. However, antibodies that predominantly bind antigens in cancer cells rather than in normal tissues have not yet been identified. Despite this lack of absolute specificity for cancer cells, antibodies with preferential cancer cell reactivity have very high tumor recognition. Therefore, the concept of targeted NPs conjugated with antibodies that bind to tumor-specific antigens is promising and widely used. Figure 7.6(b) shows schematically a

NP with a silica coating and a PEG shell that also contains antibodies conjugated to the silica coating. NPs can be loaded with additional drugs, contrast agents, radioisotopes, fluorescent tags, enzymes, etc., for multifunctional tasks.

7.2.6 Targeting

We distinguish between physical, passive, and active targeting of NPs for diagnostics and therapeutics. Physical targeting refers to NPs being distributed throughout the body by blood flow after administration to the vasculature. No specific binding site exists, but particles can penetrate leaky vessels or become trapped in stenosis regions, depending on their size. Active targeting is the opposite of physical targeting. In the latter case, NPs are covered with targeting ligands to dock at specific sites in tissues. Either the antigen-antibody interaction or the ligand-receptor interaction is used for active targeting. Passive targeting exploits differences in the vasculature and in the physiological properties of tumors versus healthy tissue that affect uptake behavior and retention times. This effect is known as the EPR effect and passive targeting is based on this effect.

7.2.7 Size

In nanomedicine, size matters. We differentiate between core size, including coating and shell, and the *hydrodynamic diameter* (HD). The core size determines the physical properties of NPs, especially their magnetic and optical properties. Coating and shell are responsible for biocompatibility, colloid stability, and biofunctionality. The HD is the size of NPs in an aqueous solution where the shell is covered by attached water molecules and eventually plasma proteins. HD size determines biodistribution and circulating half-life in the body. Longer circulating half-lives make NPs more interesting for blood pool studies. Currently, synthetic skills are very advanced and offer complete control over all sizes, shapes, and shells. NPs can be synthesized with spherical or rodlike shapes; cores can be solid, porous, or cage-like; shells perform several tasks that have been mentioned previously, and more are highlighted below.

7.2.8 Biocompatibility

Biocompatibility is the antonym of toxicity. Toxicity is easier to define than biocompatibility. Toxic are chemicals or materials that are capable of causing injury, cell death, or dysfunction of any part of the body. Toxicological hazard is the potential of a compound or material to elicit an adverse biological response, taking into account the nature of the response and the dose required to elicit it [9]. In this context, biocompatibility can be defined as "biological performance that in a specific application is judged to be appropriate for that specific situation" [10]. This implies that materials or chemicals that come into contact with biological tissue do not have any adverse effects on the tissue, such as inflammation or lesions. Conversely, materials that react by biodegrading on contact with tissues should not release or leak toxic or radioactive substances unless intended to do so. Biocompatibility has also been defined more broadly as the "ability of a material to function in a particular situation with an appropriate host response" [11].

7.3 Magnetic nanoparticles for diagnostics and therapeutics

Gadolinium (Gd) and magnetite (Fe_3O_4) are the most commonly used contrast agents for magnetic resonance imaging (MRI). So we will discuss the magnetic properties of these two materials in more detail. Magnetite is also used for hyperthermia applications. We first want to clarify the atomic origin of Fe and Gd magnetism. Next, we study cooperative effects leading to ferromagnetism on the macroscale and superparamagnetism on the nanoscale.

7.3.1 Magnetic properties of gadolinium

Gd has a half-filled 4 f shell surrounded by a screening shell of $5s^2p^6$ electrons and three electrons in the $5d^16s^2$ orbitals. The outer electrons determine the ionicity and contribute to the bonding of Gd in oxides (Gd_2O_3) or in chelate molecules. The inner 4 f electrons are essentially unaffected by the chemical bond and retain their atomic levels. In the ground state, the spins of the seven electrons in the 4 f shell of Gd line up to a total spin $S = 7/2$ and an orbital moment $L = 0$ (Fig. 7.7(a)). Therefore, the total angular moment $J = L + S$ in this case equals S: $J = S$.

The local magnetic moment per atom is given by [12]:

$$m = g_J \sqrt{J(J+1)}\mu_B,$$ (7.1)

where

$$g_J = \frac{3}{2} + \frac{S(S+1) - L(L+1)}{2J(J+1)}$$ (7.2)

(a)   (b)   (c)

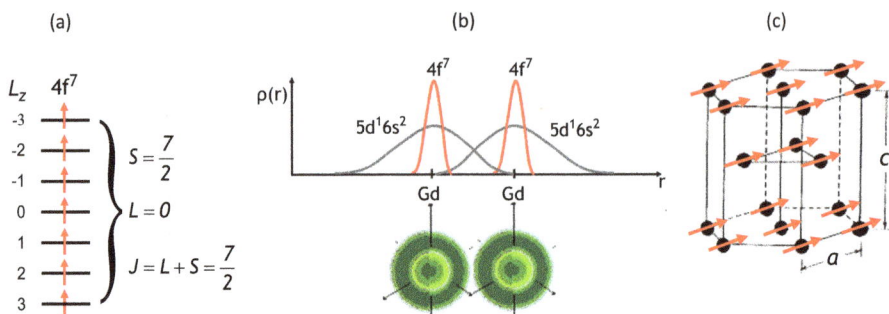

Fig. 7.7: Magnetism of Gadolinium. (a) Level scheme of the 4 f shell occupied by seven electrons with spin ½ adding up to 7/2; the orbital moment L is cancelled. (b) 4 f wave functions from neighboring atoms do not overlap, in contrast to 5d6s wave functions. Magnetic interaction between 4 f electrons is mediated by the polarization of $5d^1 6s^2$ electrons. (c) Ferromagnetic structure of hexagonal Gd metal.

is the Landé[2]-factor, and μ_B is the Bohr[3] magneton. In case of Gd with $J = S = 7/2$, the local moment is $m_{Gd} = 8\ \mu_B$. This Gd magnetic moment m_{Gd} is stable, temperature independent, and independent on whether Gd is in a gaseous, molecular, or solid-state. In metal form, Gd is a ferromagnet below the Curie[4] temperature of 292.5 K [13]. Ferromagnetism implies that all magnetic moments of different Gd atoms within the hexagonal crystal structure line up in parallel and form an ordered magnetic state (Fig. 7.7 (c)). The outer $5d^1 6s^2$ conduction electrons are magnetically polarized by the core 4 f electrons, mediating the ferromagnetic order in Gd (Fig. 7.7(b)). While the local 4 f moment is stable, their orientation in space is increasingly random as the temperature goes up. Above the Curie temperature, Gd is paramagnetic, meaning that the parallel alignment of the local Gd magnetic moments is lost. However, when a magnetic field H is applied, the local magnetic moments easily turn into the magnetic field direction, inducing partial order. This phenomenon is referred to as magnetic susceptibility χ_m. The magnetic susceptibility χ_m is defined by the change of magnetization per increment of magnetic field:

$$\chi_m = \frac{dM(T)}{dH},\qquad(7.3)$$

where the magnetization $M(T)$ is defined by:

---

2 Alfred Landé (1888–1976), German-American theoretical physicist.
3 Niels Henrik David Bohr (1885–1962), Danish theoretical physicist and Nobel laureate in Physics (1922).
4 Pierre Curie (1859–1906), French physicist and Nobel laureate in physics (1903).

$$M(T) = \frac{1}{V} \sum_I m_i. \qquad (7.4)$$

$M(T)$ is the sum over all local magnetic moments m_i within a volume V, normalized by V. The magnetic susceptibility χ_m has a maximum at the Curie temperature but drops off at higher temperatures T.

In Gd^{3+} chelates the Gd^{3+} ions are in an isolated atomic state. Any magnetic interaction is suppressed by the large separation between the molecules. Hence, Gd^{3+}-chelate molecules are in a paramagnetic state at all temperatures. In a magnetic field, the 4 f magnetic moments align partially in the field direction. As soon as the magnetic field is turned off, all magnetic moments relax back into a random state.

7.3.2 Magnetic properties of magnetite

In nanomedicine, magnetite is used for MRI contrast enhancement and hyperthermia. Magnetite (Fe_3O_4) is one of three known iron oxides. The other two are maghemite (γ-Fe_2O_3) and hematite (α-Fe_2O_3) [14]. Magnetite, the only oxide we want to discuss here in some detail, has an inverse spinel crystal structure with 24 Fe ions and 32 oxygen ions per unit cell. The 24 Fe ions are grouped in 8 Fe^{3+} ions on crystal lattice sites with tetrahedral local symmetry (T) and another 2 × 8 Fe ions on sites with local octahedral symmetry (O). Fe ions on T sites exhibit 3+ valence, those on O sites show valence fluctuation: 8 Fe ions have a valence 3+, the remaining 8 Fe ions have a valence 2+. With a half-filled 3d shell, Fe^{3+} has a total spin S = 5/2 and orbital moment L = 0. The resulting magnetic moment is 5 μ_{Bohr}. In contrast, Fe^{2+} has a sum spin S = 4/2 and a crystal field quenched orbital moment, yielding a magnetic moment of 4μ_{Bohr}/atom. The eight Fe^{3+} ions on T sites and the equivalent ones on O sites have antiparallel spins: their magnetic moments are compensated by antiparallel or antiferromagnetic ordering. Only the remaining 8 octahedral sites occupied by Fe^{2+} contribute to the magnetism of magnetite. In total, magnetite is referred to as a *ferrimagnet* with a Curie temperature of 850K and magnetic moment per formula unit of 4μ_{Bohr}. The spin structure is shown in Fig. 7.8. Maghemite (γ-Fe_2O_3) is also a ferrimagnet. In contrast to Fe_3O_4, there are fewer Fe ions on O sites (8+5), and they all have 3+ valence: 8 for compensating the magnetic moments of Fe^{3+} on T sites, and 5 more with spin 5/2 and ferromagnetic order. Often γ-Fe_2O_3 and Fe_3O_4 coexist. A distinction is possible but complicated and elaborate for tiny nanoparticles. Hematite, in contrast, is an antiferromagnet and not useful as CA in MRI. Maghemite theoretically should have a magnetic moment of 5 μ_{Bohr} per formula unit. However, it is difficult to find experimental confirmation in the literature. Often manganese ferrite ($MnFe_2O_4$) NPs are used in nanomedicine for their confirmed higher magnetic moment of 5 μ_{Bohr} per formula unit as compared to 4 μ_{Bohr} in magnetite.

(a)

Fe^{2+}, 3d^6,
S=4/2

e_g

t_{2g}

e_g

Fe^{3+}, 3d^5,
S=5/2

t_{2g}

(b)

8Fe^{3+}
S=5/2

(T)

(O)

S=5/2   S=4/2

8Fe^{3+} 8Fe^{2+}
Fe$_2$O$_3$ FeO

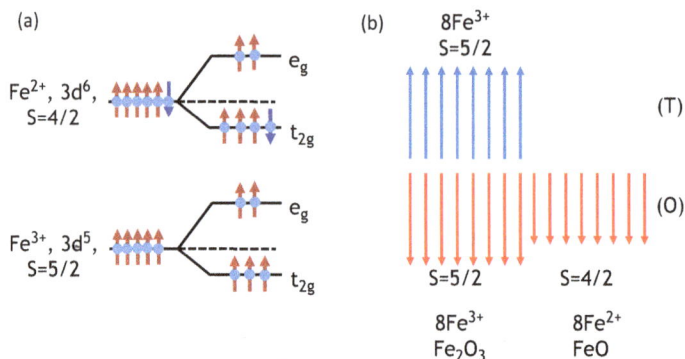

Fig. 7.8: Spin structure of ferrimagnetic magnetite. (a) Electronic shells of Fe^{2+} and Fe^{3+}. The degenerate 3d shell is split by local crystal electrical fields into t_{2g} and e_g levels. The resulting total spin for Fe^{2+} is S =4/2 and S =5/2 for Fe^{3+}. (b) Schematic occupation of Fe^{2+} and Fe^{3+} with their respective total spin angular momentum on crystal lattice sites with local tetrahedral (T) and octahedral (O) site symmetry. Note that Fe ions on octahedral sites have fluctuating valence. Fe^{3+} magnetic moments on O and T sites compensate each other by antiparallel orientation. Only Fe^{2+} ions on T sites contribute to the macroscopic magnetism of magnetite.

7.3.3 Superparamagnetism

Despite the highly complex magnetic spin structure, magnetite behaves in practice like a normal, rather soft, ferromagnetic material. In fact, it is the oldest ever known and, according to legend, discovered by shepherds of the Greek city of Magnesia, which gave its name to magnetism. A universal property of ferromagnets is their magnetic hysteresis when magnetizing in an external magnetic field H. Magnetic hystereses indicate the presence of a magnetic domain state during reversal, which is the hallmark of ferromagnetic order. Magnetization causes stray fields. In order to reduce stray fields and keep the magnetic flux inside ferromagnets, domains separated by domain walls are formed, as shown schematically in the left panel of Fig. 7.9. When such a ferromagnet is exposed to an external magnetic field H, a magnetic hysteresis can be measured, i.e., the magnetization $M(H)$ as a function of the field strength H. Magnetic hystereses are characterized by three essential properties:
1. saturation magnetization M_s at high field;
2. remanent magnetization M_r at zero field;
3. coercive field H_c at $M = 0$ when the magnetization changes sign in the ascending or descending branch of the hysteresis.

These three characteristic points are indicated in the left panel of Fig. 7.9. When ferromagnets are reduced to the size of nanoparticles, domains can no longer form since domain formation would cost too much (exchange) energy. Then each nanoparticle has a single ferromagnetic domain, which is indicated schematically in the right panel of Fig. 7.9 by a series of arrows within a nanoparticle. All the magnetic moments in a nanoparticle act together as one gigantic magnetic moment m_{macro}, called the magnetic moment of a macrospin. This macrospin is surrounded by a magnetic dipole field. Macrospins can only form in the ferromagnetic state of NPs, that is, at temperatures below the Curie temperature. Magnetite NPs have macrospins at a body temperature of 310 K, whereas Gd NPs do not, since the Curie temperature for solid Gd particles is at least 19 K below body temperature.

Next we consider an ensemble of MNPs, each macrospin having a total magnetic moment m_{macro}. In zero external magnetic field, the macrospins are randomly oriented, resulting in zero magnetization. When turning on a magnetic field, the macrospins will respond to the field and partially orient in the field direction, yielding a finite magnetization $(M(T,H) = \chi_m(T)H > 0)$. The temperature and field dependence of the magnetization of an ensemble of MNPs can be described by the Langevin[5] function $L(x)$:

$$M(T,H) = \frac{N}{V} m_{macro} L(x) \qquad (7.5)$$

with

$$L(x) = coth(x) - \frac{1}{x} \qquad (7.6)$$

where

$$x = \mu_0 \frac{m_{macro}H}{k_B T} \qquad (7.7)$$

is the ratio of the Zeeman[6] energy $\mu_0 m_{macro} H$ to the thermal energy $k_B T$, and μ_0 is the magnetic permeability of the vacuum. Because of the resemblance with a classical paramagnetic system, the magnetization of nanoparticles is called *superparamagnetism*. However, it is the magnetism of tiny ferromagnetic beads, randomly distributed in space, yielding a paramagnetic response when exposed to a magnetic field. Fig. 7.9 shows in the lower right panel schematically the magnetization of magnetic nanoparticles in an external field that lacks a hysteretic behavior. Further details on the superparamagnetism of MNPs can be found in the review article [15]. If superparamagnetism is based on iron-oxide magnetite, this system is referred to as *superparamagnetic iron-oxide* (SPIO) nanoparticles.

When superparamagnetic nanoparticles come close to each other, they couple via magnetic dipole-dipole interaction and try to align chain-like. Suspended in a

---

5 Paul Langevin (1872–1946), French physicist.
6 Pieter Zeeman (1865–1943), Dutch physicist, Nobel Prize laureate in Physics 1902.

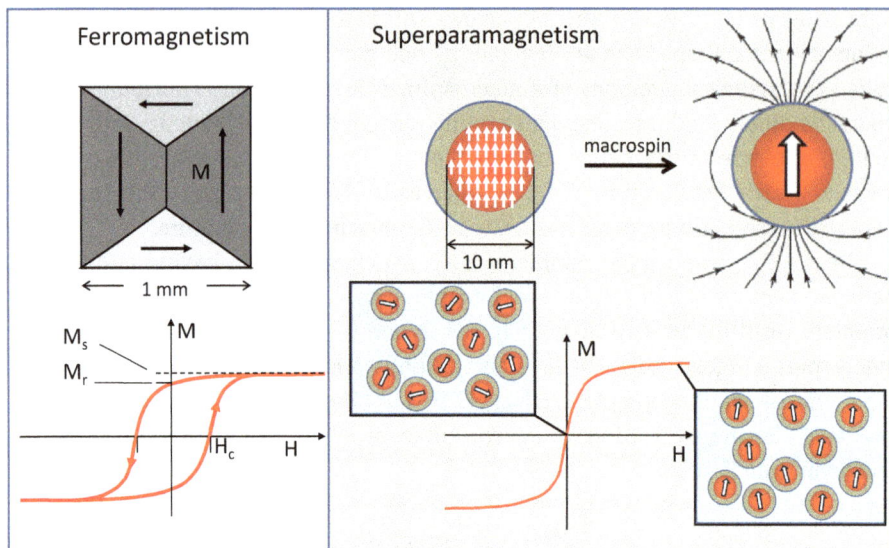

Fig. 7.9: Left panel: ferromagnet in a domain state. The magnetization of a ferromagnet is characterized by a saturation magnetization M_s in high external magnetic fields, a remnant magnetization M_r at zero field, and coercive fields H_c at the point where the magnetization changes sign on the ascending or descending branch of the hysteresis. Right panel: In sufficiently small nanoparticles, there are no magnetic domains. The tiny grains are in a single domain ferromagnetic state represented by a macrospin. An ensemble of magnetic nanoparticles (MNPs) will be randomly oriented without an external field, yielding zero magnetization. Turning on a magnetic field, the MNPs will partially align in the direction of the field, resulting in a finite magnetization. A Langevin function can describe the magnetization versus field, resembling the magnetization curve of isolated paramagnetic ions. Therefore the term *superparamagnetism* is used to describe the magnetic behavior of these nanoparticles.

liquid (ferrofluid), the MNPs tend to agglomerate. Agglomeration can be suppressed by sufficiently thick shells surrounding the NPs. The shells keep them at a distance where the rapidly decaying magnetic dipole-dipole interaction becomes negligible.

> **!** Magnetic nanoparticles can be made of transition metal oxides (Fe_3O_4) or elemental metals (Co, Ni, Fe, Gd, . . .). Fe_3O_4 is preferred because of its biocompatibility. All other magnetic nanoparticles need a protective shell to avoid toxicity effects.

7.3.4 Blocking temperature

As indicated in Fig. 7.9, the magnetic moments within MNPs are aligned parallel. The direction in which they line up is called the *easy axis*. This axis is determined by the magnetocrystalline anisotropy and is an intrinsic material property. If all moments in a MNP decided to orient themselves in another direction, for instance, in the direction opposite to the original one, they would have to overcome a potential barrier, indicated in Fig. 7.10(a). The potential barrier is given by the magnetic anisotropy energy density K times the magnetic volume V_m of the particle:

$$E_{ani} = KV_m \tag{7.8}$$

In thermal equilibrium, it is a question of temperature and time whether the magnetization within a particle will indeed succeed and overcome the barrier. From a statistical mechanics point of view, a flip will take place on average after the relaxation time τ:

$$\tau = \tau_0 \, exp \left(\frac{KV_m}{k_B T} \right), \tag{7.9}$$

where τ_0 (10^{-9} s – 10^{-10} s) is the attempt period. This equation shows that the attempt time, i.e., the average time between two flips, increases exponentially with the ratio of anisotropy energy versus thermal energy. However, the flipping time may be quite short for small particles and high temperatures.

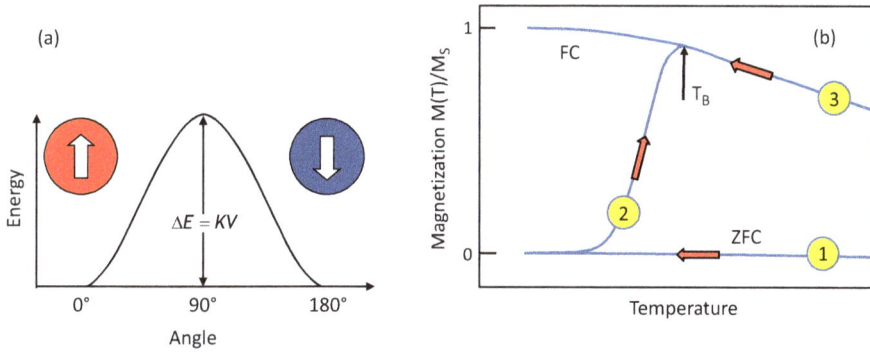

Fig. 7.10: (a) Potential barrier for flipping of magnetic NPs. (b) Path (1): Magnetization measurements during cooling in zero applied magnetic field (ZFC). Path (2): Warming up after the system has been cooled in zero field. Path (3): Magnetization measurement during cooling in an applied field (field cooling, FC). Magnetization paths (2) and (3) meet at the blocking temperature T_B.

MNPs appear to be static when during the time of observation τ_{ob} no flip occurs. This may be the case below a certain temperature T_B, called *blocking temperature*. For temperatures $T > T_B$, the ensemble of particles frequently flips, such that the time-averaged magnetization drops to zero. Hence the blocking temperature T_B separates regions of static versus dynamic magnetization. Rephrasing the equation for the relaxation time, we find for the blocking temperature:

$$T_B = \frac{KV_m}{k_B ln(\tau_{ob}/\tau_0)}, \tag{7.10}$$

which shows that the blocking temperature scales linearly with the volume of the MNPs.

Figure 7.10(b) schematically shows how the blocking temperature T_B can be determined with the help of magnetization measurements. The magnetization measurement protocol is as follows. First, the MNPs are exposed to a small magnetic field and cooled from high to low temperatures, referred to as *field cooling* (FC) measurement. In this case, the average magnetization of all NPs steadily increases with decreasing temperature and saturates at the lowest temperature. In a second measurement, the same MNPs are first cooled to low temperature in zero field. Then the magnetization stays zero for all temperatures. This magnetization path is referred to as the *Zero Field Cooling* (ZFC) state. Now, starting at the lowest temperature, the same field is turned on that was used for the FC experiment. When increasing the temperature, the magnetization will steadily increase and meet the magnetization curve of the FC curve at a certain temperature, which is called the blocking temperature T_B. Again, magnetic NPs below T_B are blocked with respect to their magnetization orientation; above T_B, the magnetization is free to rotate. For $T < T_B$, the magnetic reversal shows a hysteresis with a finite coercive field H_c; for $T > T_B$ the magnetic reversal shows no hysteresis and $H_c = 0$.

The blocking temperature increases with increasing NP size. According to [16], the blocking temperatures for Fe_3O_4 NP with diameter 3.8 nm, 4.5 nm, and 5.5 nm are 100 K, 150 K, and 250 K, respectively. From this trend, we can conclude that SPIO NPs with a diameter of 10 nm have a blocking temperature above body temperature, i.e., above 310 K. For contrast enhancement in MRI, MNPs should be designed to have a T_B above body temperature. In contrast, MNPs used for hyperthermia should have a T_B at or below body temperature.

7.4 MR imaging contrast

7.4.1 Basics of MRI contrast

MRI contrast relies on differences in spin-tissue relaxation time $T1$ and spin-spin relaxation time $T2$. $T1$ is also referred to as the longitudinal relaxation time or the time for M_z magnetization recovery; $T2$ is the transverse relaxation time or the M_{xy} magnetization dephasing time (Fig. 3.4 in Volume 2). $T1$ depends on how fast the spin excitation energy can be transported off to the surrounding molecules, whereas $T2$ is affected by magnetic field inhomogeneities generated by neighboring spins, field gradients in magnets, or magnetic dipole fields emanating from magnetic contrast agents (CAs).

In MRI, the contrast is not as rigid as in XRT: MRI contrast can be fine-tuned with the help of the parameters TR (time to repeat) and TE (time to echo) to generate $T1$ or $T2$ weighted images, or for emphasizing differences in the proton density. MRI "intensity" is not comparable with the intensity of a photon or particle beam. MRI intensity is an inductive voltage generated by a precessing in-plane magnetization M_{xy}, referred to as *free induction decay* (FID). The larger the z-component of the magnetization M_z after relaxation time T1, the larger is the signal from FID after time TE. Larger signals produce brighter pixels on MRI images after Fourier transformation (see Chapter 3, of volume 2 for further details). T1 weighted images are taken with short TR and short TE, whereas T2 weighted images require long TR and long TE.

Shortening T1 by using a CA results in faster M_z recovery and a stronger FID signal that produces brighter pixels. CAs that lower T1 are called positive contrast agents. Shortening T2 has the opposite effect. A shorter T2 relaxation time gives a weaker FID signal, resulting in a darker pixel on the image. CAs that tend to shorten T2 are referred to as negative contrast agents. Gd^{3+} chelates (Chapter 3, Section 3.10.2 and Fig. 3.41 in volume 2) belong to the class of positive contrast agents, while magnetite (Fe_3O_4) NPs are negative contrast agents. We will return to the discussion of these differences later. CAs usually affect both T1 and T2. To assess their main properties, the reciprocal relaxation times $R_i = 1/T_i$, called relaxivities, are quoted:

$$R_i = r_i x_{CA} + R_{i0}; \; (i = 1, 2). \tag{7.11}$$

Here r_i is the specific relaxivity (units: $mMol^{-1} s^{-1}$) and R_{i0} is the baseline relaxivity without CA. For a positive CA, the effect on T1 should be larger than on T2. This implies that the ratio r_2/r_1 is small. For negative CAs, the ratio r_2/r_1 becomes large. Theoretical considerations show that the relaxivities of protons in an aqueous environment affected by MNPs in high fields typical for MRI conditions can be expressed by [17]:

$$r_{1,2} \sim \frac{1}{d_{NP} D} \gamma^2 m^2_{macro} x J(\omega, \tau_D). \tag{7.12}$$

Here d_{NP} is the NP diameter, D is the diffusion constant of solvent and solute, γ is the gyromagnetic ratio of protons, m_{macro} is the magnetization of a macrospin in the core of a MNP, x is the concentration of NPs in solution, and $J(\omega, \tau_D)$ is the spectral density function, which is described in Section 3.5 of Volume 2 Eq. (7.12) shows that the relaxivity is influenced by the NP size and the diffusion constant, aside from the magnetization dependence. Here ω refers to the Larmor frequency of protons in the external field, and the relaxation time:

$$\tau_D = \frac{d^2_{NP}}{4D} \tag{7.13}$$

is the interaction time of protons with NPs. It is possible that multiple encounters of the protons with MNPs are required for the protons to relax.

7.4.2 Gd^{3+} chelates as positive contrast agent for MRI

As discussed in Section 7.3.1(a), Gd^{3+} ions have a stable atomic magnetic moment of 8 μ_B in the 4f – shell. When exposed to a magnetic field, the Gd moments partially align with the field direction and oppose the randomizing thermal energy. In zero field, the Gd moments relax to $M = 0$. This spin lattice relaxation is the same process as that for protons in aqueous medium exposed to high magnetic fields in an MRI scanner. However, in the case of Gd^{3+} ions, the magnetic moment is 5300 times larger than the magnetic moment of a proton. When Gd^{3+} chelates are administered, these large Gd moments act as an additional local magnetic field, helping proton spins to realign after *TR* and *TE* pulses, effectively shortening the *T1* relaxation time. Gd^{3+} ions in the chelate molecules have the advantage of being in close contact with the surrounding protons. In contrast, in the core of magnetic NPs, the magnetic moments are some distance away from protons in the aqueous environment, and hence the direct magnetic contact is weaker.

Gd^{3+} chelates are currently the only clinically approved *T1* contrast agents. Nonetheless, Gd-based chelates have a number of shortcomings that limit their usefulness [18]. After intravenous injection, due to their low molecular weight, the molecules are rapidly released from the blood pool into the extracellular space with a circulating half-life of only 5 min. From there, the molecules are excreted via the kidneys with a clearance half-life of 80 min. Therefore, chelate-based CAs do not provide full volume distribution. They are not useful as blood pool CAs. Due to their partial volume distribution, targeted contrast can be missed and/or contrast may be misinterpreted. To make matters worse, the *T1* relaxivity r_1 drops almost by a factor of 3 in magnetic fields from 1 to 3 T. Because of the short circulation lifetime, poor r_1 at high fields, and toxicity concerns, the research focus has shifted to *T2* contrast agents and specifically to superparamagnetic iron oxide Fe$_3$O$_4$ NPs, which are discussed next.

7.4.3 Fe$_3$O$_4$ as negative contrast agent for MRI

What is the difference between Gd^{3+} chelates and Fe$_3$O$_4$ superparamagnetic (SPIO) nanoparticles? Both are characterized by a paramagnetic response (susceptibility) in an external magnetic field. The main difference is the total moment per particle. With Gd^{3+} chelates we have 8 μ_B per molecule. The sum of all magnetic moments in a magnetite nanoparticle with a diameter of 10 nm, the Fe moments add up to a total magnetic moment of 2.8×10^4 μ_B, much larger than in the Gd-chelates. The magnetic dipole fields of the SPIO nanoparticles are strong and random, and strongly perturb the local proton spins. In addition, contact with surrounding protons in the water occurs indirectly via dipolar magnetic fields. Therefore, SPIOs give negative contrast in distinction to Gd^{3+} chelates. Negative contrast is sometimes not as favorable for imaging as positive contrast. Negative contrast darkens the pixel and makes it less distinguishable from the anatomical background. But it is beneficial to improve contrast in $T2$-weighted images. In addition, SPIOs are preferred CAs due to their superior biocompatibility. In fact, SPIOs are currently the only magnetic nanoparticles which are clinically approved.

For magnetite NPs with a diameter below 10 nm, the mass magnetization of 92 emu/g in bulk becomes reduced to about 50 emu/g. The units of mass magnetization is explained in the Infobox I. This shows that for very small NPs, the ferro- or ferrimagnetic state is not preserved. Indeed, it can be observed that spin tilting and spin disordering occur at the surface, decreasing the overall mass magnetization of the NPs below 10 nm. The effect can be used to fine-tune the r_2 relaxivity and consequently the effect on $T2$: smaller magnetite NPs have a lower r_2 than larger ones. r_2 ranges from $100 \ \text{mMol}^{-1} \text{s}^{-1}$ for 5 nm NPs to about $200 \ \text{mMol}^{-1} \text{s}^{-1}$ for 10 nm and larger NPs [17, 18].

Infobox I: Electromagnetic units

To compare the magnetization of different materials, the magnetization measured in units of A/m ($1 \ \text{A/m} = 10^{-3} \ \text{emu/cm}^3$) is usually converted to electromagnetic units per gram, also called *mass magnetization* in units of emu/g. This is easily done by normalizing the volume magnetization in units of emu/cm^3 with the mass density ρ:

$$\frac{\text{emu}}{\text{cm}^3 \, \rho} = \frac{\text{emu}}{\text{g}} . \tag{7.14}$$

Bulk metallic Fe with density of 7.93 g/cm^3 has a magnetization of 1.7×10^3 emu/cm^3 or 217 emu/g regardless of the size. Bulk magnetite has a magnetization of 4.76×10^5 A/m $= 4.76 \times 10^2$ emu/cm^3 and a density of 5.17 g/cm^3, yielding 92 emu/g or roughly 100 emu/g.

7.4.4 Size of magnetic nanoparticles

The size of NPs matters in nanomedicine, as recognized already in Section 7.2.7. This is especially true for magnetic NPs: the size determines the biodistribution of NPs in the body, the circulating half-life, and the relaxivity r_2. MNPs have been classified according to their size [19]:

1. Ultrasmall superparamagnetic iron oxide NPs (USPIO; <50 nm);
2. Superparamagnetic iron oxide NPs (SPIO; ~100 nm);
3. Micrometer-sized paramagnetic iron oxide particles (MPIO; several μ m).

Usually, MNPs are coated in a protective organic or inorganic shell and colloidally stabilized by PEG. None of the NPs would fit through the glomerular filter, requiring an overall HD size of less than 10 nm. SPIOs are recognized by macrophages in the bloodstream and selectively deposited in the liver, spleen, and bone marrow. This has a subtractive effect on MRI diagnostics. When a normal liver structure is destroyed by liver disease, such as a primary liver tumor or liver metastases, the region of Kupffer cells is depleted. Therefore, normal SPIO uptake is hampered, and strong contrast between normal and diseased parts of the tissue can be observed in $T2$-weighted images. USPIOs have longer plasma circulation times because, unlike larger SPIOs, they are not trapped by macrophages. By virtue of their rapid distribution in the intravascular space and long circulatory lifetime, USPIOs are considered ideal for blood pool CAs. Therefore, USPIOs can be used for angiography, passive targeting of tumors, and eventually functional MRI (fMRI) when primed to cross the blood-brain barrier (BBB).

7.4.5 Coating of magnetic nanoparticles

In general, the core of MNPs provides magnetic sensitivity and contrast enhancement. The coating has a number of additional tasks to fulfill:
1. biocompatibility;
2. colloidal stabilization;
3. agglomeration prevention via magnetic dipole interaction;
4. provision of reversible binding sites for drug-delivering molecules;
5. binding of targeting ligands.

Fe oxide NPs are inherently biocompatible and would not need a coating. The body stores Fe in the form of iron ferrihydrite crystals encapsulated in a hollow protein sphere called ferritin (Fig. 8.35 of Volume 1). Ferritin is stored in the liver and spleen for supply to hemoglobin when needed. However, when magnetite is used as CA, a coating is preferred for all the benefits that a coating offers. The most commonly used coatings are PEG and dextran. Both improve colloidal stability and blood circulation time.

7.4.6 Alternative magnetic nanoparticles

In addition to magnetite, a number of alternative magnetic nanoparticles have been produced and tested in animal studies, such as Fe, FeCo, and FePt alloys, mixed oxides $CoFe_2O_4$, $MnFe_2O_4$, and $NiFe_2O_4$, hydroxyapatite (HAP, $Ca_5(PO_4)_3(OH)$) doped with Co^{2+}, Fe^{2+}, Fe^{3+}, and Gd^{3+} ions [20]. HAP is a bio-conformal mineral that is also found in bones and teeth (see Fig. 3.11 of Volume 1). It could be shown that Gd^{3+}-doped HAP has a high magnetic moment along with hemocompatible properties, which could make it a good alternative to magnetite as MRI-CA. Reviews of the use of MNPs as CAs for MRI are presented in [19, 21].

7.5 Magnetic hyperthermia

7.5.1 Relaxation mechanisms

Magnetic hyperthermia refers to a moderate and local temperature increase produced by the relaxation of MNPs excited in an oscillating (AC) magnetic field with frequencies up to MHz. Hyperthermia is used to treat cancer because moderate heating to around 44 °C leads to cell death (apoptosis). To understand the physical principles, we revisit the relaxation mechanism of MNPs in solution and without interparticle interactions.

We have already seen that MNPs can overcome the orientational potential barriers at temperatures $T > T_B$ with a relaxation time τ (Fig. 7.10). When an external field is switched off, the magnetization of the MNPs relaxes to zero within the same relaxation time τ. The question is how the relaxation takes place. There are three possible mechanisms leading to orientational relaxation, sketched in Fig. 7.11:

1. Néel[7] relaxation;
2. Brown[8] relaxation;
3. Hysteretic relaxation.

The Néel relaxation assumes a coherent rotation of the magnetization within the particle without any physical rotation of the particle itself. The Brown relaxation assumes a physical rotation of the particle without rotation of the magnetization inside [22]. Hysteretic relaxation requires the formation of domains that propagate through the NP to complete reversal. These are three extreme cases. Hysteretic relaxation can be ruled out for very small single-domain MNPs, but domain formation and domain spreading can occur for particles larger than 50 nm. The Néel relaxation requires MNPs with negligible magnetocrystalline anisotropy. Brown relaxation is the opposite case and requires high magnetocrystalline anisotropy. For temperatures below

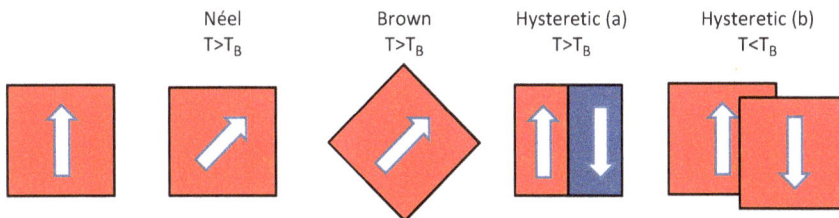

| Néel | Brown | Hysteretic (a) | Hysteretic (b) |
| $T > T_B$ | $T > T_B$ | $T > T_B$ | $T < T_B$ |

Fig. 7.11: Different relaxation processes of magnetic nanoparticles depending on size and temperature: Néel, Brown, hysteretic domain formation above T_B, and hysteretic magnetization reversal below T_B.

---

7 Louis Eugène Felix Néel (1904–2000), French physicist, Nobel prize in physics (1970).
8 William Fuller Brown (1904–1983), American physicist.

the blocking temperature T_B, the magnetic reversal is hysteretic, regardless of how the magnetic reversal actually takes place.

7.5.2 Relaxation in an AC magnetic field

Exposing MNPs to an alternating (AC) magnetic field, the magnetization in MNPs will be flipped back and forth. In the case of Brownian motion, the magnetization and the entire particle is sterically rocked. Since the particle is embedded in an aqueous liquid, we expect the most effective energy dissipation and thus heating of the local environment for this type of motion. The Brown relaxation time τ_B depends on the viscosity η of the fluid, the effective hydrodynamic volume of the particle $V_{HD} = \pi d_{HD}^3/6$, which is larger than the magnetic volume V_m, and the thermal energy $k_B T$ [19]:

$$\tau_{Brown} = \frac{\pi \eta d_{HD}^3}{2 k_B T}. \tag{7.15}$$

The theoretical background of the Brown relaxation time is the *Debye relaxation model*, which is explained in Infobox II and here applied to magnetic NPs.

In case of Néel relaxation, the damping is first due to spin-lattice relaxation within the MNPs, and in a second step, the lattice heat is transferred to the surrounding fluid. Therefore the Néel relaxation time depends on the magnetic volume and is given by [23]:

$$\tau_{N\acute{e}el} = \tau_0 \sqrt{\frac{\pi k_B T}{4 K V_m}} \, exp\left(\frac{K V_m}{k_B T}\right). \tag{7.16}$$

In general, both relaxation processes contribute, but with different weights depending on the particle size. In general, the effective relaxation time is composed of Néel and Brownian relaxation:

$$\frac{1}{\tau_{eff}} = \frac{1}{\tau_{Brown}} + \frac{1}{\tau_{N\acute{e}el}}. \tag{7.17}$$

The power dissipation to the surrounding fluid can be determined from a Debye[9] relaxation model according to [23]:

$$P = \frac{1}{2} \omega \mu_0 \chi_0 H_0^2 \frac{\omega \tau}{1 + (\omega \tau)^2}. \tag{7.18}$$

---

9 Peter Debye (1884–1966), Dutch physicist and Nobel prize winner in chemistry (1936).

Here H_0 is the AC magnetic field amplitude, and ω is the angular frequency of the AC field. χ_0 is the equilibrium susceptibility, determined from the initial slope of the Langevin equation for the NP magnetization, and expressed by:

$$\chi_0 = \frac{\mu_0 m_{macro}^2 V_m}{k_B T}. \tag{7.19}$$

Here V_m is the volume of the MNP and m_{macro} is its magnetization. The expected temperature rise ΔT after exposure time Δt before reaching equilibrium can be expressed as:

$$\Delta T = \frac{P \Delta t}{C}, \tag{7.20}$$

where C is the average heat capacity of MNPs and fluid.

Numerous experiments have tested the dependencies of the power dissipation P as a function of frequency, susceptibility, magnetocrystalline anisotropy, magnetic and hydrodynamic particle volume and shape, particle size distribution, and as a function of fluid viscosity [24–26]. All these studies implicitly assume that the blocking temperature is below the probing temperature. However, this may not be the case for particles with a diameter of 10 nm and larger. If the blocking temperature is above the measurement temperature, the particles are in a blocked state. In this case, the magnetization curve opens, showing remnant magnetization and finite coercive fields. An open magnetic hysteresis entails a hysteretic energy loss equal to the area enclosed by the hysteresis. This hysteretic energy loss adds to the energy losses already considered by the Néel and Brown relaxations.

Depending on magnetocrystalline anisotropy and size, the relaxation of magnetic nanoparticles can be Néel-like, Brown-like, or hysteretic. !

Infobox II: Debye relaxation model i

Debye original designed his model to describe the relaxation of electric dipoles in fluids exposed to an oscillating electric field $E(t)$. But it can also be applied to the relaxation of paramagnetic dipoles in an AC magnetic field.

Debye's dipole theory is based on the assumption that the dipoles attached to atoms or molecules can rotate in a solution of viscosity η; this viscosity slows down the dipole movement and determines the relaxation time τ. When an electric field E is applied, the polarization P of the dipoles always tends towards the equilibrium value:

$$P(t \rightarrow \infty) = \varepsilon_0 (\varepsilon_{stat} - \varepsilon_\infty) E$$

as $t \rightarrow \infty$. Here ε_∞ is the contribution of the polarization effective only at high frequencies $\omega \gg 1/\tau$, ε_{stat} is the static dielectric constant, and ε_0 is the vacuum dielectric constant.

The time dependence of the polarization $P(t)$ exhibits a time delay characterized by the relaxation time τ:

$$\dot{P} = \frac{1}{\tau}(P(t) - P(t \rightarrow \infty))$$

The approach of the polarization $P(t)$ towards equilibrium has a real part and a dissipative viscous part, both depending on the frequency. Expressed in terms of the complex dielectric constant $\varepsilon^*(\omega)$, the frequency dependence of the orientational polarization is:

$$Re\ \varepsilon^*(\omega) = \varepsilon_\infty + (\varepsilon_{stat} - \varepsilon_\infty)\frac{1}{1 + (\omega\tau)^2}$$

$$Im\ \varepsilon^*(\omega) = (\varepsilon_{stat} - \varepsilon_\infty)\frac{\omega\tau}{1 + (\omega\tau)^2}$$

The frequency response of the real part $Re\ \varepsilon^*(\omega)$ shows a step at $\omega = \tau$ from ε_{stat} to ε_∞. Simultaneously, the imaginary part $Im\ \varepsilon^*(\omega)$ passes through a maximum at the same frequency. Both curves are shown below as function of $\omega\tau$ on a logarithmic scale. The dielectric loss increases significantly for frequencies $\omega > \tau^{-1}$ compared to lower frequencies.

In the case of magnetic nanoparticles, the Debye relaxation time $\tau = 4\pi\eta r_{HD}^3/k_BT$ has to be replaced by the Brown relaxation time: $\tau_{Brown} = \pi\eta d_{HD}^3/2k_BT$.

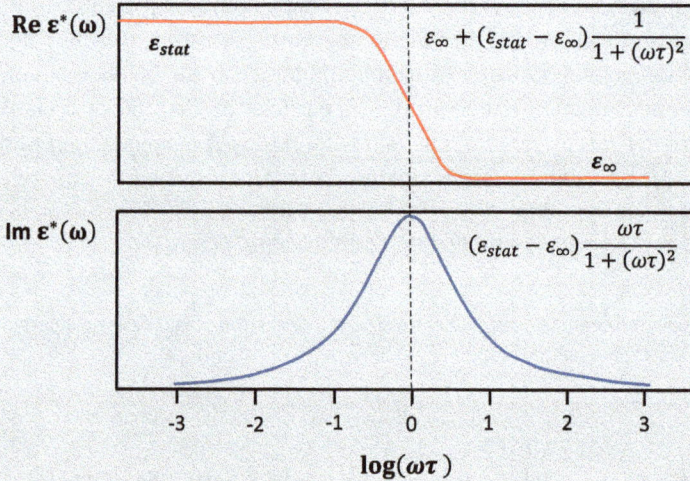

Real and imaginary part of the dielectric function.

7.5.3 Power loss density

Magnetic hyperthermia necessitates magnetic nanoparticles that preferentially accumulate in cancerous tissue. This can be accomplished either through physical or active targeting. In the latter case, the MNPs have to be conjugated with antibodies that recognize the antigens of cancer cells. The MNPs can be injected intravenously or directly into the tumor tissue. It is crucial to achieve a homogeneous distribution of MNPs across the tumor volume for a uniform temperature topography that avoids hot or cold spots. Many technical challenges need attention for successful hyperthermia treatment. First, the concentration of MNPs should be kept as low as possible to avoid adverse biomedical effects. Next, optimized parameters for the applied power are to be identified in a vast parameter space: magnetic material selection, particle size, shape and distribution, field amplitude, and AC frequency. Not all parameters are independent. The particle size is related to the relaxation mechanism and the relaxation mechanism determines the efficiency of heat transfer. The particle size also determines whether the relaxation process occurs above or below the blocking temperature. After considering all these issues, the main goal is to maximize the specific heating power of MNPs in an AC magnetic field to reach the target temperature with a minimal concentration of MNPs. The critical physical quantity is the power loss density. Rewriting the expression in eq. (7.18), we find for the power loss density in Watt/m³:

$$P = \left(\mu_0 M_s H_0\right)^2 \left(\frac{\omega V}{2k_B T}\right)\left(\frac{\omega\tau}{1+(\omega\tau)^2}\right). \tag{7.21}$$

According to this equation, for low frequencies, $P \sim \omega^2$, while for $\omega\tau \gg 1$, P becomes frequency independent with the saturation value:

$$P = \left(\mu_0 M_s H_0\right)^2 \left(\frac{V}{2k_B T\tau}\right). \tag{7.22}$$

With constant field amplitude H_0 and frequency ω, the strong size dependence of the relaxation time leads to a sharp maximum of the power loss density within a narrow size range of MNPs [23, 25]. Therefore one can select the particle size or the frequency and then adjust all other parameters accordingly. The *specific loss power* (SLP) is also known as *specific absorption rate* (SAR).

To convert the power loss density in Watt/m³ to units of Watt/g, the power density must be divided by the density of the nanoparticles. Values for SLP vary widely in the literature from 10 to 100 Watt/g for magnetic field amplitudes of 10 kA/m and AC frequencies of around 400 kHz. A record SLP was recently reported for 18 nm MnFe$_2$O$_4$ NPs embedded in an amphiphilic copolymer shell. The magnetic nanoparticles loaded into these polymer nanospheres had an SLP of 580 Watts/g at a frequency of 435 kHz and a field amplitude of 4 kA/m [27]. With such a high SLP, the concentration of MNPs to be administered in vivo can be drastically reduced.

Interest in MHT increased after a 2001 report by Jordan and co-workers on clinical trials with very promising treatment results in prostate cancer [28]. MNPs were dispersed in ferrofluids in these studies and injected into the tumor volume. Since then, numerous in-vitro and animal studies have been carried out, but approval for clinical applications has so far only been granted in Germany. The remaining problems of magnetic hyperthermia are seen in targeting specific tumor areas, while sparing neighboring cells. Adjacent tissues can be affected by hyperthermia if MNPs are not completely cleared from healthy tissues. Diffusion of MNPs from tumor areas into surrounding tissues and heat conduction are also potential problems. If these problems are solved, MNPs have the potential for multifunctional use in diagnosing and treating various diseases.

7.5.4 Magneto-mechanical cell destruction

Using magnetic nanoparticles, a completely different approach has recently been pursued to destroy cancer cells. In this approach, magnetic particles exert mechanical force on cell membranes in weak magnetic fields and at low frequencies [29]. This is in contrast to magnetic hyperthermia, which uses high fields and high frequencies of several hundred kHz [25]. We have already noticed that a multidomain ferromagnetic state collapses into a single domain state when the lateral extension is reduced to the nanometer length scale. A spin-vortex state arises in circular disks of a certain diameter and aspect ratio, which is a magnetic spin swirl. The vortex state's critical diameter and aspect ratio are determined by the exchange interaction, magnetocrystalline anisotropy, and shape anisotropy. For a $Ni_{0.8}Fe_{0.2}$ alloy, known as soft magnetic permalloy, the critical diameter is about 1 μm. The magnetic vortex state in a zero external magnetic field is shown in Fig. 7.12(a). In the vortex state, the magnetic flux within the disk is completely closed, no stray fields emanate from the disks, and the magnetization M integrated over the disk volume is zero. This has key advantages, as these vortex disks in the residual state do not

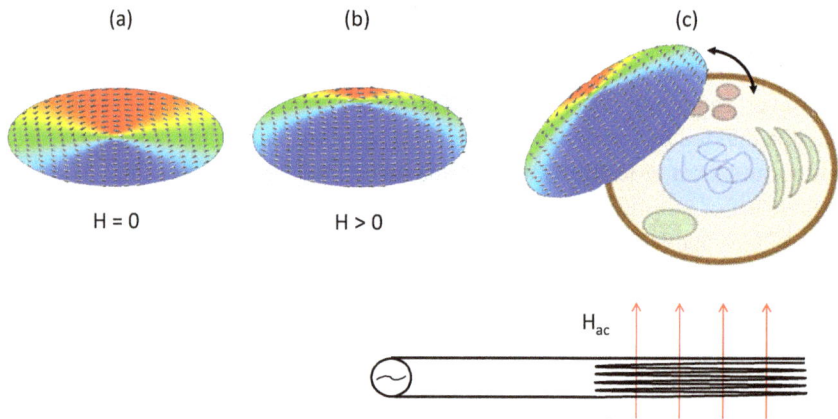

Fig. 7.12: Magnetic microdisks used for magneto-mechanical destruction of cancer cells. (a) Magnetic spin vortex state in a circular disk; (b) application dc magnetic field moves the vortex core out of the center; (c) magnetic disks immersed in the extracellular space of tumor tissue can be rocked in an AC magnetic field supplied by a coil, schematically indicated in the lower part. The low frequency oscillation exerts mechanical force onto cellular membranes that initiates apoptosis.

tend to agglomerate in aqueous solution due to the lack of magnetic dipole-dipole interaction, in contrast to magnetic single-domain nanoparticles.

When a magnetic field H is applied to the magnetic disk, the vortex core moves from the center in the direction perpendicular to the applied field direction, as shown in Fig. 7.12(b). This causes an asymmetry in the spin distribution generating a finite magnetization. For small fields, the magnetization is linearly proportional to the external field: $M = \chi_{vort} H$, where χ_{vort} is the magnetic susceptibility of the spin vortex. At the same time, the magnetic field exerts a torque $\vec{T}$ on the disk $\vec{T} = \vec{M} \times \vec{H}$, which causes the disk to rotate in the direction of the field. This effect can be used to rock the magnetic disks in the extracellular space of tumor volumes back and forth by an external alternating magnetic field H_{AC}. It turns out that already small field amplitudes of less than 100 Oe and low frequencies of about 10–20 Hz have a dramatic effect on the integrity of cell membranes, initiating apoptosis within a short exposure time of only 10 min [29]. The mechanism of induced cell death is not entirely clear. The damage can cause disruption of the membrane or ion channels. In any case, the mechanical vibration of the magnetic disks attached to the cancer cells appears to be converted into a transmembrane ion current that triggers apoptosis. For cancer cell targeting, the microdisks are first coated with a gold layer to improve biocompatibility and then conjugated with antibodies that bind to specific antigens overexpressed on the surface of the targeted cancer cells.

7.6 Metal nanoparticles for diagnostics and therapeutics

Contrast problems are common in x-ray and CT imaging. X-ray contrast depends solely on lateral differences in electron density. The biomolecular structure of tissue can vary greatly. However, if there is no difference in local electron density, attenuation based x-ray radiography will not record any difference. Bones are clearly distinguishable due to their high Ca content with an above-average electron density. However, contrasting soft tissues and organs is more difficult. For imaging of the gastrointestinal tract, it is common to first swallow a higher-contrast Ba_2SO_4 slurry before taking an x-ray picture. Similar problems arise when taking x-rays of the lungs, which are also obstructed by the thoracic skeleton in the foreground (see Section 8.7 of Volume 2 for discussions), of internal organs (kidneys, liver, spleen, etc.), or when imaging blood vessels (angiography). In all of these cases, contrast enhancement through the use of nanoparticles with a high atomic number can be a solution.

7.6.1 X-ray radiography contrast

Currently, the most commonly applied contrast-enhancing agents are small nonionic iodine (Z = 53)-containing molecules that exist in various configurations differing in viscosity, osmolality, and the number of iodine ions per formula unit. The molecular structure of one of the iodine contrast agents (CAs) is shown in Fig. 8.20 of Volume 2. There are a number of issues associated with these CAs. Large amounts must be injected to compensate for the low iodine concentration. These small molecules have a short circulating half-life in the body due to rapid renal clearance, which, in turn, can cause kidney problems. To overcome the adverse effects of iodinated CAs, various alternative NPs have been developed in recent years, the main idea of which is briefly presented. More detailed overviews can be found in the references [30, 31].

In general, when engineering NPs for x-ray contrast enhancement, they should fulfill at least part of the following conditions:
1. high contrast efficacy;
2. excellent biocompatibility;
3. proper size;
4. long in-vivo circulation time;
5. long colloidal stability in the body;
6. reasonable cost effectiveness.

High contrast efficacy[10] implies the use of high-Z, high-density materials to reduce the total amount of NPs needed for the required contrast. NP size and long in vivo circulation times are interrelated requirements, as discussed in Section 7.2. Nanoparticles smaller than 10 nm are rapidly cleared by the kidneys, resulting in a shortened circulation time. At the same time, such small NPs can pose a potential hazard to the glomerular filtration system. From a size of 100 nm, they can be eliminated by monocytes and deposited in the liver and spleen. The correct size is therefore important and is somewhere between 20 and 80 nm. However, the size effect can also be used to preferentially image the liver and spleen with particles larger than 100 nm that the RES recognizes. Vice versa, if those NPs do not yield a high contrast of liver and spleen, this would indicate damaged macrophages and Kupffer cells resulting from diseases in those organs.

High contrast efficacy is achievable using NPs containing a core of high Z metal ions, such as Au (Z = 79), Bi (83), or lanthanide ions ranging from Gd (64) to Yb (70). The high Z atoms should be nonoxidizing and nonsolvable. Since some of the metals, except Au, are highly oxidizing, the metal core has to be protected by a tight and impermeable shell surrounding the heavy metal core. This is usually accomplished by a silica coating and PEG shell that feature additional biocompatible properties, as indicated before and schematically illustrated in Fig. 7.5.

---

10 Efficacy and efficiency, what is the difference? In medical sciences, these terms are used in the following way.

Efficacy is the ability to produce a desired or intended result. Efficiency means getting the same desired result in the most economic way.

7.6.2 K-edge contrast

In standard x-ray radiography, resonant K-edge absorption of x-ray photons does not play any role. The K-edges of soft materials have very low energies (283 eV for carbon K-edge) compared to the mean photon energy of the bremsstrahlung of about 100 keV. Even the K-edges of Ca (4 keV) and I (33.2 keV) are too low to be of any relevance in radiography. However, K-edge absorption becomes more significant for nanoparticles containing heavy metal cores, such as Au (80.7 keV) or Bi (90.5 keV) and even for the lanthanide ions, K-edge absorption is significant (61.4 keV for Yb). For these CAs, K-edge absorption contributes markedly to the total attenuation of the x-ray beam [32]. Using this fact, NPs can be designed containing heavy metal alloys with K-edges covering a wide energy range. Furthermore, *K-edge contrast imaging* can be applied by taking two sequential pictures, one with weighted average x-ray photon energy above an absorption edge and another one below the same absorption edge. How the weighted average photon energy can be adjusted and how dual-energy images can be analyzed is explained in Section 8.7.3 of Volume 2. Presently, Au NPs are favored for K-edge contrast imaging because of their superior biocompatibility, nontoxic and nonreactive chemical properties, and comparatively easy synthesis.

7.6.3 Bimodal contrast

NPs containing high-Z magnetic materials such as Gd can also be used for MRI contrast enhancement. NP containing Au can also be used for near-infrared plasmonic resonance for photothermal therapy. These are just two examples of NPs with bimodal properties: x-ray and MRI diagnostics for Gd-containing NPs; x-ray diagnostics together with photothermal therapy using Au NPs. Along the same lines, multimodal NPs with sensitivity to multiple imaging modalities and functionalities have been developed, such as Au NPs coated with Gd chelates or Fe_3O_4 for x-ray contrast, MRI contrast, and photothermal therapy. Multimodal NPs are discussed further below.

7.6.4 Alternative concepts

There are a variety of NPs that can be used to replace iodinated CAs. Many have been tested in vivo, but only a few have been approved for clinical use. The remaining main problem is cost-efficiency. Nanoparticles containing Au or lanthanide ions tend to be much more expensive than iodine-based CAs, which are currently mass-produced at a lower cost. For a normal CT, the cost can be 10–20 times higher if NP-CAs with high Z-metal content is used instead of iodine-containing molecules, prohibiting their routine use. But for research and for patients with iodine intolerance, the NP route is a viable alternative.

7.7 Plasmon resonance

7.7.1 Mie scattering

The optical properties of metal NPs are of great interest in various fields of science. This is due to the absorption of electromagnetic radiation, which depends on the size and aspect ratio of the NPs and thus determines their color variation. It turns out that the optical properties of metal NPs are also of interest for nanomedical applications for three reasons: 1. enhancement of the fluorescence yield of fluorophores; 2. use for hyperthermia, and 3. use for photochemical therapy.

When light (electromagnetic waves, EM waves) strikes molecules or particles, the electrons in the particles are excited to dipole radiation. If the particle diameter d is much smaller than the wavelength of light ($d \ll \lambda$), the emitted EM wave radiates homogeneously in all directions (4π). This type of radiation is called Rayleigh[11] scattering and is responsible for the blue sky during daytime and red sky in the morning and evening. With increasing particle size, the light scattering becomes preferentially directed in the forward direction. This scattering is known as Mie scattering. Both Rayleigh and Mie scattering are compared in Fig. 7.13.

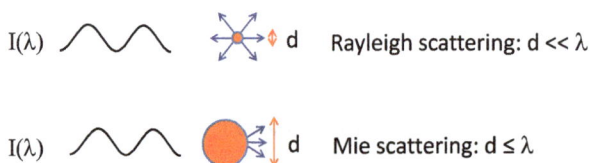

Fig. 7.13: Rayleigh and Mie scattering of light at particles of various size.

Suppose the particles consist of a metal containing a "free" electron gas. The incident EM wave can set these electrons into "plasma" oscillations: the free negative electrons oscillate against the rigid positive ionic background like in a tiny antenna. This has a decisive consequence for the interaction of EM radiation with metal particles: metal particles strongly scatter *and* absorb EM waves. Hence the refractive index has to be treated as a complex and wavelength dependent number. The absorption and scattering cross-sections of EM waves at spherical metal nanoparticles was worked out by Mie[12] [33]. Here we outline a simplified version of the main results.

In a static case, the dipolar polarization $\vec{P}$ of matter in an electric field $\vec{E}$ is expressed by:

---

11 John William Rayleigh (1842–1919), English physicist and Nobel prize winner 1904 in physics.
12 Gustav Mie (1868–1957), German physicist.

$$\vec{P} = \alpha\vec{E}, \tag{7.23}$$

where α is the polarizability characterizing the material, in our case, the metal NP. For a metal sphere of radius r in an oscillating EM field, the polarizability can be expressed as [34]:

$$\alpha = \frac{\varepsilon_p - \varepsilon_m}{\varepsilon_p + 2\varepsilon_m} r^3 = Qr^3. \tag{7.24}$$

Here ε_p and ε_m are the dielectric permittivities of metal particles (p) and the surrounding medium (m), respectively. This leads to the scattering and absorption cross-sections of EM waves at metal particles of radius r [34]:

$$\sigma_{scat} = \frac{8\pi}{3} k^4 r^6 |Q|^2 \tag{7.25}$$

$$\sigma_{abs} = 4\pi k r^3 Im|Q|. \tag{7.26}$$

Here $k = 2\pi/\lambda$ is the wavenumber of the EM radiation. Note that the absorption cross-section scales with r^3, while the scattering cross-section scales with $(r^3)^2$. Thus the NP size determines which cross-section dominates. The scattering cross-section contains the typical k^4 dependence that is characteristic for Rayleigh scattering. Absorption and scattering together contribute to the total extinction (attenuation) cross-section in the forward direction that determines the color appearance of metal NPs in transmission.

The Q-factor in eqs. (7.25) and (7.26) is strongly wavelength dependent and determines the frequency w of the absorption maximum. According to the Drude[13] model (see Infobox III), the dielectric response of a free electron gas is [35]:

$$\varepsilon_p = 1 - \left(\frac{w_p}{w}\right)^2 \tag{7.27}$$

with

$$w_p = \sqrt{\frac{n_e e^2}{\varepsilon_0 m^*}}. \tag{7.28}$$

Here w_p is the plasma frequency of the free electron gas, specific for each metal. n_e is the free electron density, e is the electrical charge, ε_0 is the vacuum permittivity, and m^* is the effective electron mass in the specific metal.

---

13 Paul Karl Ludwig Drude (1863–1906), German physicist.

7.7.2 Surface plasma resonance

In first approximation, an absorption maximum is expected for the condition $\varepsilon_p + 2\varepsilon_m = 0$ in the denominator of the polarizability, eq. (7.24). From this and eq. (7.27), we estimate the frequency for the *surface plasma resonance* (SPR) peak to be at:

$$\omega_{max} = \frac{\omega_p}{\sqrt{2\varepsilon_m + 1}}. \tag{7.29}$$

The last equation can be roughly estimated as $\omega_{max} = \omega_p/2$, or $\lambda_{max} = 2\lambda_p$.

Figure 7.14 shows the absorption maxima for spherical Au NPs of different sizes, reproduced from [36]. Au particles have an absorption maximum at $\lambda_{max} = 525$ nm but low absorbance in the red regime from 600 to 700 nm. Therefore Au particles appear red for particle sizes below 100 nm. At a particle size of 100 nm the absorbance peak is shifted to 600 nm, while low absorbance occurs in the 400–500 nm wavelength regime. These particles, therefore, appear bluish. Ag particles show an absorption maximum in the UV range at $\lambda_{max} = 375$ nm, shifting to longer wavelengths with increasing particle size. The absorbance shift with increasing size is generally observed for the SPR peak position. Although the size dependence of the SPR frequency shift is poorly understood [37], it is an important property for nanomedical treatments. Most likely, the shift can be related to the effective electron density variations in the surface layer of NPs.

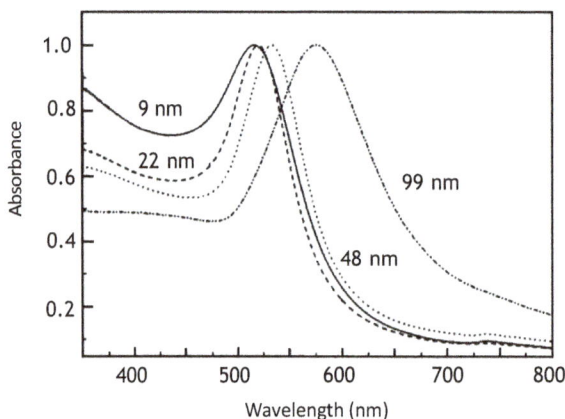

Fig. 7.14: Absorbance of Au NPs with different size as function of incident wavelength. All curves are normalized to the maximum absorbance (reproduced from [36] by permission of ACS).

! Metal nanoparticles exhibit various colors depending on their size and aspect ratio. The perceived color is the result of subtractive color mixing: scattered wavelengths mix which have not been absorbed by the nanoparticle.

Infobox III: Plasma frequency (Drude model)

The Drude model describes the dc electric conductivity of metals. Drude proposed that electrons are free to move in metals like an ideal gas, being accelerated by the external electric field E and scattered at positive ion cores. The scattering is the "friction" that is responsible for the relaxation in time τ and steady state properties of the metal. Later, it turn out that the classical Drude model is not the correct description of metal conductivity and was replaced by the quantum-mechanical Bloch-Bethe-Sommerfeld model. However, for deriving the Plasma frequency, the Drude model is still sufficient.

Now we expose the metal electrons to an AC electric field:

$$E(\omega) = E_0 \exp(i\omega t).$$

We set up an equation of motion, for simplicity in 1D, where $x(t)$ describes the time displacement of the electrons from equilibrium:

$$\ddot{x} + \frac{1}{\tau}\dot{x} = -\frac{e}{m^*}E(\omega) \tag{7.30}$$

in the ac electric field. m^* is the effective electron mass. We seek a solution for the differential equation in terms of an oscillating displacement amplitude $x(\omega)$ of the free electrons in phase with the ac electric field:

$$x(\omega) = x_0 \exp(i\omega t).$$

Yielding:

$$i\omega\dot{x} + \frac{1}{\tau}\dot{x} = -\frac{e}{m^*}E_0. \tag{7.31}$$

Solving for $\dot{x}$ yields:

$$\dot{x} = -\frac{eE_0\tau}{m^*(1-i\omega\tau)}. \tag{7.32}$$

Now we use the relationship that connects current density j with the displacement amplitude and conductivity σ:

$$j = -ne\dot{x} = -\frac{ne^2 E_0\tau}{m^*(1-i\omega\tau)} = \sigma(\omega)E_0. \tag{7.33}$$

Here n is the electron density. The dielectric function of the conduction electrons and the conductivity are related according to:

$$\varepsilon(\omega) = 1 + \frac{4\pi i}{\omega}\sigma(\omega). \tag{7.34}$$

For very high frequencies ($\omega\tau \gg 1$), the electrons can no longer follow the external ac field, and the dielectric constant approaches:

$$\varepsilon(\omega) = 1 - \left(\frac{\omega_p}{\omega}\right)^2 \tag{7.35}$$

with the plasma frequency:

$$\omega_p = \sqrt{\frac{4\pi ne^2}{m^*}}. \tag{7.36}$$

The plasma frequency is the eigenfrequency of the free electron gas in the metal and is characteristic of each metal. The color of metals (Au, Ag, Al, Cu, etc.) is determined by ω_p.
For $\omega > \omega_p$, metals are transparent, which, for instance, is the case for x-rays.
The graph indicates the free electron oscillation in metal against the positive ionic background.

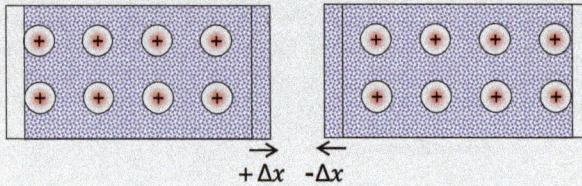

$$+\Delta x \quad -\Delta x$$

While there is still some uncertainty regarding the SPR frequency shift as a function of spherical NP size, there is a clear dependence of the SPR frequency on its aspect ratio. Spherical NPs show strong quadrupole contributions in addition to the dipolar extinction [30]. By changing the shape from spherical to prolate, the quadrupole contribution vanishes. In Au nanorods, the absorption maximum becomes a clear function of the aspect ratio (AR), as seen in Fig. 7.14 [38]. Actually, there are two resonances: one fixed and one that changes with the AR. The changing one is due to the *longitudinal* SPR in the direction parallel to the rod, while the fixed one refers to the *transverse* SPR in the direction normal to the rod axis. The tunability of the SPR by the AR is a very interesting property that can be exploited in photothermal therapy. It is also important that these SPR peaks coincide with the water window of tissues, which extends from 650 to about 1300 nm. In this wavelength range, light has the greatest penetration depth into the tissue, see Fig. 7.15.

Au particles with optical tunability in the infrared regime have also been fabricated in form of shells coating silica cores. It turns out that here the SPR is solely determined by the ratio of shell to core thickness, independent of the particle size [39]. This has great advantages for biomedical applications: first the particle size can be adapted to the required biodistribution in the body; and second, reshaping of particles under heat load is not relevant.

The cross-sections quoted above (eqs. 7.25, 7.26) indicate that the scattering cross-section increases more rapidly with increasing radius than the absorption cross section. This is indeed the case, as seen in Fig. 7.16 for Au nanospheres [40]. At a diameter of 80 nm, the scattering cross-section approaches the one for absorption; for NP size larger than 100 nm, scattering completely dominates absorption. This dependence guides nanomedical applications: smaller-sized NPs are preferentially used for absorptive applications; larger-sized NPs are more beneficial for imaging because of their strong scattering.

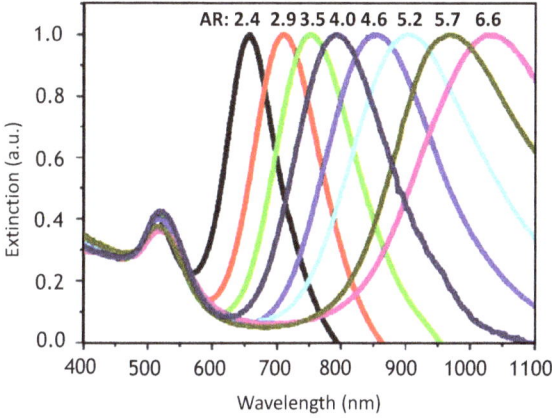

Fig. 7.15: Longitudinal and transverse surface plasmon resonance peaks in Au nanorods as function of aspect ratio (AR). All spectra are normalized to maximum extinction. The colors of the curves have no relation to the actual color appearance of the nanoparticles in an aqueous solution. The latter ones vary from bluish for AR = 2.4 to reddish for AR = 6.6 (reproduced from [38] by permission of Elsevier Publishing Inc.).

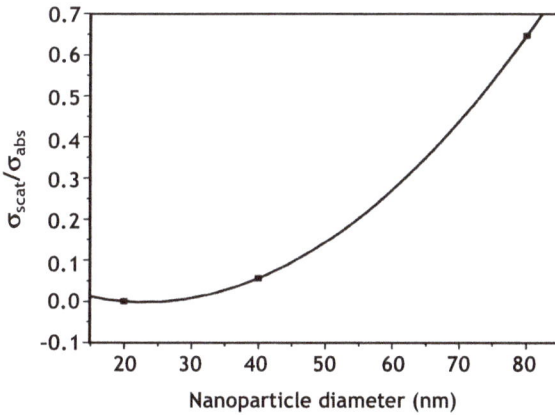

Fig. 7.16: Ratio of scattering cross-section to absorption cross-section. For NPs of about 20–30 nm in diameter, absorption completely dominates over scattering. With increasing diameter, the scattering cross-section becomes comparable to the absorption cross-section (adapted from [40] by permission of ACS).

7.7.3 Photothermal therapy

Some basics of photothermal therapy (PTT) by laser exposure have already been discussed in Section 6.4.7. PTT, with the help of metal NPs, is much more effective than without. The most preferred NPs for PTT are Au and Ag nanorods due to their relative chemical insensitivity and near-infrared (NIR) absorption bands. Figure 7.14 shows the absorbance peak of Au NPs as a function of wavelength. The peak position can be varied by the NP's size and aspect ratio. On the other hand, the laser wavelength is selected based on the required penetration depth. According to Tab. 6.1, skin treatment is best performed with a YAG or CO_2 laser, while deeper tissues can be reached with an Ar laser. Once the laser wavelength is chosen, the size and shape of Au or Ag NPs can be tailored for the maximum extinction of surface plasmons. However, nanorod particles have an intrinsic shape stability problem. Nanorods are thermodynamically unstable because of their high surface energy. To lower the surface energy, high aspect ratio NPs tend to reshape towards sphericity. Surface diffusion is enhanced, and reshaping of the NP is accelerated at a high heat load while in plasmonic resonance. Reshaping, in turn, shifts the plasma resonance energy and also the photothermal conversion efficiency. We notice that using metal nanorods for plasmonic therapy is not trivial.

> **Infobox IV: What are absorption, absorbance, extinction, and attenuation? And what is their difference?**
>
> As explained in Chapter 5 of Volume 2, attenuation is the loss of beam intensity by interaction with a medium via absorption, elastic scattering, inelastic scattering, or reflection. Attenuation merely expresses the fact that the transmitted beam intensity is lower than the incident intensity.
>
> For instance, the sun light intensity becomes attenuated by clouds. Attenuation, extinction, and absorbance have the same meaning in physics. It is a question of taste which term to use. In chemistry and biochemistry, often the term "absorbance" is preferred, whereas in physics this term is not used at all.
>
> Absorption is the loss of beam intensity in a medium by removal of incident particles and converting them in different particles or different forms of energy. Example is the photoelectric absorption of photons. For instance, in the retinal of our eyes, photoelectric absorption causes a cis-trans isomerization of the retinal that initiates vision.

Au nanoparticles can be fabricated with a solid Au core (Au NP) or by coating a spherical core of dielectric material, forming an Au nanoshell (Au NS). In the latter case, the SPR is determined by the outer diameter and the shell thickness. The tunability of the SPR appears to be the result of hybridization between the inner cavity and outer surface plasmon resonance. A red shift of 300 nm has been achieved upon variation of the shell thickness between 20 and 5 nm [41].

The therapeutic benefit of Au NPs or NSs results from resonant light absorption at the respective plasmon energy. Photon excitation of the electron gas in the metal NPs is followed by a relaxation on the femtosecond timescale, mainly due to electron-electron scattering, leading to a rapid surface temperature increase. Electron-phonon[14] coupling then causes heating of the crystal lattice on the picosecond time scale [42].

---

14 Phonons are quantized lattice vibrations.

The steps leading from light absorption to dissipative thermal energy of Au NPs is illustrated in Fig. 7.17. NPs in contact with tissue dissipate their thermal energy to the surrounding medium conductively and by infrared radiation. When Au NPs embedded in a cellular medium are illuminated by laser light, the large temperature difference between hot Au NP surface and cooler surrounding tissue leads to cell death. The heating rate and cooling rate must be in equilibrium. Otherwise, the NPs would melt. Thermal equilibrium can be better controlled with a cw laser than a pulsed laser. As mentioned before, intense photothermal heating may cause melting or reshaping of the NPs that changes, in turn, its optical properties irreversibly.

In conclusion, Au NPs have excellent properties for hyperthermia, including nontoxicity and tunable plasma resonance for optimal absorption. The only limitation is the penetration depth of laser light, which is on the order of 10 mm in the infrared regime. For most applications this is sufficient. Alternatively, hyperthermia with MNPs can be used, which is not depth limited.

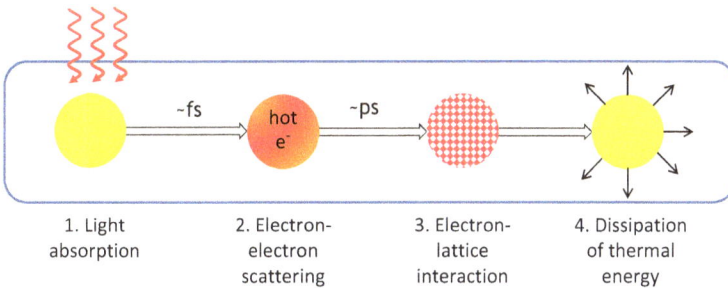

| 1. Light absorption | 2. Electron-electron scattering | 3. Electron-lattice interaction | 4. Dissipation of thermal energy |

Fig. 7.17: Resonant light absorption by Au nanoparticles in tissue and time scale of subsequent processes: 1. light absorption at plasmonic resonance; 2. electron-electron scattering heats up the electron gas in the metal; 3. electron–lattice interaction heats up the crystal lattice of the nanoparticle; 4. dissipation of thermal energy by radiation and heat conduction to the surrounding tissue.

7.8 Imaging and spectroscopy

7.8.1 Dark field imaging

As stated in Section 7.7.2, Au NPs scatter strongly light, the details depend on size, shape and geometry. The large scattering cross-section for particles larger than 50 nm make them very promising for imaging carcinoma, provided that they are conjugated to antibody receptors and targeted to cancer cells. Upon attachment to antigens, the NPs reveal the location of cancerous cells by their strong and colorful scattering properties. The SPR scattering from Au NPs can particularly well be imaged by a dark field microscope, which collects only scattered light against a dark background [43]. The principle of a dark field microscope is shown in Fig. 7.18 in comparison to a normal bright field microscope. The main difference is the aperture which blocks direct light from the white lamp. So the objective lens images only those rays that have been scattered at the sample. Figure 7.19 shows an image taken with a dark field microscope from Au nanorods accumulated in cancer tissue. In comparison, Au nanorods in healthy tissue are dispersed at low density but are not agglomerated. Cancer imaging is made possible by using Au NPs conjugated to antibodies which specifically and homogeneously bind to the surface of the cancer cells with greater affinity than to the noncancerous cells. The difference can be as high as a factor of 6 [43]. The gold nanoparticles show color in red due to SPR in the NIR region.

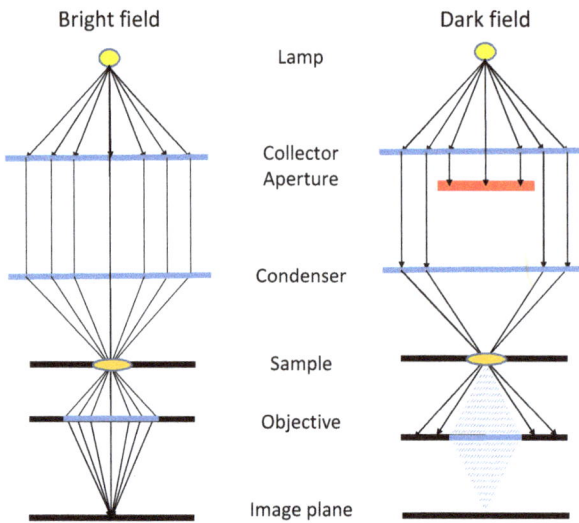

Fig. 7.18: Comparison of a bright-field and a dark field microscope. In a dark field microscope, particles in the sample, which scatter light, appear bright against a dark background.

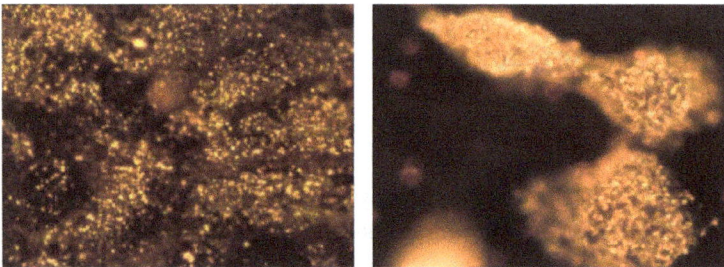

Fig. 7.19: Dark field images of Au nanorods conjugated with antibodies in healthy tissue (left) and cancerous tissue (right). In healthy tissue, the Au NPs are randomly distributed, whereas the NPs are agglomerated in cancer cells. The Au nanorods show a red color due to SPR in NIR region (reproduced from [43] by permission of ACS).

7.8.2 Surface enhanced Raman scattering

Metal NPs exhibit even more remarkable optical properties than discussed so far upon excitation of their surface plasmon resonance. At SPR conditions, a considerable electromagnetic field enhancement occurs close to the NP surface. Due to this overlap, the fluorescence emission of fluorophores can be enhanced by a factor of 50 [44]. However, it is important to keep a proper distance of a few nanometer between fluorophore and NP surface. The enhancement effect can be used for imaging chemical sensors and/or biosensors with specific resonances overlapping with the SPR when conjugated to metal NPs. Similarly, Raman scattering of biomolecules experiences a dramatic enhancement under SPR conditions and can also be used for imaging [45, 46].

Raman scattering is a spectroscopic method for studying vibrational excitations of molecules. When incident light in the infrared to UV region is scattered at atoms or molecules, the outgoing wave may have the same frequency or maybe blue or red-shifted. In the first case, we talk about Rayleigh scattering. Rayleigh scattering is an elastic scattering process without any energy transfer to the specimen. In the graph of Fig. 7.20, Rayleigh scattering is symbolized by an upscattering of the sample to a virtual and nonresonant energy level E_i, indicated by a dashed line. The upscattering is followed by down-scattering to the initial energy state. The energy of the incident and scattered waves are identical: $E_i = E_f$. Elastic or Rayleigh scattering is the dominant process when scattering light at molecules or nanoparticles.

In contrast to Rayleigh scattering, Raman[15] scattering is an inelastic scattering process and, under normal circumstances, comparatively rare. On average, only about 1 ppm scattering events are of Raman type. Some energy $\Delta E = \pm \hbar \Omega$ is transferred between photons and internal excitations of the sample. In Raman spectroscopy, these small energy shifts are analyzed. Raman spectroscopy has to be distinguished from *photoelectron emission spectroscopy* (PES). In PES, the entire photon energy is transferred to potential and kinetic energy of electrons, and the remaining kinetic energy of the emitted electron is then analyzed. In the case of Raman scattering, the energy transfer is only partial. Raman scattering can be best described in a quantum picture. Let us assume a monochromatic incident photon with energy $E_i = \hbar \omega$. This photon modulates the molecule such that it becomes excited to a vibrational or rotational energy level of energy $\Delta E_{vib} = \pm \hbar \Omega$. Because of energy conservation, the final photon energy is $E_f = \hbar \omega - \hbar \Omega$. The excitation of a molecule leads to a red-shifted *Stokes*[16] line. Vice versa, the photon may also take up the vibrational energy of a molecule during the scattering process. Then the final photon energy is blue-shifted (anti-Stokes line):

---

15 Chandrasekhara Venkata Raman (1888.1970), Indian physicist and Nobel prize winner 1930 in physics.
16 George Gabriel Stokes (1819–1903), Irish English physicist and mathematician.

$E_f = \hbar\omega + \hbar\Omega$. Alternatively, these internal energy levels can be excited directly and resonantly by EM waves in the terahertz frequency regime (THz). In principle, momentum conservation should be considered in these scattering processes. However, as the vibrational excitation energies are by a factor of 10^3 smaller than the incident light energy, momentum conservation can be neglected.

Rayleigh scattering and Raman scattering are fast processes on the time scale of pico- to femtoseconds. In contrast, fluorescence is a much slower process on the time scale of nanoseconds, and the luminescence is even slower. First, molecules are excited to a higher energy level. Often a nonradiative transition takes place to an intermediated vibrational level, followed by a transition to the ground state. The fluorescence emission is recognized by colored light after exposing the specimen to white light or UV radiation, sometimes even long after exposure. Fluorescence spectra are characteristic of the (bio-)molecule exposed to light but independent of the wavelength of the exciting radiation as long as $\lambda_{excit} < \lambda_{fluo}$. Fluorescence spectroscopy is similar to Raman spectroscopy and can be used for discriminating malignant skin tumors from benign ones. The schematics of these different spectroscopic transitions are illustrated in Fig. 7.20.

> **!** Surface enhanced Raman scattering allows probing vibrational energy levels in molecules that distinguishes normal and malignant tissues. Histological samples can be scanned to gain topographic images of vibrational excitations in tissues.

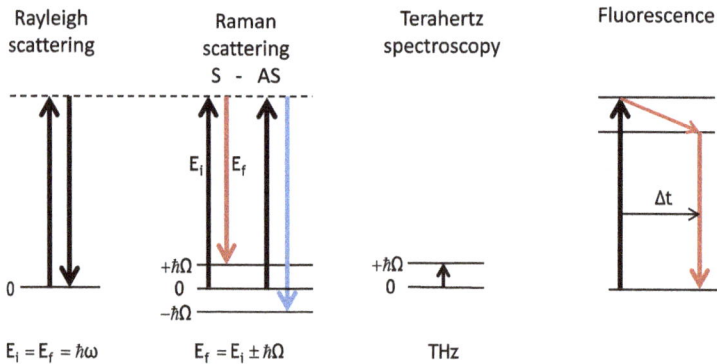

Fig. 7.20: Comparison of light interactions with matter: elastic Rayleigh scattering, inelastic Raman scattering, direct level excitation with terahertz radiation, and fluorescence transition. Fluorescence light is not immediate but occurs with some time delay Δt. S = Stokes, AS = anti-Stokes.

Raman spectroscopy is a widely used method in condensed matter physics and physical chemistry to study vibrational excitations of solids and molecules. It is also used in medicine, particularly for histological studies of healthy versus pathological cells. The Raman spectrum of molecules is a fingerprint of these molecules and can be used to identify and characterize them. Scanning a monochromatic beam across the surface of a specimen with the Raman analyzer tuned to a specific and characteristic Raman transition, the distribution of this molecule in the specimen can be imaged [45].

As already stated, Raman scattering is a rare process under normal circumstances, which occurs on the order of 1 ppm with respect to ordinary Rayleigh scattering. However, the Raman scattering cross-section becomes drastically enhanced in the presence of Au or Ag NPs. When these NPs are excited to SPR, the strong electric field at their surfaces enhances the Raman cross-section, which scales with the square of the electric field amplitude. This enhancement process is termed *surface-enhanced Raman scattering* (SERS). It is more effective for nanorods than nanospheres as nanorods act like local antennas. The enhancement factor can be as high as 10^4–10^5 for adsorbed molecules on nanorods [44]. The current status and future perspectives of Raman imaging in medicine are discussed in [45–47].

7.9 Multimodality of theranostic nanoparticles

One and the same nanoparticle may feature several beneficial properties for diagnostics and therapeutics. The best examples are Au NPs and magnetite nanoparticles. Because of their strong light scattering properties, Au NPs can be used for hyperthermia, CT contrast agents, and dark-field imaging. Magnetite NPs are used for hyperthermia and for MRI contrast enhancement. To further increase the multimodality of nanoparticles, different materials can be combined. An example is shown in Fig. 7.21 from Ref [2]. Here the nanoparticle has a gold shell deposited on a dielectric silicon-oxide core. The shell diameter and thickness are selected on the one hand for surface plasmon resonance in the near infrared, and on the other for enhanced light scattering in dark-field imaging. In the Au nanoshell are embedded magnetite nanoparticles for MRI contrast enhancement. At a certain distance to the Au shell surface fluorophores emitting in the green spectral range are deposited. The nanoshell is then functionalized with proteins to bind antibodies for targeting cancer cells. Moreover, the nanoparticle is coated with PEG to support biodistribution in the body and to hinder protein adsorption and opsonization. Finally, antibodies are attached to the surface to target cancer cells. Aside from protective coatings, this nanoparticle has seven functionalities: targeting with antibodies; therapy with PTT (and eventually magnetic hyperthermia); diagnostic with CT and MRI contrast enhancement, SERS, fluorescence, and dark field imaging. The panels (b) and (c) of Fig. 7.21 show the specific relaxivities provided by magnetite NPs embedded in the Au nanoshell and the enhanced fluorescence intensity in the infrared regime.

While the development of NPs for cancer diagnostics and therapy has progressed tremendously over the past years, clinical studies are mostly missing. In animal studies, accumulation of targeted NPs in cancer tissue has been confirmed and regression of tumor volume upon application of hyperthermia has been observed. However, long term survival tests are still largely missing; therefore, it is presently

(a)

- Au nanoshell
- Fe_3O_4 NPs
- ICG fluorophore
- Streptavidin functionalization
- PEG
- Antibody

(b)

$r_2 = 390$ mM^{-1}s^{-1}

T_2^{-1}(sec^{-1})

[Fe] (mM)

(c)

Enhanced

FL Intensity (cps)

Unenhanced

Wavelength (nm)

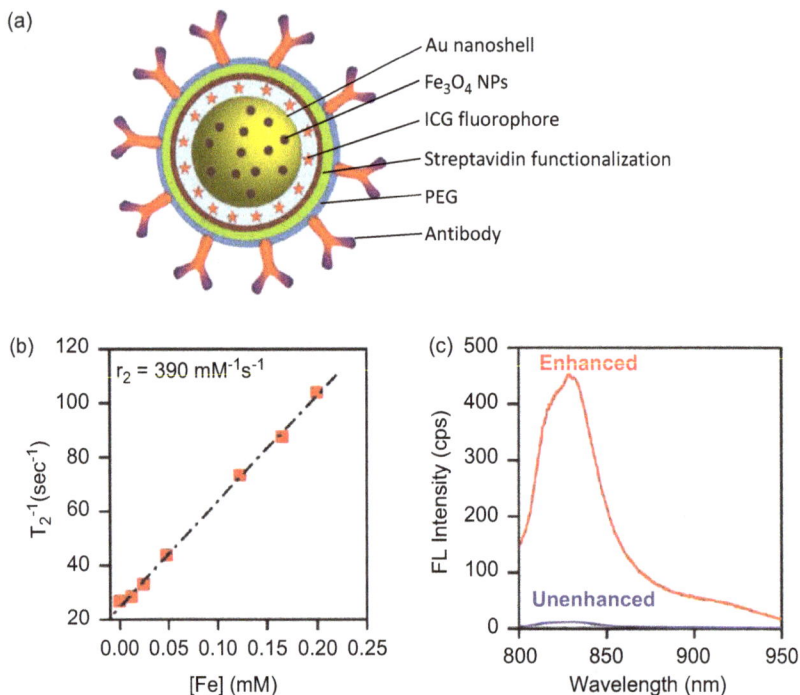

Fig. 7.21: Multimodal nanoparticle for targeting, imaging, and photothermal therapy. (a) Au nanoshell contains magnetite nanoparticles with adjustable density. The nanoshell is coated with an Indocyanine green (ICG) fluorophore; the nanocomplex is then functionalized with the protein streptavidin for binding antibodies. The nanocomplex is finally passivated with PEG to reduce nonspecific binding. (b) Transverse relaxation rate (T_2^{-1}) as a function of Fe concentration of the nanocomplex; r_2 is the specific relaxivity obtained from the slope. (c) Fluorescence (FL) spectra of enhanced ICG from nanocomplexes and unenhanced control (adapted from [2] by permission of ACS).

difficult to judge the success rate of a cancer treatment based on nanomedical methods. On the other hand, nanoparticles can also be used in other areas of medicine but oncology. For instance, gold-coated nanoshells activated by SPR can fuse together arteries, and functionalized NPs may bind pathogens and toxins in the bloodstream. This emerging field holds many promises.

Despite the impressive progress towards design and production of innovative nanoparticles for a variety of medical applications, there are still concerns with respect to their biocompatibility, pharmacokinetic distribution in the body, and metabolism. A state-of-the-art summary of theranostic nanoparticles, their applications, and their diverse administration routes for diagnosis and therapeutics are discussed in [48, 49].

7.10 Summary

S7.1   Nanomedicine is a multidisciplinary field with contributions from physics, materials science, biochemistry, polymer sciences, and many others.

S7.2   Nanomedicine uses nanoparticles for diagnostics and therapeutics.

S7.3   Nanoparticles, which have diagnostic and therapeutic capabilities, are called theranostic nanoparticles.

S7.4   Using nanoparticles, the pathway through the body has to be considered with respect to glomerular filteration, opsonization, and deposition in the liver.

S7.5   The reticuloendothelial system (RES) is part of the immune system and consists of scavenger cells that help filtering out dead and toxic cells.

S7.6   Enhanced permeation and retention (EPR) effect of nanoparticles is the result of a leaky and raptured blood vessel system in the tumor volume combined with a porous tissue.

S7.7   For keeping NPs in the bloodstream, their hydrodynamic diameter (HD) should be between 10 and 50 nm.

S7.8   PEG coating consists of a biocompatible and water soluble polymer of varying chain length. PEG shells hinder protein adsorption, which suppresses opsonization and therefore bypasses RES.

S7.9   We distinguish between physical targeting, passive targeting, and active targeting of NPs for diagnostics and therapeutics.

S7.10 We differentiate between core size, size including coating and shell, and the hydrodynamic diameter (HD).

S7.11 Superparamagnets are nanoparticles containing single domain ferrogmagnets.

S7.12 The magnetization of superparamagnets fluctuates above the blocking temperature, but is stable below the blocking temperature.

S7.13 The blocking temperature depends on the crystal anisotropy and particle size.

S7.14 Magnetic contrast agents can be positive or negative. Positive contrast agents shorten $T1$, negative contrast agents shorten $T2$.

S7.15 Magnetic nanoparticles can be used as contrast agent in magnetic resonance imaging (MRI) and for hyperthermia.

S7.16 High x-ray contrast efficacy requires the use of high Z elements.

S7.17 The size and shape of gold nanoparticles allows tuning the frequency of the maximum light absorption in the near infrared water window.

S7.18 The therapeutic benefit of Au nanoparticles and nanoshells results from resonant light absorption at the respective plasmon frequency.

S7.19 Larger Au nanoparticles of about 100 nm diameter scatter strongly light, which can be utilized for dark field imaging.

? Questions

Q7.1 What are the three main tasks of nanoparticles in medicine?

Q7.2 When nanoparticles enter the body, they are recognized by the immune system. What is it called, and how does it act on the nanoparticles?

Q7.3 The size of nanoparticles is an important parameter from a physiological and a physical point of view. Please explain both.

Q7.4 What types of targeting possibilities do nanoparticles have?

Q7.5 What is a Curie temperature?

Q7.6 What is a macrospin?

Q7.7 Why is the magnetic moment of Gd $8\mu_B$ and not $7\ \mu_B$, as one might expect from $m = gS\mu_B$ with $S = 7/2$ and a Landé-factor of 2 for spin only magnetic moment?

Q7.8 What is the blocking temperature of nanoparticles?

Q7.9 For contrast enhancement of MRI, MNPs should be designed such that T_B is above body temperature. In contrast, for hyperthermia the MNPs should have a T_B at or below body temperature. Why?

Q7.10 What are the three relaxation mechanisms for magnetic nanoparticles used for hyperthermia?

Q7.11 What is meant by positive contrast in MRI as compared to negative contrast?

Q7.12 Why do magnetite nanoparticles provide negative contrast?

Q7.13 What is the reason that Gd^{3+} chelates act as positive contrast agent?

Q7.14 What is K-edge contrast imaging?

Q7.15 Compare Rayleigh scattering and Mie scattering of electromagnetic waves. Which type of scattering prevails as function of wavelength?

Q7.16 What is the preferential direction of Mie scattering?

Q7.17 How does the optical absorbance of metal nanoparticles vary with size and aspect ratio?

Q7.18 Why are smaller sized NPs beneficial for absorptive applications, whereas larger sized NPs are more beneficial for imaging?

Q7.19 Describe briefly the essential points of photothermal therapy (PTT).

Q7.20 How can the accumulation of Au nanoparticles in tumorous tissue be made visible?

Q7.21 What are theranostic nanoparticles?

⌇ Attained competence checker

	+	0	–
I know what nanoparticles can be used for			
I can distinguish between antigen and antibody			
I know what the RES is and I can tell its function			
I know what the optimal radius of nanoparticles is to keep them in the vascular system			
I know what PEG stands for			
I can distinguish between positive and negative contrast agents			
I know that the most favorable magnetic nanoparticles are Fe_3O_4			
I can distinguish between Néel, Brown, and hysteretic relaxation of magnetic nanoparticles			
I realize that metal nanoparticles scatter light strongly			

I know how the color of metallic nanoparticles can be tuned

I realize the importance of metal nanoparticles for PTT

I can state the main ideas of the Debye relaxation and the Drude relaxation

I know the characteristics of theranostic nanoparticles

Exercises

E7.1   **Gd magnetic moment**: Show that Gd has a magnetic moment of $8\,\mu_B$

E7.2   **NP macrospin**: What is the magnetic moment of a magnetite nanoparticle of diameter 10 nm?

E7.3   **Magnetic induction of a magnetite nanoparticle**: What is the magnetic induction generated by a magnetite NP of diameter 10 nm.

E7.4   **Blocking temperature**: The blocking temperature T_B of magnetite nanoparticles increases with the NP diameter d: $T_B = 100$ K for $d = 3.8$ nm, $T_B = 150$ K for $d = 4.5$ nm, and $T_B = 250$ K for $d = 5.5$ nm. What is the blocking temperature for magnetite nanoparticles with diameter of 10 nm?

E7.5   **Positive and negative contrast agents**: Explain and rationalize why Gd-chelates is a positive contrast agent and magnetite is a negative one.

References

[1]   Nanomedicine, An ESF–European Medical Research Councils (EMRC) Forward Look Report; 2005.

[2]   Bardhan R, Lal S, Joshi A, Halas NJ. Theranostic nanoshells: From probe design to imaging and treatment of cancer. Acc Chem Res. 2011; 44: 936–946.

[3]   Kolosnjaj-Tabia J, Lartiguea L, Javed Y, Luciania N, Pellegrino T, Wilhelma C, Alloyeaub D, Gazeaua F. Biotransformations of magnetic nanoparticles in the body. Nano Today. 2016; 11: 280–284.

[4]   Rao L, Xu JH, Cai B, Liu H, Li M, Jia Y, Xiao L, Guo SS, Liu W, Zhao WZ. Synthetic nanoparticles camouflaged with biomimetic erythrocyte membranes for reduced reticuloendothelial system uptake. Nanotechnol. 2016; 27: 085106.

[5]   Longmire M, Choyke PL, Kobayashi H. Clearance properties of nano-sized particles and molecules as imaging agents: Considerations and caveats. Nanomedicine (Lond). 2008; 3: 703–717.

[6]   Ranganathan R, Madanmohan S, Kesavan A, Baskar G, Krishnamoorthy YR, Santosham R, Ponraju D, Rayala SK, Venkatraman G. Nanomedicine: Towards development of patient-friendly drug-delivery systems for oncological applications. Int J Nanomedicine. 2012; 7: 1043–1060.

[7]   De Crozals G, Bonnet R, Farre C, Chaix C. Nanoparticles with multiple properties for biomedical applications: A strategic guide. Nano Today. 2016; 7: 435–463.

[8]   Scott AM, Wolchok JD, Old LJ. Antibody therapy of cancer. Nat Rev Cancer. 2012; 12: 278–287.

[9]   Biological evaluation of medical devices – Guidance on the conduct of biological evaluation within a risk management process, ISO TR 15499, 2012.

[10] Black J. Biological performance of materials: Fundamentals of biocompatibility. 4th. Boca Raton, London, New York: CRC Press, Taylor & Francis; 2006.

[11] Williams DF editor, Definitions in biomaterials: Proceedings of a Consensus Conference of the European Society for Biomaterials. Amsterdam, Boston, Heidelberg, London, New York, Chester, England: Elsevier; 1987.

[12] Blundell S. Magnetism in condensed matter. Oxford master series in condensed matter physics. Oxford, New York, Athens: Oxford University Press; 2007.

[13] Graham CD. Magnetic behavior of gadolinium near the curie point. J Appl Phys. 1965; 36: 1135.

[14] Benitez MJ, Mishra D, Szary P, Badini Confalonieri GA, Feyen M, Lu AH, Agudo L, Eggeler G, Petracic O, Zabel H. Structural and magnetic characterization of self-assembled iron oxide nanoparticle arrays. J Phys Condens Matter. 2011; 23: 126003. p. 1–12.

[15] Bedanta S, Barman A, Kleemann W, Petracic O, Seki T. Magnetic nanoparticles: A subject for both fundamental research and applications. J Nanomater. 2013; 952540.

[16] Fang M, Ström V, Olsson RT, Belova L, Rao KV. Particle size and magnetic properties dependence on growth temperature for rapid mixed co-precipitated magnetite nanoparticles. Nanotechnol. 2012; 23: 145601.

[17] Koenig SH, Kellar KE. Theory of 1/T1 and 1/T2 NMRD profiles of solutions of magnetic nanoparticles. Magn Reson Med. 1995; 34: 227–233.

[18] Estelrich J, Sánchez-Martín MJ, Busquets MA. Nanoparticles in magnetic resonance imaging: From simple to dual contrast agents. Int J Nanomed. 2015; 10: 1727–1741.

[19] Bin Na H, Song IC, Hyeon T. Inorganic nanoparticles for MRI contrast agents. Adv Mater. 2009; 21: 2133–2148.

[20] Laranjeira MS, Moc A, Ferreirad J, Coimbrae S, Costag E, Silvag AS, Ferreirah PJ, Monteiroa FJ. Different hydroxyapatite magnetic nanoparticles for medical imaging: Its effects on hemostatic, hemolytic activity and cellular cytotoxicity. Colloids Surf B. 2016; 146: 363–374.

[21] Chaughule RS, Purushotham S, Ramanujan RV. Magnetic nanoparticles as contrast agents for magnetic resonance imaging. Proc Natl Acad Sci India Sect A Phys Sci. 2012; 82: 257–268.

[22] Brown WF. Thermal fluctuations of a single-domain particle. J Phys Rev. 1963; 130: 1677.

[23] Rosensweig RE. Heating magnetic fluid with alternating magnetic field. J Magn Magn Mater. 2002; 252: 370.

[24] Wang X, Gub H, Yang Z. The heating effect of magnetic fluids in an alternating magnetic field. J Magn Magn Mater. 2005; 293: 334–340.

[25] Fortin JP, Wilhelm C, Servais J, Ménager C, Bacri JC, Gazeau F. Size-sorted anionic iron oxide nanomagnets as colloidal mediators for magnetic hyperthermia. J Am Chem Soc. 2007; 129: 2628–2635.

[26] Ilg P, Kröger M. Dynamics of interacting magnetic nanoparticles: Effective behavior from competition between Brownian and Néel relaxation Phys. Chem Chem Phys. 2020; 22: 22244–22259.

[27] Liu XL, Guang Choo ES, Ahmed AS, Zhao LY, Yang Y, Ramanujan RV, Xue JM, Fan DD, Fan HM, Ding JD. Magnetic nanoparticle-loaded polymer nanospheres as magnetic hyperthermia agents. J Mater Chem B. 2014; 2: 120–128.

[28] Jordan A, Scholz R, Maier-Hau K, Johannsen M, Wust P, Nadobny J, Schirra H, Schmidt H, Deger S, Loening S, Lanksch W, Felix T. Presentation of a new magnetic field therapy system for the treatment of human solid tumors with magnetic fluid hyperthermia. J Magn Magn Mater. 2001; 225: 118–126.

[29] Kim DH, Rozhkova EA, Ulasov IV, Bader SD, Rajh T, Lesniak MS, Novosad V. Biofunctionalized magnetic-vortex microdiscs for targeted cancer-cell destruction. Nat Mater. 2010; 9: 165–171.

[30] Liu Y, Ai K, Lu L. Nanoparticulate x-ray computed tomography contrast agents: From design validation to in vivo applications. Acc Chem Res. 2012; 45: 1817–1827.
[31] Lusic H, Grinstaff MW. X-ray-computed tomography contrast agents. Chem Rev. 2013; 113: 1641–1666.
[32] Lee N, Choi SH, Hyeon T. Nano-sized CT contrast agent. Adv Mater. 2013; 25: 2641–2660.
[33] Mie G. Contributions to the optics of diffuse media, especially colloid metal solutions. Ann Phys. 1908; 25: 377–445.
[34] Kelly KL, Coronado E, Zhao LL, Schatz GC. The optical properties of metal nanoparticles: The influence of size, shape, and dielectric environment. J Phys Chem B. 2003; 107: 668–677.
[35] Ashcroft NW, Mermin ND. Solid state physics. New York, Chicago, San Francisco: Holt, Rinehart, and Winston; 1976.
[36] Link S, El-Sayed MA. Size and temperature dependence of the plasmon absorption of colloidal gold nanoparticles. J Phys Chem B. 1999; 103: 4212–4217.
[37] Peng S, McMahon JM, Schatz GC, Gray SK, Sun Y. Reversing the size-dependence of surface plasmon resonances. Proc Natl Acad Sci USA. 2010; 107: 14530–14534.
[38] Huang X, El-Sayed MA. Gold nanoparticles: Optical properties and implementations in cancer diagnosis and photothermal therapy. J Adv Res. 2010; 1: 13–28.
[39] Jain PK, El-Sayed MA. Universal scaling of plasmon coupling in metal nanostructures: Extension from particle pairs to nanoshells. Nano Lett. 2007; 7: 2854–2858.
[40] Jain PK, Lee KS, El-Sayed IH, El-Sayed MA. Calculated absorption and scattering properties of gold nanoparticles of different size, shape, and composition: Applications in biological imaging and biomedicine. J Phys Chem B. 2006; 110: 7238–7248.
[41] Prodan E, Radloff C, Halas NJ, Nordlander P. A hybridization model for the plasmon response of complex nanostructures. Science. 2003; 302: 419–422.
[42] Link S, Hathcock DJ, Nikoobakht B, El-Sayed MA. Medium effect on the electron cooling dynamics in gold nanorods and truncated tetrahedral. Adv Mater. 2003; 15: 393–396.
[43] Huang X, El-Sayed IH, Qian W, El-Sayed MA. Cancer cell imaging and photothermal therapy in the near-infrared region by using gold nanorods. J Am Chem Soc. 2006; 128: 2115–2120.
[44] Bardhan R, Grady NK, Cole JR, Joshi A, Halas NJ. Fluorescence enhancement by Au nanostructures: Nanoshells and nanorods. ACS Nano. 2009; 3: 744–752.
[45] Mikla VI, Mikla VV. Raman spectroscopy in medicine. Med Imaging Technol. 2014; 8: 129–141.
[46] Keating ME, Byrne HJ. Raman spectroscopy in nanomedicine: Current status and future perspective. 2013; 8: 1335–1351.
[47] Ou J, Zhou Z, Chen Z, Tan H. Optical diagnostic based on functionalized gold nanoparticles. Int J Mol Sci. 2019; 20: 4346.
[48] Zhang P, Lia Y, Tang W, Jie Zha J, Jing L, McHughc KJ. Theranostic nanoparticles with disease-specific administration strategies. NanoToday. 2022; 42: 101335.
[49] Anani T, Rahmati S, Sultana N, David AE. MRI-traceable theranostic nanoparticles for targeted cancer treatment. Theranostics. 2021; 11: 579–601.

Further reading

Howard AK, Vorup-Jensen T, Peer D editors, Nanomedicine. Berlin, Heidelberg, New York: Springer Verlag; 2016.
Jain KK editor, Handbook of nanomedicine. Berlin, Heidelberg, New York: Springer Verlag; 2012.

Hehenberger M. Nanomedicine. Boca Raton, London, New York: CRC Press, Taylor and Francis; 2016.

Nanomedicine, Wikipedia.

Chang H, Rho WY, Son BS, Kim J, Lee SH, Jeong DH, Jun BH. Plasmonic nanoparticles: basics to applications (I). Adv Exp Med Biol. 2021; 1309: 133–159.

Blundell S. Magnetism in Condensed Matter. In: Oxford master serie in condensed matter physics. Oxford, New York, Athens: Oxford University Press; 2007.

Part F: **Measuring and statistics**

8 Elements of medical statistics: measurements, population, distribution, reliability

Overview on important definitions

Systematic error	$\Delta x_{s,i}$
Random error	$\Delta x_{r,i}$
Number of measurement points	n
Absolute error	Δx
Relative error	$\Delta x/x$
Mean value	$\bar{x}$
Variance	s^2
Standard error	s
Reliability of the mean value	$\Delta\bar{x} = s/\sqrt{n}$
Coefficient of variation	$CV = s/\bar{x}$
Expectation value	μ
Standard deviation	σ
Area under the curve	$AUC = \sigma\sqrt{2\pi}$
Coefficient of variation (standard distribution)	$c_v = \sigma/\mu$

8.1 Introduction

There is no doubt that statistical methods are of the utmost importance in medical research and clinical trials. Therefore, knowledge of basic statistical concepts is essential for daily practice and the publication of research results. In this short chapter, the most important definitions and concepts are presented and explained by examples.

We need at least two items to measure physical quantities: a suitable measuring device such as a ruler and an internationally recognized system of units, usually the MKS or SI system. The ruler should be calibrated and provide reliable results. Measuring physical observables under these conditions can range from trivial to challenging. In any case, measurements of physical properties always contain certain errors. The error can be due to a lack of sensitivity of the instrument used, a faulty device or an incorrectly designed experimental setup to obtain the required information. Even if all systematic sources of error have been considered and eliminated, statistical errors still have to be treated using statistical methods. When performing a statistical analysis of a population with certain characteristics, it is important to select a sufficiently large representative group from the larger population with the same specified characteristics.

Statistical methods are used in all science branches, such as economics, social sciences, physical sciences, and engineering. In medical sciences, statistical analysis is particularly important as the success of specific diagnostic and therapeutic methods can only be judged by the positive outcome in one group compared to a control group. In the following paragraphs, we discuss some important concepts

https://doi.org/10.1515/9783111168739-008

and terms of error analysis and provide examples that will guide to more advanced texts. The following treatment is very basic. Therefore, this chapter does not replace a full course on medical statistics, but hopefully, it will provide sufficient information for understanding statistical issues and concepts.

8.2 Precision and accuracy

Before we go any further into measuring methods, we first want to clarify the often confusing terms *accuracy* and *precision*. We illustrate the concepts with a game of darts as shown in Fig. 8.1. In panel (a), all the darts thrown are fairly well centered on the inner ring and the spread of dart hits is small. We call this first example precise and accurate. In panel (b), we observe a similarly narrow spread of arrow hits, but they are all shifted to the right. In this round, the darts are thrown accurately, but miss the target and are therefore imprecise. In the third example shown in panel (c), the spread of the darts is wide but centered; hence the distribution of the darts is accurate but imprecise. The last example in image (d) shows a wide and off-center dart distribution, i.e., the hits are inaccurate and imprecise. In most cases considered in medical physics, especially when reaching a target volume for cancer treatment, we strive for the highest precision and the greatest possible accuracy. Precision is a matter of the random spread (scatter) of points, which must be kept as small as possible. The inaccuracy is determined by systematic shifts in the distribution to an off-center target position, which in turn should be as small as possible. It is the task of medical physicists and engineers to strive for ever-improved methods to increase precision and accuracy in medical practice.

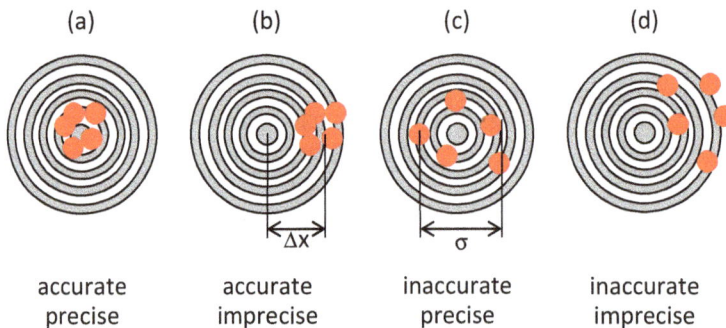

(a)	(b)	(c)	(d)
accurate	accurate	inaccurate	inaccurate
precise	imprecise	precise	imprecise

Fig. 8.1: Game of darts with five throws. In panel (a), all throws are precise and accurate; in panel (b), the throws are precise but miss the target by the shift Δx and are therefore inaccurate; in panel (c), the throws are centered around the target but widely scattered over an area of width σ. Hence they are accurate but imprecise, and in (d) the throws are inaccurate and imprecise.

8.3 Random errors and systematic errors

Usually it is not sufficient to measure a physical observable only once. All measurements are affected by systematic and random or stochastic errors. In order to figure out how big the error of a single measurement is, we have to repeat the same measurement many times.

Any measurement of a physical quantity x contains mistakes. The true value of x, which we want to call x_t, may be never known. A measurement of x yields the measured value x_i. The measured value x_i deviates from the true value x_t by an unknown amount Δx_i:

$$x_i = x_t \pm \Delta x_i. \tag{8.1}$$

The deviation or error $\pm \Delta x_i$ is composed of a systematic error $\pm \Delta x_{s,i}$ and a random error $\pm \Delta x_{r,i}$ such that

$$x_i = x_t \pm \Delta x_i = x_t \pm (\Delta x_{s,i} + \Delta x_{r,i}). \tag{8.2}$$

The sign ± expresses the fact that the deviation from the true value can be positive or negative. Systematic errors are due to faulty measurement device or effects that have not been properly considered, such as thermal expansion when measuring extensions above or below room temperature. Systematic errors cannot be eliminated by repeating the measurement. They need to be removed or accounted for in a proper data analysis. However, stochastic errors fluctuate and can be treated by statistical methods only. Figure 8.2 shows a sequence of 12 individual measurements of the same physical quantity. In the left panel, the measurements scatter about the unknown true value. The scatter of the points in the right panel is identical, but shifted by a constant amount Δx_s, due to a systematic error. Systematic errors may have dramatic consequences. For instance, missing the target volume by a few millimeters in EBRT of cancerous tissue would have fatal consequences for the malignant and healthy tissue. Similarly, missing the target with a laser beam in ophthalmology would be disastrous for the patient. But how do we become aware of systematic errors and how can we avoid them? This topic goes beyond the discussion of this

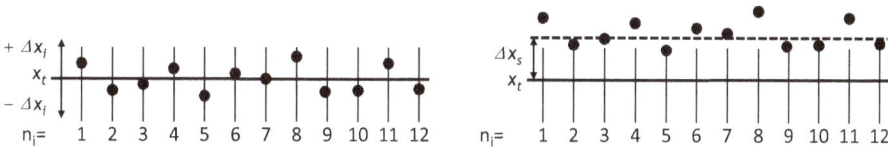

Fig. 8.2: Illustration of random errors (left) and random plus systematic errors (right) in a sequence of measurements of some physical parameter x. The center solid black line represents the unknown "true" value of x. The dashed line in the right panel indicates a constant shift of the data points due to a systematic error.

chapter and is part of the so-called metrology. In short, systematic errors can only be identified by repeated calibration of instruments against standards.

8.4 Mean, variance, and confidence interval

For the following, we assume that any systematic error has been eliminated. Thus, we can treat the data taken by statistical methods. An estimate of the true value of x can be gained by taking the average of all measured values x_i, i.e., summing up all measured results and dividing the sum by the number n of measurements taken:

$$\bar{x} = \frac{1}{n} \sum_{i=1}^{n} x_i \tag{8.3}$$

$\bar{x}$ (pronounced x bar) is called the average or the *mean value* of x, in short, the *mean*. The number of measurement points n is also called the *number of degrees of freedom*. In statistics, unlike in physics, the number of degrees of freedom is the number of independent measurements that can be freely taken without violating any constraints that may be imposed on them.

Next, we want to determine the width of the scattered points, i.e., whether the distribution of measured points is narrow, as illustrated in Fig. 8.3 (left), or whether the distribution is wide, as indicated in the right panel of Fig. 8.3. For that, we calculate the *variance s^2*, which is the sum of squared deviations from the mean, divided by $n-1$:

$$s^2 = \frac{1}{n-1} \sum_{i=1}^{n} (x_i - \bar{x})^2 = \frac{1}{n-1} \sum_{i=1}^{n} (\Delta x_i)^2 \tag{8.4}$$

The normalization factor is $n-1$ instead of n because one degree of freedom has already been used for calculating the mean $\bar{x}$. The difference between $n-1$ and n is significant only for small n. For small n, the variance of measured points would be underestimated when dividing by n instead of $n-1$.

The smaller the variance s^2, the smaller is the scatter of the measured points, i.e., the more narrow is the distribution. With a bit of algebraic manipulation, we obtain for the variance the expression:

$$s^2 = \frac{1}{n-1} \left[\sum_{i=1}^{n} x_i^2 - \frac{\left(\sum x_i\right)^2}{n} \right] = \frac{1}{n-1} \left[\sum_{i=1}^{n} x_i^2 - n\bar{x}^2 \right] \tag{8.5}$$

These expressions are easier to compute than the sum in eq. (8.4), as the sum of squares $\sum x_i^2$ has to be determined only once. The square root of the variance is called the *standard error s* of $\bar{x}$:

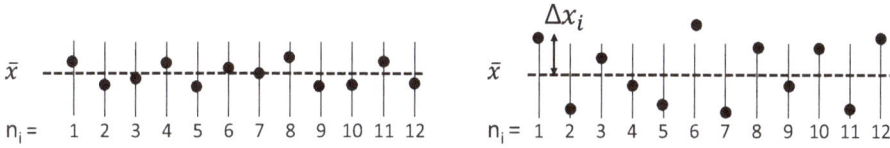

Fig. 8.3: Left panel: narrow distribution of measurement points about the mean. Right panel: wider distribution with larger variance.

$$s = \left| \sqrt{s^2} \right| \tag{8.6}$$

Experimental values are usually quoted as the mean value of a measurement series $\bar{x}$ together with the standard error s:

$$x = \bar{x} \pm s. \tag{8.7}$$

The total extension of a standard error $2s$ defines the length of error bars for any measurement of the physical quantity x.

Returning to the mean value $\bar{x}$, it should be stated that also $\bar{x}$ has an error that can be estimated. The mean value becomes more reliable the higher the number n of measurements (sampling number) is. The average error of the mean value, i.e., the deviation of the mean value from the true value $\Delta\bar{x} = \bar{x} - x_t$ decreases with the reciprocal root of the number of measurements taken:

$$\Delta\bar{x} = \bar{x} - x_t = \frac{s}{\sqrt{n}}. \tag{8.8}$$

$\Delta\bar{x}$ is also known as the *reliability of the mean value*, the *confidence interval* (CI), or the standard error of the mean. Note that the standard error of the mean is smaller than the standard error of the measurement value by the factor $1/\sqrt{n}$. Increasing the sampling number n is beneficial for decreasing the standard error and for increasing the reliability of the mean value. In fact, for $n \to \infty$, the mean value approaches the true value of the physical parameter x.

An example may illustrate the concept of CI. Measurements of the length of the Eustachian tube made of 100 subjects yield a mean value of 3.6 cm with a standard deviation of $s = \pm 1$ mm. The standard deviation of the mean value, i.e., the CI, is $\Delta\bar{x} = s/\sqrt{n} = 1/\sqrt{100} = \pm 0.1$mm. In order to improve the CI by one order of magnitude, 10 000 instead of 100 subjects need to be tested. Then $\Delta\bar{x} = s/\sqrt{n} = 1/\sqrt{10000} = \pm 0.01$mm. Note that increasing the sampling number n does not change the mean of the distribution and its variance. But it considerably reduces the CI.

Often the ratio:

$$CV = \frac{s}{\bar{x}} \tag{8.9}$$

is quoted. This ratio is called *coefficient of variation* (CV) or *relative standard error* and is a measure of how narrow or wide a distribution is. Small numbers below 1 reflect a narrow distribution, and large numbers above 1 indicate a wide distributions. The following example illustrates the importance of this ratio.

Example: Measurements of the speed of cars on a freeway may yield the following sets of data. In a first test containing 10 cars, the speed in kilometer per hour was: 1×80, 2×90, 4×100, 2×110, and 1×120. The mean is $\bar{x} = 1000/10 = 100$ km/h, the variance is $s^2 = 133$ (km/h)^2, and the standard error $s = \pm 11.5$ km/h. In another test containing 16 cars the outcome was: 1×40, 2×60, 3×80, 4×100, 3×120, 2×140, and 1×160. Here mean $\bar{x} = 1600/16 = 100$ km/h is the same, but the variance is $s^2 = 1066$ (km/h)^2, and the standard error amounts to $s = \pm 32.6$ km/h. For the first distribution, we have a CV of 0.11, whereas in the second distribution the CV is 0.33, i.e., in the second case, the CV is considerably larger, indicating a broader distribution than in the first test.

Finally, note that none of the evaluated parameters, such as mean, variance, standard error, CI, or CV, are observables. These parameters are solely evaluated by statistical means after repeated measurements of an observable.

8.5 Absolute and relative errors

Next, we distinguish between *absolute* errors and *relative* errors. The absolute error Δx of a variable x is independent of the value of x and is given in absolute numbers and with the same units as the variable x. For instance, the length of a board maybe $1\,\text{m} \pm 1\,\text{mm}$, $2\,\text{m} \pm 1\,\text{mm}$, $3\,\text{m} \pm 1\,\text{mm}$, etc. The absolute error Δx is usually expressed in terms of a one-standard deviation s.

The *relative error* Δx scales with x such that the ratio $\Delta x / x$ remains constant. The relative error is expressed in percent and is a pure number without units. For instance, if the relative error of a measured length is $\pm 0.1\%$, length measurements then yield $1\,\text{m} \pm 1\,\text{mm}$, $2\,\text{m} \pm 2\,\text{mm}$, $3\,\text{m} \pm 3\,\text{mm}$, etc., according to eq. (8.7).

Systematic errors often result in absolute errors Δx, whereas for random errors, the ratio $\Delta x / x$ stays constant. The difference may be illustrated by the following example.

Example: Manufacturers usually express the precision of an instrument in terms of a relative error such as $\pm 1.5\%$. If the systolic pressure of a patient measured with a pressure gauge has been determined to be 150 hPa and the relative error is quoted as $\pm 1.5\%$, then the expected absolute error of a single measurement is ± 2.25 hPa. However, if the absolute measurement error of, for instance, a clinical thermometer is given by the manufacturer as ± 0.1 C, then the relative error for a temperature measurement of 40 C is 0.25%.

8.6 Error propagation

We consider a function $f = f(x, y, z)$ of the independent variable (x, y, z). Then the squared total deviation $(df)^2$ is the sum of the squared partial differentials times the squared deviations:

$$(df)^2 = \left(\frac{\partial f}{\partial x}\right)^2 (\partial x)^2 + \left(\frac{\partial f}{\partial y}\right)^2 (\partial y)^2 + \left(\frac{\partial f}{\partial z}\right)^2 (\partial z)^2 \tag{8.10}$$

Here df is the standard error of f and $(\partial x), (\partial x), (\partial x)$ are the standard errors of the variables (x, y, z). If the function and the variables are normal-distributed (see 8.10), then we can call the deviations also standard deviations. Equation (8.10) is known as the formula of the *error propagation.*

If the measurement value f is the sum of several different and independent observables x, y and z with coefficients a, b, c, yielding:

$$f = ax + by + cz, \tag{8.11}$$

then the absolute error of f follows, according to (8.10), from the algebraic sum of the errors for the individual observables x, y, and z:

$$\Delta\bar{f} = \sqrt{(a\Delta\bar{x})^2 + (b\Delta\bar{y})^2 + (c\Delta\bar{z})^2}, \tag{8.12}$$

where $\Delta\bar{x}$, $\Delta\bar{y}$, and $\Delta\bar{z}$ are the CIs of x, y, and z.

If the measurement value f is the product of the observables x, y, z:

$$f = x \cdot y \cdot z, \tag{8.13}$$

then the absolute error of f is:

$$\Delta\bar{f} = \sqrt{(yz\Delta\bar{x})^2 + (xz\Delta\bar{y})^2 + (xy\Delta\bar{z})^2}, \tag{8.14}$$

and the relative error is:

$$\frac{\Delta\bar{f}}{f} = \sqrt{\left(\frac{\Delta\bar{x}}{x}\right)^2 + \left(\frac{\Delta\bar{y}}{y}\right)^2 + \left(\frac{\Delta\bar{z}}{z}\right)^2}. \tag{8.15}$$

If the measurement value is expressed as the ratio of the observables x and y:

$$f = \frac{x}{y}, \tag{8.16}$$

then the absolute error is:

$$\bar{\Delta f} = \frac{1}{y^2} \sqrt{(y\Delta\bar{x})^2 + (x\Delta\bar{y})^2}, \tag{8.17}$$

and the relative error follows from:

$$\frac{\bar{\Delta f}}{f} = \sqrt{\left(\frac{\Delta\bar{x}}{x}\right)^2 + \left(\frac{\Delta\bar{y}}{y}\right)^2}. \tag{8.18}$$

Examples of relative and absolute errors of physical quantities composed of several observables are discussed in the exercise section.

8.7 Error of measuring devices and sensitivity

Most measurement devices these days have digital displays. When measuring an observable, such as the voltage U, one can observe that the last digit in the display can jump back and forth between two values. For example, the true value of a voltage U to be measured can be 23.17 V. However, the voltmeter used to measure U only has an accuracy of ±0.02 V. Then, the digital display jumps between the values 23.16 and 23.18 V. The result of the measurement can then be quoted as $U = (23.17 \pm 0.01)$ V with an error of $\Delta U = \pm 0.01$ V.

As a rule of thumb, the quoted result of a measurement should not contain more digits than evaluated via the standard error s or the CI of the mean $\Delta\bar{x}$. If the error is, for instance, $\Delta U = \pm 0.01$V, then quoting the measurement value as $U = (23.17 \pm 0.01)$ V is reasonable, but quoting the result as $U = (23.17432 \pm 0.01)$ V would not make sense, even if a calculator may show more digits.

The *sensitivity* of a measuring device is defined as the ratio of display value to entrance value. In case of analog displays with a needle movement proportional to the measurement value, the sensitivity of the instrument is clearly visible. If a voltage measurement of 1 mV causes the needle to move by 1 mm, the sensitivity of the voltmeter is 1 mm/1 mV or 1 m/V. In a galvanometer, the needle may rotate by 10°, yielding a sensitivity of 10°/1 mV. In the case of digital instruments, the sensitivity is determined by the last digit. An instrument that shows three digits after the decimal point can measure millivolts but not microvolts. The sensitivity is therefore quoted as 1 mV. A voltmeter that shows 3.5 digits after the decimal has a sensitivity of 0.5 mV = 500 μV, etc.

The *accuracy class* is defined by the maximum error occurring at the maximum range of the display. If a voltmeter has a maximum range of 1000 V with an accuracy $U = (1000 \pm 15)$ V, the sensitivity of the instrument is $15/1000 = 0.015 = 1.5\%$. This sensitivity is then quoted as an accuracy class of 1.5. The expected measurement error when measuring 500 V is then $U = (500 \pm 7.5)$ V, such that $\Delta U/U$ remains constant.

8.8 Population and sampling

When we repeat the measurement of a physical quantity several times, such as the length or the weight of a metal block, the variance of the measurements will be relatively small, assuming that the measurements are correctly and carefully executed. We have a completely different situation when we measure the same physical property of many different but comparable objects, such as the length of 1000 nails that a machine has produced in the same run. Then these 1000 nails belong to a *population* of nails and each individual nail in this population will have a slightly different length due to production reasons. Here we neglect the measurement error of the length of an individual nail and assume that this error is small compared to the length variation within the population of nails. It is of interest to determine again the variance and the standard deviation since a large standard deviation may indicate a faulty production machine.

Both types of error analysis are relevant in medical physics: (1) random errors of repeated measurements of one and the same object or person and (2) fluctuations of a physical quantity within a population. Examples of the first kind may be the measurement of the bone density of a particular individual, the body temperature or the systolic–diastolic blood pressure difference. Examples of the second kind are, for instance, measurements of the vital lung capacity of male persons at age 20, or the accommodation width of female persons at age 40, or the intraocular pressure drop after taking certain drugs in a group of people affected by glaucoma.

The examples given for the second kind of measurement contain subtle differences, noticeable only after carefully scrutinizing the procedures, as explained in the following.

To determine the vital lung capacity of 20-year-old male subjects, it would be impossible to perform this test for everyone in this group. Therefore, a vital capacity measurement requires a subset of randomly selected male subjects (often referred to as a cohort). This test is then referred to as a random but hopefully *represrentative sampling* of the population of 20-year-old males. The number of subjects participating in this measurement is referred to as the *sample size n*.

The situation is different in the case of the intraocular pressure drop. Here the group of people receiving the drug is finite and all members of that group will participate in the test. Therefore, the test is done with the group's total *population* being exposed to the drug.

In summary, we distinguish between three types of measurement statistics and error analysis:

1. Repeated measurements of an observable taken from a single object;
2. Single measurement of an observable taken from all members of a defined group, called *population*;
3. Single measurement of an observable taken from an unspecified random subset of a defined population, called *random sampling or representative sampling*.

8.9 Frequency, histogram, and distribution

Next we consider how measurements of an observable x are distributed among individuals in a population. The number of individuals in the population with a particular value or quality is called the *frequency of occurrence* or simply the frequency of that quality. As an example, we examine the age size of 10-year-old boys. The population of 10-year-old boys is far too large to be fully measured and analyzed. Therefore, we consider a random but representative sample of size $n = 30$ boys. The results of the length measurements $x_i (i = 1 - 30)$ in units of centimeters are shown in Tab. 8.1 in ascending order. For convenience, measurement results have been rounded to the nearest whole number.

Tab. 8.1: Results of size measurements of 30 boys at age 10.

127	131	132	132	133	134	135	136	136	136
137	137	138	139	139	140	140	141	141	142
142	143	144	145	146	147	148	149	151	152

These 30 data points are represented in a *histogram* of frequencies, where the x-axis is subdivided into intervals of equal size Δx, and the y-axis shows the frequency within the intervals. Here the size of intervals is chosen to be $\Delta x = 4.99$ cm, extending from 130 to 134.99 cm, from 135 to 139.99 cm, etc. Intervals need to have an adequate width or extension. They should neither be too small nor too big to obtain a meaningful distribution of frequencies. The distribution of data listed in Tab. 8.1 is plotted in Fig. 8.4. One boy has a size between 125 and 129.99 cm, five boys have sizes between 130 and 134.99 cm. We immediately recognize that half of the boys or 50% have a size below 140 cm, the other 50% are taller than 140 cm. The median is marked in Fig. 8.4 by a red line. Next we calculate the mean, the variance, and the standard error for this distribution.

We determine the mean $\bar{x}$ by taking the sum of all values listed in Tab. 8.1 and divide the sum by the number of data points (sample size) n: $\bar{x} = 139.76$ cm. For the variance we obtain $s^2 = (587\,155 - 30 \times (139.76)^2)/29 = 38.39$ cm^2, and for the standard error we find $s = \pm 6.2$ cm. For this symmetric distribution, the median is identical with the mean.

Often histograms are asymmetric or they may show two maxima instead of one. In Fig. 8.5 various types of histograms are compared schematically. A *symmetric distribution* like in panel (a) with a single maximum is called a *single mode distribution*. Panel (b) shows a *bimodal distribution* with two maxima, and panels (c) and (d) display *skewed distributions*. For the symmetric distribution, the location of the maximum is equivalent to the mean, which, in turn, is identical with the median. For bimodal and

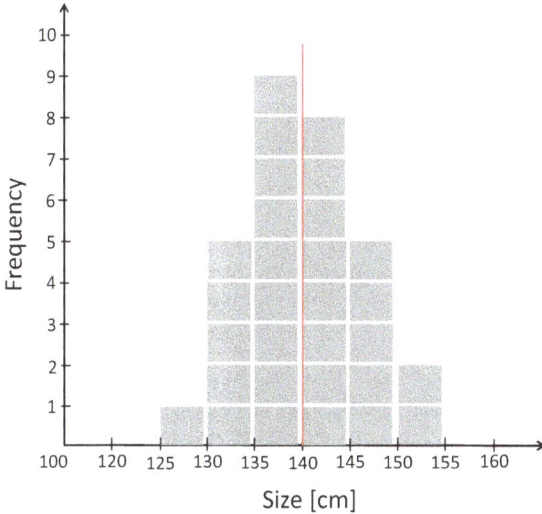

Fig. 8.4: Histogram of size measurements according to the values listed in Tab. 8.1. Each gray square box represents one measurement (or the size of one boy), organized according to the size intervals Δx shown on the abscissa.

asymmetric distributions, quoting a mean is meaningless. But quoting *quantiles* is useful: the first *quartile* contains 25% of all data, the *median* contains 50% of all values, and the last quartile covers 75%. If a distribution is symmetric, the mean and median are identical, as we already noted, and the distribution usually can be described by a *standard distribution,* which is the topic of the next section.

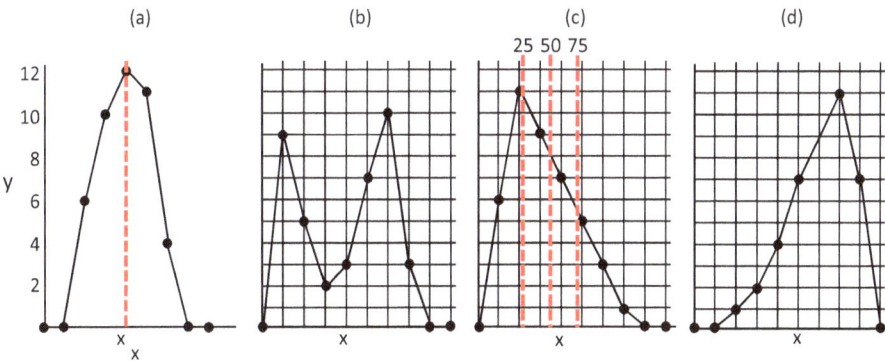

Fig. 8.5: Different distributions of measured values: (a) single mode distribution; (b) bimodal distribution; (c) and (d) show skewed distributions. The vertical dashed lines in panel (c) indicate the quantiles for 25%, 50%, and 75% of all plotted values.

8.10 Standard distribution

The standard distribution, also called *normal distribution, standard normal distribution*, or *Gaussian*[1] *distribution*, is a mathematical model that describes rather well symmetric histograms of stochastic values. The mathematical expression for the standard distribution is:

$$h(x) = \frac{1}{\sqrt{2\pi}\sigma} \exp\left(-\frac{(x-\mu)^2}{2\sigma^2}\right)$$ (8.19)

Here x is the measured value, μ is the *expectation value*, σ^2 is the variance, and σ is the *standard deviation* from the expectation value. The standard distribution is plotted in Fig. 8.6. The expectation value is a mathematical construct originating from probability theory. Consider the following example. The probability of rolling a six with an ideal dice is $p = 1/6$. p is the expectation value. In the case of 100 throws, the expectation value will be $100 \times p = 100/6 = 16.66$. After 100 throws, the pip number 6 should appear 16.66 times. With an increasing number of throws, the average value will increasingly approach the expectation value.

The standard distribution has a maximum at $x = \mu$ with the normalized height of:

$$h_{max} = h(x = \mu) = \frac{1}{\sigma\sqrt{2\pi}}.$$ (8.20)

This shows that the maximum of the normal distribution scales inversely with the standard deviation σ. While the total *area under the curve* (AUC) remains constant, the height drops with increasing width. Integrating over the total area of the standard distribution yields the value:

$$AUC = \sigma\sqrt{2\pi}.$$ (8.21)

Therefore, the prefactor $1/\sigma\sqrt{2\pi}$ is the normalization factor of the normal distribution. The coefficient of variation (CV) of the standard distribution is defined by the dimensionless ratio:

$$CV = \frac{\sigma}{\mu}$$ (8.22)

For a fixed mean μ, the CV is small for high and narrow distributions, and big for low and wide distributions.

The inflection point is reached at one standard deviation to the left and right of the maximum with the values:

$$h(x = \mu \pm \sigma) = \frac{1}{\sigma\sqrt{2\pi e}}.$$ (8.23)

---

1 Johann Carl Friedrich Gauss (1777–1855), German mathematician, astronomist, and physicist.

The area from $(x-1\sigma)$ to $(x+1\sigma)$ indicated by a gray shade in Fig. 8.6 is called the *one-sigma area*. The 1σ-area contains 68.3% of all values. An additional $2 \times 13.6\%$ of all values are in the 2σ – area (ocher color), thus covering 95.5% of the total area. The remaining blue area contains only 4.5% of the total area. In statistical analysis, usually the 1σ and more rarely the 2σ deviation from the expectation value is quoted.

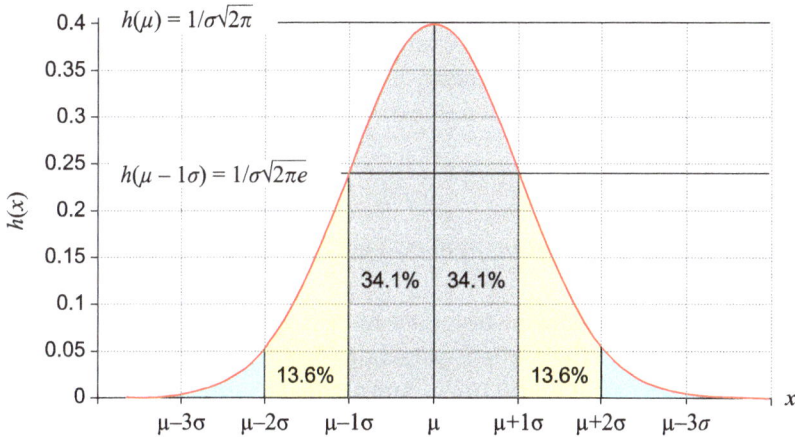

Fig. 8.6: Plot of the standard distribution with a single maximum at $x = \mu$ and height $h(\mu) = 1/\sigma\sqrt{2\pi}$ with σ set to 1. The 1σ standard deviation at $x = \mu \pm \sigma$ is shaded gray and contains 68.3% of the total area. (Adapted from wikimedia@ creative commons).

The normal distribution has three important properties:
1. The mean marks the maximum of the distribution where most of the data points are taken;
2. The mean divides the normal distribution into two equal parts: 50% of all collected data is equal to or less than the mean, and 50% is greater than the mean.
3. The area between the standard deviations, the $\pm 1\sigma$ area, contains 68.3% of all data points.

Other important distributions are the binomial distribution and the Poisson distribution. Both distributions are mainly used to analyze counting statistics. They are discussed in Section 8.5.1 of Volume 2 and therefore not repeated here.

8.11 Assessing histograms by standard distribution

Now we want to use the standard distribution for describing a symmetric histogram. The standard distribution is a smooth and continuous mathematical curve of the variable x, while a histogram is a discontinuous plot of measured values, subdivided into equal intervals Δx. The histogram becomes smoother by increasing the sample size n and decreasing the width of the intervals Δx. Nevertheless, it will always remain a discontinuous distribution, unless we take the limits $n \to \infty$ and $\Delta x \to 0$. If we try to describe a symmetric histogram by a standard distribution, we aim for the best fit of a standard distribution to a histogram. The fit parameters are μ and σ, whereas the sample size n is fixed.

A good estimate for the expectation value μ of the histogram is the mean value $\bar{x}$. For the standard deviation σ we use the standard error s as the best estimate, and for the variance, we set $s^2 = \sigma^2$. Furthermore, if the total number of measurements (sample size) is n, then the height at $x = \mu$ is accordingly:

$$h(x = \mu) = \frac{n}{\tilde{\sigma}\sqrt{2\pi}}. \tag{8.24}$$

Here the tilt indicates that the standard deviation σ is to be given in terms of the interval size: $\tilde{\sigma} = \sigma / \Delta x$.

As an example, we reconsider the histogram of the stature-for-age of 10-year-old boys, which we discussed in the previous paragraph. For this distribution, we have a sample size $n = 30$, a mean of $\bar{x} = 139.76$ cm, and a standard error $s = 6.2$ cm. As the interval chosen was $\Delta x = 5$ cm, we have for $\tilde{\sigma} = \sigma / \Delta x = 1.24$. Figure 8.7 shows a plot of the standard distribution together with the histogram from Fig. 8.4 using the parameters quoted before.

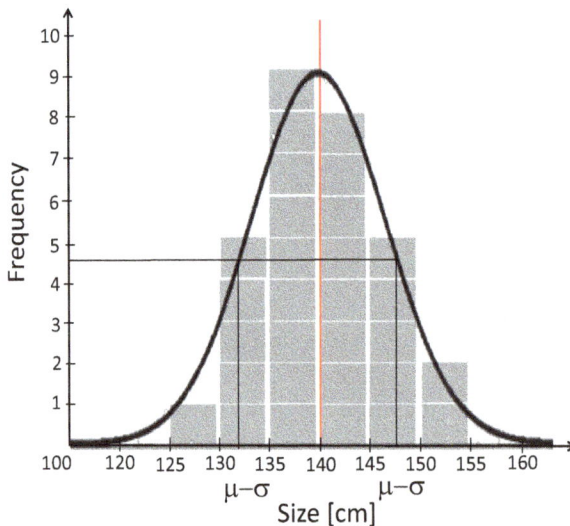

Fig. 8.7: Histogram and standard distribution. The solid line represents the best fit to the histogram using a standard distribution with the mean value $\mu = \bar{x} = 139.76$ cm and a standard deviation of $\sigma = s = 1.24$ in units of the interval size. The sample size n is 30.

The standard normal distribution with the estimated values of $\mu = \bar{x}$ and $\sigma = s$ agrees rather well with the histogram. So far so good. However, we also want to estimate how well a standard normal distribution can describe a distribution of measured values or whether the measured distribution deviates from a normal distribution.

For the standard normal distribution, there are fixed probability values for each point x of the distribution. The ones at $x = \mu \pm 1\sigma$, $\mu \pm 2\sigma, \mu \pm 3\sigma$ are the most prominent check points and they are independent of the expectation value μ. The corresponding probabilities $P(x)_{standard}$ are listed in Tab. 8.2. If a measurement is normal-distributed, then at a standard deviation of -1σ we expect that 16% of all measurement points have values that are smaller than the predicted $\mu - 1\sigma$ value, etc.

Now we evaluate the measured probabilities $P(x)_{measured}$ at the checkpoints of the histogram: $x = \bar{x} \pm ks$ ($k = 1, 2, 3$). For this evaluation, we use again the size distribution of the 30 boys listed in Tab. 8.1. For the $-3s$ value, we have $x = \bar{x} - 3s = 140 - 3 \times 6.2 = 121.4$. At this point, there are zero measurement points. For the $-2s$ value at $x = \bar{x} - 2s = 128$, we have just one measurement point, which has a probability of $1/30 = 3.3\%$. At $\bar{x} - 1s = 133.8$, we have five points with a probability of $5/30 = 16.6\%$, etc. Now we plot the standard normal deviations $x = \mu \pm k\sigma$ on the Y-axis against the measured values of the histogram $x = \bar{x} \pm ks$ on the X-axis where the expected probabilities have actually been reached. If a standard normal distribution can represent the histogram, the plot should show a straight line, as indeed is the case for the considered histogram of the boy's size distribution plotted in Fig. 8.8(a). This type of plot is called *normal plot*. However, if the measured probabilities occur earlier or later than expected by the normal distribution, the plot will deviate strongly from a straight line. Hence, the normal distribution is not a good description. Figure 8.8(b) shows the normal plot for the distribution provided in Fig. 8.5(c). Clearly, this skewed distribution cannot be described by a normal standard distribution.

Tab. 8.2: Probabilities P(x) for integrated values reached at -3σ up to $+3\sigma$ according to the standard normal distribution and probabilities actually determined from the histogram plotted in Fig. 8.4.

$x = \mu \pm k\sigma$	$P(x)_{standard}$	$x = \bar{x} - ks$	$P(x)_{measured}$
-3σ	0.001	121.4	0
-2σ	0.023	127.6	0.033
-1σ	0.159	133.8	0.166
0	0.5	140	0.5
$+1\sigma$	0.841	146.2	0.833
$+2\sigma$	0.977	152.4	1.0
$+3\sigma$	0.999	158.6	1.0

(a)   (b)

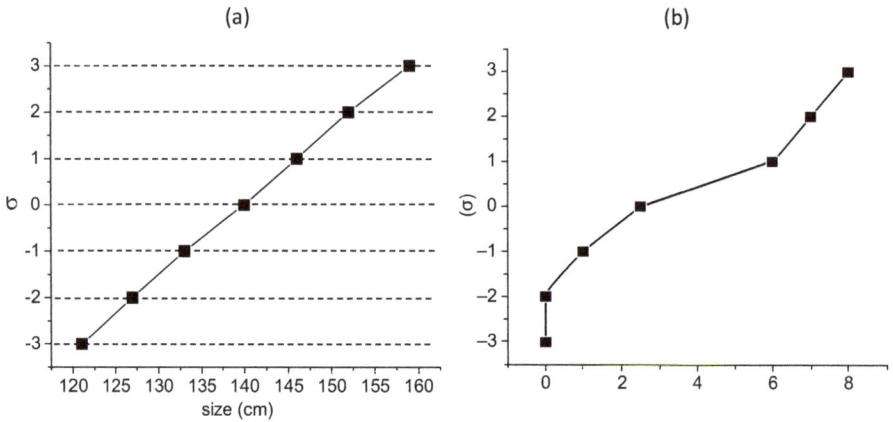

Fig. 8.8: Normal plots for two distributions: (a) a histogram that can be described by normal standard distribution and (b) for a histogram that deviates strongly from a normal distribution. In panel (a), the standard normal deviations of the first column in Tab. 8.2 are plotted against the measured standard deviations according the third column. In panel (b), the same analysis is performed for the histogram shown in Fig. 8.5(c).

8.12 Normal probabilities

A common task of normal distributions is the evaluation of probabilities. Often the question is asked: what is the probability that a certain quantity is larger or smaller than a given value? Re-considering the boy's size distribution, we may ask what the probability is that boys are taller than 150 cm. Hence, we seek the probability $P(x > 150)$, which is equivalent to finding the probability for: $P(x \leq 150) = 1 - P(x > 150)$. $P(x \leq 150)$ is the probability for finding boys that are smaller or equal to 150 cm.

Evaluating $P(x \leq 150)$ is simpler because integration over the normal distribution always starts at $-\infty$ and goes up to the variable x. However, the integration is performed in a normalized fashion with the normal distribution expressed as:

$$h(z) = \frac{1}{\sqrt{2\pi}} \exp\left(-\frac{z^2}{2}\right) \tag{8.25}$$

where

$$z = \frac{x - \mu}{\sigma} \tag{8.26}$$

The probability that a variable x lies between minus infinity and z is now:

$$P(Z) = \frac{1}{\sqrt{2\pi}} \int_{-\infty}^{z} \exp\left(-\frac{z^2}{2}\right) dz \tag{8.27}$$

The integrated area of the standard normal distribution is shown in Fig. 8.9. The yellow part refers to probability values $P(X \leq Z)$, and the blue area is the complement part of the standard normal distribution.

$P(Z)$ values for the integral of the normal distribution can be found in tabulated form at many places [1]. An abridged version of the probabilities $P(Z)$ for negative z-values is reproduced in Tab. 8.3. Probabilities for positive z-values follow from $1 - P(Z) = P(-Z)$.

Returning to the boy's size distribution and the evaluation of the probability $P(x>150)$, we find by normalization:

$$z = \frac{x - \mu}{\sigma} = \frac{150 - 140}{6.2} = \frac{10}{6.2} = 1.61$$

According to the Tab. 8.3, $P(z \leq 1.61) = 1 - P(-z) = 0.9463$. Therefore, the complement of the standard normal value is $P(z > 1.61) = 1 - P(z \leq 1.61) = 0.0537$. Hence, the probability for boys being taller than 150 cm is 5.4%.

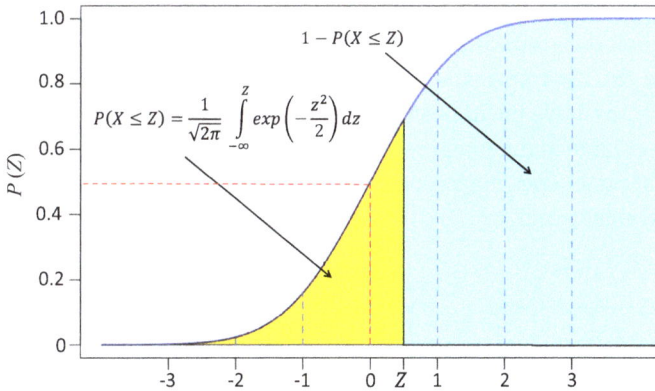

Fig. 8.9: Probability density function of the standard normal distribution.

Tab. 8.3: Negative z-table for z-values left of the mean. $P(z)$ values define the area under the standard normal curve to the left of the respective z-value.

z	P(z)	z	P(z)	z	P(z)
−3.0	0.0013	−2.0	0.0228	−1.0	0.1587
−2.8	0.0026	−1.8	0.0359	−0.8	0.2119
−2.6	0.0047	−1.6	0.0548	−0.6	0.2743
−2.4	0.0082	−1.4	0.0808	−0.4	0.3446
−2.2	0.0139	−1.2	0.1151	−0.2	0.4207
				0.0	0.5000

8.13 Linear regression and correlation

Linear regression analysis is used to study linear relationships between two observables, i.e., whether a straight line can represent two variables. The term "regression" has historical reasons and doesn't really make sense. A better term would be
"linear correlation," but the traditional term is hard to change. A simple example
for a linear relationship is the distance y and the time t of a uniformly moving body
expressed via $y = vt + b$, where v is the velocity of the body and b is the starting
point at $t = 0$. Measurement of both independent observables, y and t, should verify
or falsify the hypothesis of a linear relationship. If the linear relationship does not
hold in our example, we will conclude that the velocity of the moving body was not
uniform on average. In any case, the task of a linear regression is to test whether
the data can be represented by a linear relationship, via the general expression
$y = ax + b$. If yes, we also want to know how reliable such a representation is. This
task requires determining the coefficients a and b and stating a CI for both. The parameter a yields the slope and b the value of the abscissa at $x = 0$. In Fig. 8.10, two
possible outcomes of such measurement and corresponding linear regressions are
sketched. In the left panel, the scattering of measurement points is small and a linear regression through the data points is justified with $b = 0$ and $a > 0$ (positive
slope). In the right panel, we have the opposite situation. The scattering of the measurement points is rather large and a linear regression yields the values $a = 0$, $b > 0$.
The example in the right panel does not support the conclusion of a linear relationship between the observables y and x.

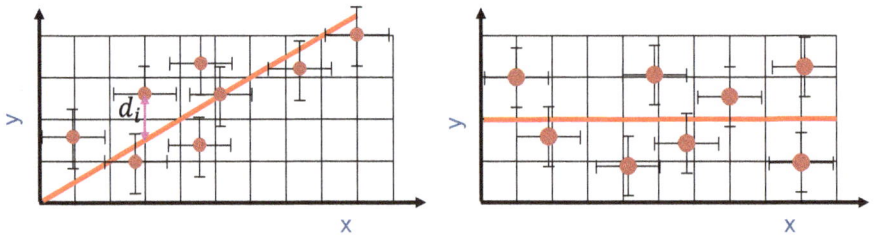

Fig. 8.10: Left and right panels show a scatter plot of data points. All data points are plotted
together with their error bars. The straight red lines represent a best fit via linear regressions.

Measurements of the quantities x and y yield the pair of values (x_i, y_i). If (x_i, y_i) are
functionally related and assuming here a linear relationship, then x_i is the *predictor variable* and y_i is the *response variable*. Measuring for fixed x_i the response y_i
several times will result in different values y_i. Hence, y_i is a random variable to
the fixed value x_i.

Now we consider a linear regression model with only one independent variable
x_i. The linear regression model is expressed as

$$\hat{y}_i = ax_i + b. \tag{8.28}$$

Here $\hat{y}_i$ is the predicted response value of the measured response variable y_i to the predictor variable x_i. a, b are the best fit parameters to be determined, and n is the number of measurement points. Finding the best fit parameters is done by minimizing the quadratic deviations between measurement point y_i and the predicted value $\hat{y}_i$ for each variable x_i. The deviation corresponds to the vertical differences:

$$d_i = y_i - \hat{y}_i = y_i - (ax_i + b), \tag{8.29}$$

as indicated in the left panel of Fig. 8.10. For the best fit, the sum of the quadratic deviations

$$Z = \sum_{i=1}^{n} (d_i)^2 \tag{8.30}$$

should have a minimum value. Therefore the first derivatives of Z with respect to a and b should be zero:

$$\frac{dZ}{da} = \frac{dZ}{db} = 0. \tag{8.31}$$

From this condition, we obtain the equations for a and b:

$$a = \frac{n \sum_i x_i \cdot y_i - \sum_i x_i \sum_i y_i}{A} \tag{8.32}$$

and

$$b = \frac{n \sum_i (x_i)^2 \sum_i y_i - \sum_i (x_i \cdot y_i) \sum_i x_i}{A}, \tag{8.33}$$

where

$$A = n \sum_i (x_i)^2 - \left(\sum_i x_i \right)^2, \tag{8.34}$$

and n is the number of measurement points. In fact, only four sums need to be evaluated in order to determine the coefficients a and b:

$$c = \sum_i x_i; \ d = \sum_i y_i; \ f = \sum_i x_i \cdot y_i; \text{ and } g = \sum_i (x_i)^2 \tag{8.35}$$

In short notation, the coefficients follow from the determinants:

$$a = \begin{vmatrix} f & c \\ d & n \end{vmatrix} / A; \qquad b = \begin{vmatrix} g & f \\ c & d \end{vmatrix} / A; \qquad A = \begin{vmatrix} g & c \\ c & n \end{vmatrix}. \tag{8.36}$$

The result of this calculation is a linear curve that must include the relation:

$$\bar{y} = a\bar{x} + b, \tag{8.37}$$

where $\bar{x}$ and $\bar{y}$ are the mean values of x and y. Thus, the abscissa may also be determined via:

$$b = \bar{y} - a\bar{x}, \tag{8.38}$$

where $\bar{x} = c/n$ and $\bar{y} = d/n$.

Having determined the coefficients a and b, we want to know how well the data points are represented by the linear function. For this, we consider again the sum of the squared deviations and divide it by the degrees of freedom associated with the n measurement points, yielding the variance:

$$\sigma^2 = \frac{Z}{n-2} = \frac{\sum_i (d_i)^2}{n-2}. \tag{8.39}$$

The degrees of freedom is $n-2$ since two degrees of freedom are already used for evaluating the parameters a and b. The standard deviation is simply the positive square root of the variance:

$$s = \left| \sqrt{\sigma^2} \right|. \tag{8.40}$$

Example: To illustrate the procedures, we discuss a simple example: temperature T and length L of a metal rod are related as:

$$L(T) = \alpha T + L_0, \tag{8.41}$$

where α is the thermal expansion coefficient. The results of a fictive measurement with a total number of measurement points $n = 10$ are listed in Tab. 8.4.

The measurement points together with the least square fit (red line) are plotted in Fig. 8.11. From Tab. 8.4 we obtain the following values:

$a = 2.043$; $b = -1.66$; $\bar{x} = c/n = 55$; $\bar{y} = d/n = 110.7$; $\sum d_i = 0$; $Z = 318.91$; $\sigma^2 = 39.86$; $s = 6.31$. From these values, we conclude that the variance is reasonably small and that the hypothesis of a linear relationship between temperature and length of the metal rod is supported by the measured data points.

Tab. 8.4: List of the results of length measurement of a metal rod at incremental temperatures from 10 to 100 °C.

N	Temp. x_i	Length y_i	$x_i y_i$	x_i^2	$\hat{y}_i$	d_i	d_i^2
1	10	22	220	100	18.77	3.23	10.43
2	20	39	780	400	39.2	−0.2	0.04
3	30	58	1740	900	59.63	−1.63	2.65
4	40	83	3320	1600	80.06	2.94	8.64
5	50	101	5050	2500	100.49	0.51	0.26
6	60	115	6900	3600	120.92	−5.92	35
7	70	144	10 080	4900	141.35	2.65	7.02
8	80	150	12 000	6400	161.78	−11.78	138.76
9	90	185	16 650	8100	182.21	2.79	7.78
10	100	210	21 000	10 000	202.64	7.36	54.17
$n = 10$	$c = 550$	$d = 1107$	$f = 77\ 740$	$g = 38\ 500$		0	$Z = 318.91$
Total	$c^2 = 302\ 500$						

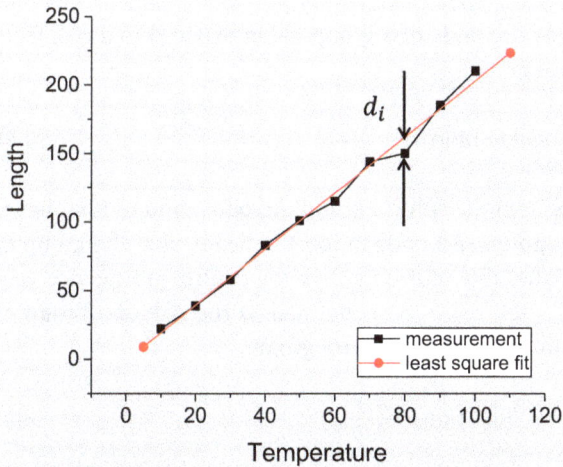

Fig. 8.11: Plot of measured length versus temperature for a metal rod in arbitrary units. Black points are the measured values, and red line is the result of the linear regression analysis (least squares fit).

8.14 Risk and odds

In medical literature, often the following terms are used: odds, risk, odds ratio, and risk ratio. In the following, these terms will be explained. *Odds* are calculated by dividing the number of times x an event takes place in a group of y members by the number of times it does not occur in the same group. In other words, in a group of y members, x members are affected, while y-x members are not affected. Thus the odds is defined as the ratio:

$$\text{Odds} = \frac{x}{y - x} \tag{8.42}$$

If 10 in a group of 100 patients suffer a side effect from a treatment, the odds are $10/90 = 0.11$.

In contrast, the *risk* is defined by the number of times an event occurs in a group of patients divided by the total number of patients in that group:

$$\text{Risk} = \frac{x}{y} \tag{8.43}$$

For the quoted example this means that the risk is $10/100 = 0.1$ or 10%. The odds and the risk can be significantly different if the number of events in the group is rather large. For instance, with 50 events in a group of hundred, the odds are 1 and the risk is 50%.

The *odds ratio* is defined as the odds of a group being exposed to a risk factor divided by the odds in a control group:

$$\text{Odds ratio} = \frac{x/(y - x)}{\tilde{x}/(\tilde{y} - \tilde{x})} \tag{8.44}$$

Here the tilde indicates values from the control group. An odds ratio of 1 implies no difference between these two groups. An odds ratio > 1 indicates that the group exposed to a risk factor has a higher rate of events than the control group; vice versa for an odds ratio < 1. Similarly, the *risk ratio* is defined as the risk of a group exposed to a risk factor divided by the risk of a control group:

$$\text{Risk ratio} = \frac{x/y}{\tilde{x}/\tilde{y}}. \tag{8.45}$$

8.15 Sensitivity, specificity, predictive values of diagnosis

Sensitivity and specificity are important concepts when analyzing diagnostic results. We have already defined the term "sensitivity" in connection with measurement technology as the ratio of response amplitude to input value. The sensitivity in the medical sense is based on the result of measurements carried out on two different groups as detailed below.

In common language, a good diagnostic method for a disease X should detect that disease with a high probability. Then we say that this method has a high sensitivity to detect disease X. However, this method may also detect with a similar probability disease Y. Then, the method is sensitive to both diseases X and Y but not specific to either one of them. For instance, for detecting a particular carcinoma, MRI and FDG–PET may have a similar sensitivity of about 93%. However, the specificities are very different: 50% for MRI and 93% for FDG–PET. In the following, we want to clarify and quantify this concept. We consider two groups. One with disease X and second control group of the same size but without disease X. Diagnostic measures are performed with both groups and the results yield the following values.

- The group with disease: In TP patients the disease was properly detected, called *true positive*; in FN patients the disease was not detected, called *false negative*.
- The group without disease: in FP persons, the disease was "diagnosed," called *false positive*, and in TN persons, the disease was not detected, called *true negative*. These four terms together with the diagnostic results are summarized in Tab. 8.5.

Tab. 8.5: Overview on true and false diagnostic results.

Diagnosis	Group of patients with disease	Group of persons without disease
Positive	TP (true)	FP (false)
Negative	FN (false)	TN (true)

The *sensitivity* SE of the diagnosis is now defined as the ratio:

$$SE = \frac{TP}{TP + FN}. \tag{8.46}$$

Ideally, this ratio would be SE = 1.

The specificity discriminates true positive results against false-positive results, i.e., for a healthy person, the diagnosis should yield a true negative result. Thus the *specificity* is defined as:

$$SP = \frac{TN}{TN + FP}. \tag{8.47}$$

Ideally, this ratio would also be 1. However, diagnostic methods are never perfect. Therefore, it is of interest to evaluate the probability that a diagnosed positive patient will have a disease. This probability is called *positive predictive value* and is defined as the ratio:

$$PPV = \frac{TP}{TP + FP}. \tag{8.48}$$

Vice versa, the probability of detecting a disease that does not exist is called *negative predictive value* (NPV) and is defined as the ratio:

$$NPV = \frac{TN}{TN + FN}.$$ (8.49)

PPV and NPV ratios would ideally also be 1. The lower the values are, the lower is the sensitivity and specificity of a diagnostic method.

Some further definitions are redundant and can be derived from the already defined definitions. The *false-positive rate* (FPR) is defined as:

$$FPR = \frac{FP}{FP + TN} = 1 - SP,$$ (8.50)

and the *false-negative rate* (FNR) is defined accordingly:

$$FNR = \frac{FN}{FN + TP} = 1 - SE$$ (8.51)

For an illustration of these ratios, we take a lung cancer test performed with 100 patients that showed lung cancer and a control group of 100 persons without lung cancer. The diagnostic results are as follows: 80 patients were diagnosed positive in the cancer group, 30 persons were diagnosed false positive in the control group. This yields the following values: $SE = 0.8$, $SP = 0.7$, $PPV = 0.73$, $NPV = 0.77$. These numbers have the following meaning. In the group with lung cancer, 80% are diagnosed correctly, but in 20% of the cases, lung cancer was not detected although present (false-negative result). In the healthy group, 70% of the persons are diagnosed correctly, but 30% had a false-positive result. Furthermore, there is a 73% chance that if the test is positive, the person actually shows lung cancer, and there is a 77% chance that if the test is negative, the person does not have lung cancer. If the group sizes are different, the numbers for both groups have first to be normalized before evaluating PPV and NPV, This means that the following sums should add up to 1:

$$TP + FN = TN + FP = 1.$$ (8.52)

In summary, the most important quality criteria for the significance (validity) of diagnostic laboratory tests are specificity and sensitivity:
- Sensitivity means the probability that a positive sample will also be recognized as positive (exclusion of false negatives).
- Specificity means the probability that a negative sample will also be recognized as negative (exclusion of false positives);

For a reliable test result, values close to 100% are aimed for both criteria; high sensitivity ensures that no person with a disease is accidentally overlooked; a high specificity indicates that no "false alarm" (e.g., due to cross reactivity) is triggered.

8.16 Summary

S8.1   Precision implies hitting a target with narrow distribution, but the center of the distribution may deviate from the target center.

S8.2   Accuracy implies hitting a target with the center of the distribution located at the target center.

S8.3   All measurements are affected by errors. Errors consist of random errors and systematic errors.

S8.4   Random errors can be treated by statistical methods. Systematic errors need to be recognized and removed.

S8.5   The mean is the sum of all measurements of an observable divided by the number n of measurements taken.

S8.6   Variance is defined by the sum of squared deviations from the mean, divided by $n - 1$:

S8.7   Standard error is defined as the square root of the variance.

S8.8   Confidence interval is defined as the standard error divided by the square root of n.

S8.9   Coefficient of variation is defined as the ratio of the standard deviation and the mean.

S8.10   If a measurement has a fixed absolute error, then the measurement result is quoted by its mean value plus/minus the standard deviation.

S8.11   Relative errors are quoted as ratios of standard error and mean value.

S8.12   The sensitivity of a measuring device with a digital display is given by the last digit shown.

S8.13   Error propagation analysis is required whenever the measured observable is composed of independent parameters connected by addition, multiplication, or division.

S8.14   Single measurement of an observable taken from all members of a defined group, is called *population*;

S8.15   Single measurement of an observable taken from an unspecified random subset of a defined population, is called *random sampling*.

S8.16   A histogram is a plot of number events (frequency) recorded within a defined interval of values versus the values subdivided into intervals.

S8.17   A histogram represents a frequency distribution of measured values for a certain observable.

S8.18   The histogram may have the shape of a single mode distribution, a bimodal distribution, or a skewed distribution.

S8.19   Single mode histograms may be modeled by a standard distribution.

S8.20   The standard distribution is a symmetric distribution about an expectation value, representing the maximum of the distribution. The width of the standard distribution is given by the standard deviation.

S8.21   Linear regression is used to study linear relationships between measurable observables.

S8.22   Linear regression is a method to find the best fit for data points, which are linearly related, by a straight line. If the measured data are linearly related, the slope of the best linear fit curve should be different from zero.

S8.23   Sensitivity of a diagnostic result is defined as the ratio of true positive results and the total sample number.

S8.24   Specificity is defined as the ratio of true negative results and the sum of true negative and false-negative results. True negative results are taken from a control group.

? Questions

Q8.1   What is the difference between accuracy and precision?

Q8.2   Can systematic errors be eliminated by statistical methods?

Q8.3   How is the mean value of a measurement series defined?

Q8.4   Why is the variance normalized by $n - 1$ and not by n, where n is the number of independent measurements, also called the sampling number?

Q8.5   How can the confidence interval of the mean value be improved?

Q8.6   If the quoted relative error of a measurement is 1.2%, what is the expected error of a single measurement of the weight, displaying 875 N on the scale?

Q8.7   What is the difference between sensitivity and accuracy class quoted for measurement devices?

Q8.8   Define the terms population and representative sampling.

Q8.9   A histrogram is a plot of what?

Q8.10   What types of distributions do you know? Name at least three.

Q8.11   The standard normal distribution is a mathematical concept with the following five characteristics: the distribution has only one maximum, the maximum represent the 50%-tile, symmetric distribution, 95.5% of the total area lies within the $\pm 1\sigma$ range, the expectation value corresponds to the mean value. Which statement is wrong?

Q8.12   How can a histogram be tested whether it is normal distributed?

Q8.13   What is a linear regression analysis?

Q8.14   How is a linear regression analysis performed?

Q8.15   If the slope of linear regression turns out to be zero, what does this tell you about the scattering of the measurement points?

Q8.16   How can the quality of the linear regression analysis be judged?

Q8.17   What is the difference between the terms odds and risk?

Q8.18   When a group is tested to contain 40% false-positive results, how would you judge the specificity of such a diagnostic test?

Q8.19   How is the sensitivity of a diagnostic method defined?

Attained competence checker	+	0	–	⚡
I know the difference between precision and accuracy				
I can distinguish between random and systematic errors				
I know that systematic errors cannot be eliminated by statistical methods				
I know how to evaluate the mean value of a measurement series and the variance				
I can construct a histogram from collected data				
I know what a normal distribution is and I can draw such a distribution				
I appreciate that in a normal distribution the mean equals the median				
I know what a population means in statistical terms				
I can name other distributions than the normal distribution				
I know how many data points lie within the $\pm 1\sigma$ range of a normal distribution				
I know what a linear regression analysis is good for				
I understand the difference between the terms risk and odds				
I know how the terms sensitivity and specificity are defined for diagnostic analysis				
I realize that sensitivity and specificity are statistical terms expressing probabilities				

Exercises

E8.1 **Systematic error**: How do I notice that my data are shifted by a systematic error?

E8.2 **Cube volume**: Determine the absolute and relative error of measuring the volume of a cube with a side wall of 0.75 m and a measurement precision of ±1 mm. What is the total error of the measurement result?

E8.3 **Error propagation**: Assume a function $f = ax^n$, where n is the power of x and a is a constant. What is the relative and absolute error of f?

E8.4 **Probability**: Bags of dog food have a mean weight of $\mu = 8$ kg with a standard deviation of $\sigma =$ 100 g. Assuming that the weight distribution follows a normal distribution, what is the probability that the weight X of a bag is more than 8.2 kg: $X > 8.2$ kg.

E8.5 **Sensitivity and specificity**: A certain disease has a prevalence of 0.01 or 1%, meaning that 10 people out of a population of 1000 have the disease. A diagnostic test has a sensitivity of 80%. The same diagnostic test has a false positive rate of 10%.

   a. When the test is applied, how many people are tested true positive, how many false negative on average?

   b. Confirm the sensitivity of the test.

   c. Determine the specificity of the test.

   d. Confirm the false positive rate.

E8.6   **PSA test**: Men at age above 60 have a 21% chance of developing prostate cancer. The PSA test has a diagnostic sensitivity of 19.5%, i.e., only 19.5% of the prostate cancer cases test positive. The specificity of the test is rather high at 94% [2].
   a.   How many men in a population of 1000 are diagnosed correctly?
   b.   Determine the false positive rate.
   c.   What is the false-negative rate?
   d.   What is the predictive value of the PSA test?

References

[1]   https://en.wikipedia.org/wiki/Standard_normal_table
[2]   Ankerst DP, Thompson IM. Sensitivity and specificity of prostate-specific antigen for prostate cancer detection with high rates of biopsy verification. Arch Ital Urol Androl. 2006; 78: 125–1299.

Further reading

Neter J, Wasserman W. Applied linear statistical methods. Homewood, Illinois; Georgetown, Ontario: Richard D. Irwin Inc; 1974.

Weisberg S. Applied linear regression. 4th. New York, London, Sydney, Toronto: Wiley & Sons; 2014.

Bland M. An introduction to medical statistics. Oxford, New York, Athens: Oxford University Press; 2015.

Kirkwood B, Sterne J. Essential medical statistics. Malden, Massachusetts, USA; Carlton, Australia: Blackwell Science. 2nd; 2003.

Harris M, Taylor G. Medical statics made easy. London, New York, Melbourne: Martin Dunitz-Taylor & Francis Group; 2003.

Huges IG, Hase TPA. Measurements and uncertainties. A practical guide to modern error analysis. Oxford, New York, Athens: Oxford University Press; 2010.

Useful website

The following webpage contains all important definitions: https://en.wikipedia.org/wiki/Sensitivity_and_specificity

Appendix

9 Answers to questions

Chapter 1

A1.1   Interphase and mitosis.

A1.2   During the S phase.

A1.3   P53.

A1.4   Mutation, extension, attack, penetration, angiogenesis, and invasion.

A1.5   Somewhere between penetration and angiogenesis.

A1.6   During angiogenesis with additional oxygen supply.

A1.7   S is an exponential function of the dose D.

A1.8   They are most sensitive during the M phase and least sensitive during the S phase.

A1.9   The probability to irradiate tumor cells during the most sensitive M phase is higher when the total dose is fractioned in smaller amounts over several sessions. In the interval time, healthy tissues have a chance for self-repairing.

A1.10   Direct action by a single-strand break or double-strand break of a DNA. Indirect action via the generation of free toxic radicals.

A1.11   RBE is defined as the ratio of x-ray dose to the dose of other types of radiation for the same survival fraction.

A1.12   The RBE is highest when the interaction distance between two hits matches the diameter of the DNA. This is the case for radiation with a linear energy transfer of about 100 keV/μm. At the maximum, the RBE reaches values of about 8–10.

A1.13   Fast neutrons, alpha particles, carbon ions.

A1.14   X-rays or gamma rays, protons, slow neutrons.

A1.15   OER is the oxygen enhancement ratio, meaning that the radiation effect is enhanced in the presence of oxygen within the cancerous volume.

A1.16   For low-LET radiation OER is important; for high-LET radiation, OER is not important.

A1.17   OER is important because, with the presence of oxygen, free and aggressive radicals can be produced that attack the DNA.

A1.18   The presence of oxygen supports the radiotherapy of LET radiation.

A1.19   Low dose causes tumor growth delay but no growth control. High dose provides tumor control.

A1.20   Chemotherapy enhances the radiation sensitivity.

A1.21   Repairing, reoxygenation, redistribution, repopulation.

A1.22   QA is important to avoid systematic errors in the tumor assessment and radiation planning and to avoid communication mistakes.

A1.23   TNM is a tumor categorization code specifying tumor size, proximity to nearby lymph nodes, and extent of metastasis.

https://doi.org/10.1515/9783111168739-009

Chapter 2

A2.1   4–25 MeV.

A2.2   By an electron linac. At the end of the accelerator path, the electrons hit a tungsten target for converting electrons to x-ray photons.

A2.3   80% of photons undergo Compton scattering, the remaining 20% cause pair production.

A2.4   Electrons straggle for a certain distance and then stop. Photons are first converted into Compton electrons and electron/positrons and then build up the dose by ionizing soft matter.

A2.5   For Kerma, only the transfer of photon energy into kinetic energy of electrons is important, which occurs immediately starting beneath the skin, while the dose develops only after these charged particles straggle until they stop.

A2.6   There are two reasons: electrons have no skin sparing effect, see Fig. 12.3. And second, the range of electrons is too limited. We need the photon beam as a "carrier of electrons" to deeper regions. Only after photon-electron conversion, the electrons become dose relevant.

A2.7   At roughly 1 to 3 mm below the surface, depending on the photon beam energy.

A2.8   The reason is the conversion of photons into Compton electrons and electron/positron pairs, which straggle over a certain distance and deposit kinetic energy for ionization of molecules.

A2.9   SSD refers to constant source-surface distance; the reference point is the skin surface. SAD refers to constant source axis distance, which is the constant source-target volume distance.

A2.10   The SSD simulation of the treatment plan is simpler, but moving the treatment table is required to meet the condition. SAD does not require patient movement. However, dose-to-patient simulation is more difficult for the SAD geometry than for SSD.

A2.11   GTV is the target volume to be irradiated. CTV is an extra clinically defined volume that may be affected by the proliferation of cancer cells. PTV allows for any patient or organ movement and encloses GTV and CTV.

A2.12   With IGRT, the PTV can be considerably reduced, sparing healthier tissue from high-dose radiation.

A2.13   This is only possible by multiple exposures from different angles or orientations. The target volume accumulates a high dose by exposure from all different orientations, while the tissue between surface and target is exposed only once.

A2.14   IMRT-MLC stands for intensity modulated multileaf collimation. It is presently the best way to define a tumor volume with the prescribed dose.

A2.15   Cyberknife technology overcomes most of the shortcomings of traditional XRT with a coplanar gantry. The linac is short, lightweight and can be positioned at any angle by a robotic arm. Patient and organ movement is compensated for by an infrared and x-ray tracking system. The linac produces a maximum of 8 MeV photons, staying below the photon-neutron production threshold. MLC is not required as the fine x-ray beam automatically exposes a well-defined target volume.

A2.16   MigRT stands for magnetoresonance guided radiation therapy.

A2.17   FLASH promises a similar tumor control compared to conventional radiotherapy while being better in sparing healthy surrounding tissues.

A2.18   The gamma-knife was used for treating brain tumors. The main justification is simultaneous irradiation of a target volume with high-energy gamma photons that can penetrate the skull. Gamma-knife technology requires ^{60}Co isotopes. This technology has become obsolete by the introduction of cyberknife technology.

Chapter 3

A3.1   Ionization of core electrons, Coulomb scattering at nuclei, elastic recoil scattering.

A3.2   At the end of the track.

A3.3   Yes, the range is equivalent with the position of the Bragg peak.

A3.4   The maximum energy is 230 MeV, yielding a range of 32 cm.

A3.5   About 1.1.

A3.6   PRT is a low RBE treatment.

A3.7   Very low in the beginning, increasing LET in the Bragg region. But still rather low compared to fast neutrons.

A3.8   SOPB can be formed by tuning the incident energy and the proton fluence.

A3.9   The proximal side receives much less radiation than in the case of XRT. The distal side is not exposed at all. Therefore healthy tissue is better spared, i.e., the non-target dose is significantly lower.

A3.10   Proton beams can be monitored by recording β^+ – decay, or recording the prompt γ-emission spectrum.

A3.11   The tail of the dose profile is due to fragmentation of carbon nucleons.

A3.12   Larger LET and RBE compared to PRT and a much sharper Bragg peak.

A3.13   Size of facility and high investment cost.

A3.14   Ion source, accelerator, beam delivery, and gantry.

Chapter 4

A4.1   1. Direct neutron irradiation with fast neutrons as a fission product in a re-
        actor; 2. Irradiation with neutrons produced by (p,n) conversion using an
        accelerator; 3. Thermal neutron capture by ^{10}B, producing alpha particles
        in tumor cells referred to as boron-neutron capture therapy.

A4.2   Cold neutrons, thermal neutrons, epithermal neutrons, fast neutrons, and
        high-energy neutrons.

A4.3   Fast neutrons have high LET and high RBE, while their OER is 1. These are
        favorable parameters for cancer treatment.

A4.4   Inoperable tumors, tumors with radiation resistance to XRT and PBT, ex-
        tended tumors close to the surface.

A4.5 ^{9}Be(p,n)^{9}B, ^{9}Be(d,n)^{10}B, ^{3}H(d,n)^{4}He.

A4.6   The shape is similar to the one for high-energy x-rays. The dose increases
        below the surface, then goes through a maximum, and drops off linearly with
        increasing depth.

A4.7   High energy neutrons first need to slow down to about 2 MeV before devel-
        oping their maximum LET. This occurs a few millimeters below the surface,
        depending on the initial neutron energy.

A4.8   The number of fractions can be reduced in high-LET irradiation; the dose
        per fraction is kept constant.

A4.9   The role of thermal neutrons is to be captured by ^{10}B that causes a nuclear
        reaction releasing fast α-particles. The short-range α-particles destroy can-
        cer cells.

A4.10   α-particles and Li ions.

Chapter 5

A5.1   Brachytherapy is a radiation therapy where radioisotopes are in direct con-
        tact with carcinoma.

A5.2   External beam therapy.

A5.3   Prostate, cervical, breast, and skin cancer.

A5.4   Either in the form of capsules containing radioisotopes, radioactive wires,
        or plaques.

A5.5   Very local treatment with high dose over short time, saves healthy tissue.

A5.6   Mainly gamma radiation, beta radiation is stopped in the container wall.

A5.7   137Cs, 60Co, 192Ir, 103Pd.

A5.8   Mainly by neutron activation.

A5.9   Either by hand or by using an applicator, allowing the insertion of a wire.

A5.10   Interstitial (implantation) or contact on the inner or outer skin.

A5.11   LDR is typically 2 Gy/h, HDR is typically 12 Gy/h.

A5.12   In standard dosimetry, there is one source, and the target is at a certain distance from the source. In BT, there are multiple sources; the radiation from all these sources overlap, and they go in 4π.

Chapter 6

A6.1   Light amplification by stimulated emission of radiation.

A6.2   Laser light is monochromatic, unidirectional, and the electromagnetic waves have phase coherence.

A6.3   For removing tumors close to inner or outer surfaces, lens corrections, removal of tattoos, fragmentation of stones, glaucoma surgery.

A6.4   Mainly for increasing the laser flux that enables photoablation processes.

A6.5   Mechanical switching, Q-switch, and mode coupling.

A6.6   Electro-optical and acousto-optical.

A6.7   By forming wavepackets with a broad band of wavelengths that constructively interfere at specific nodes traveling in space and time.

A6.8 CO_2, Nd:YAG, Ar+, and excimer lasers.

A6.9   In the order of 10^{12} W/cm^2, used mainly for LASIK and LASEK.

A6.10   PDT is a procedure for cancer treatment with reactive O_2 molecules in the singlet state. First, a photosensitive drug is administered. After accumulation in tumor cells, a laser is used to excitation the photosensitizer that transfers its energy to oxygen, thereby transforming triplet oxygen to singlet oxygen.

A6.11   LASIK conserves the epithelium layer of the cornea, LASEK disposes the epithelium layer.

A6.12   LASEK, LASIK, diabetic macular edema removal, proliferative diabetic retinopathy, cataract, and glaucoma treatments.

Chapter 7

A7.1   Precise targeting of local areas of tumors, support of imaging diagnostics, providing cures. Short: targeting, diagnostics, and therapeutics.

A7.2   This part of the immune system is called reticuloendothelial system and consists of monocytes and macrophages that deposit the nanoparticles in the liver and spleen for excretion.

A7.3   From a physiological point of view, the size of nanoparticles determines the RES and the EPR and therefore the lifetime of nanoparticles in the body.

From a physical point of view, the size of nanoparticles determines the blocking temperature for magnetic NPs and plasmon frequency in the case of metal nanoparticles.

A7.4 Active and passive targeting. Physical targeting by the size of particles is passive targeting. Active targeting requires ligands that can dock to specific sites.

A7.5 Curie temperature characterizes ferromagnets. Curie temperature is an ordering temperature. Below the Curie temperature, the magnetic moments in a solid are in an ordered state. Above the Curie temperature, the magnetic moments are disordered, also called paramagnet.

A7.6 Below the Curie temperature, the magnetic moments in a nanoparticle are ordered and form a tiny single magnetic domain. All individual magnetic moments in the single domain contribute to the magnetic moment of the macrospin.

A7.7 The total magnetic moment of the half-filled Gd-shell does not follow from $gS\mu B$. The correct equation is $g(S(S + 1))^{1/2}$, yielding $8\ \mu B$.

A7.8 The blocking temperature separates low-temperature behavior of magnetic nanoparticles from a high-temperature behavior. At low temperatures, nanoparticles do not reorient within a typical observation time; at high temperatures, macrospins are not fixed in space but fluctuate strongly.

A7.9 In MRI it is important that the magnetization of the NPs align in the field of the Helmholtz coils and are stable along the z-direction. This is only the case for nanoparticles whose blocking temperature is below body temperature. For hyperthermia, the magnetization of the nanoparticles should follow the external ac field. This is only possible for NPs in the superparamagnetic state above the blocking temperature.

A7.10 Néel relaxation, Brown relaxation, and hysteretic relaxation.

A7.11 In the case of positive contrast, T1 is shortened. In the case of negative contrast, T2 is shortened.

A7.12 Magnetite nanoparticles with large magnetic dipole fields contribute to the fluctuation of local fields that shortens T2.

A7.13 Gd3+ chelates are in direct contact with protons and support the external field with their large magnetic moment. This shortens the T1 relaxation time and therefore acts as a positive contrast agent.

A7.14 K-edge contrast imaging is a radiography modality by taking two sequential pictures, one with a mean x-ray photon energy above the K x-ray absorption edge of a heavy metal particle like Au, and another one taken below the K-edge of the same element.

A7.15 Rayleigh scattering occurs for objects such as molecules which are small compared to the wavelength of electromagnetic waves. Mie scattering is the scattering of EM waves at objects with a diameter smaller or equal the wavelength of EM radiation.

A7.16 Forward direction with respect to the incoming wave.

A7.17 The absorbance maximum moves to a longer wavelength with increasing particle size and an increasing aspect ratio.

A7.18 Because smaller size NPs show stronger absorption effect of EM radiation, whereas for larger size NP scattering of EM waves prevails.

A7.19 By using plasmonic excitation of metal nanoparticles with a laser, heat is generated in particles that is transferred to the tissue by thermal conductance and by radiation.

A7.20 The illuminated nanoparticles can be best made visible by a dark field microscope.

A7.21 Theranostic nanoparticles are those which serve the purpose of diagnostics, for instance by contrast enhancement, and for therapeutics, such as hyperthermia. In any case, theranostic nanoparticles serve a multimodality purpose.

Chapter 8

A8.1 Precision implies hitting a particular point with little scattering, but not necessarily the target. Accuracy implies hitting the target, but not necessarily with a narrow distribution.

A8.2 No. Systematic errors must be eliminated by analyzing the cause and fixing the problem.

A8.3 The mean value is defined by the sum over all measurement results, divided by the number of independent measurements n taken.

A8.4 The variance is normalized by $n - 1$, because one degree of freedom is already used for evaluating the mean.

A8.5 The confidence interval is proportional to the inverse root of the number of measurements n: $CI \sim 1/\sqrt{n}$. Therefore, the CI can be improved, i.e., decreased, by increasing the number n of independent measurements.

A8.6 The error of a single measurement is ± 10.5 N.

A8.7 Sensitivity of a measuring device is defined as the ratio of entrance value to display value. In the case of digital instruments, the sensitivity corresponds to the decimal of the last digit displayed. The accuracy class is defined by the maximum error occurring at maximum range of the display.

A8.8 In statistical terms, a population is the total number of people that express one or several specified characteristics. A representative sample is a randomly chosen subset of a population.

A8.9 A histogram is a plot of the frequency of occurrence versus the measurement interval of a certain observable.

A8.10 Single mode, bimodal, and skewed distributions.

A8.11 Wrong statement: 95.5% of the total area lies within the ±1σ range. Correct is: 68.3% of the total area lies within the ±1σ range.

A8.12 This test can be performed by a normal plot. A normal plot displays the probabilities actually reached at specific intervals (−3σ to +3σ) in comparison to the expected probabilities according to the standard distribution.

A8.13 A linear regression analysis tests whether a set of measurements of two observables x and y are linear related according to a linear function $y = ax + b$.

A8.14 A linear regression analysis is performed by minimizing the sum of squared deviations between measured values y_i and predicted value $\hat{y}_i$: $\sum_i (d_i)^2 = \sum_i (y_i - \hat{y}_i)^2$.

A8.15 If the slope of the linear regression turns out to be zero, it indicates that the observables x and y are randomly distributed but mot linearly related.

A8.16 The quality of the linear distribution can be judge by its variance. The variance should be as small as possible.

A8.17 Odds is he number of events occurring with a specific outcome divided by the number of events not occurring. Odds is therefore the ratio of true-positive events to false-negative events: TP/FN.

Risk refers to the number of events occurring with a specific outcome divided by number of all events. Risk is therefore the ratio of true-positive events and the total number of events and has thus the same definition as the sensitivity: TP/(TP + FN).

A8.18 False-positive results can only be tested with a control group. 40% FP implies 60% TN. Thus the specificity defined as TN/(TN + FP) = 0.6 or 60%.

A8.19 The sensitivity of a diagnostic method is defined by the ratio of TP/(TP + FN).

10 Solutions to exercises

Chapter 1

E1.1   **Solution: Base sequence**
        ATGCTGA.

E1.2   **Solution: Cell survival**
        a:

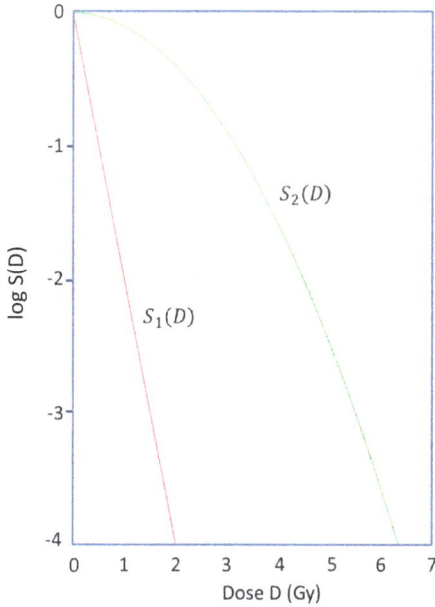

b.   The coefficient α can be evaluated from the slope to be 2. The coefficient β can be determined from the point where lög $S(D)$ touches the x-axis at $D = 6.2$. Then $\beta = -\log S/D^2 = 4/6.2^2 = 0.1$. The α/β ratio is $2/0.1 = 20$.

https://doi.org/10.1515/9783111168739-010

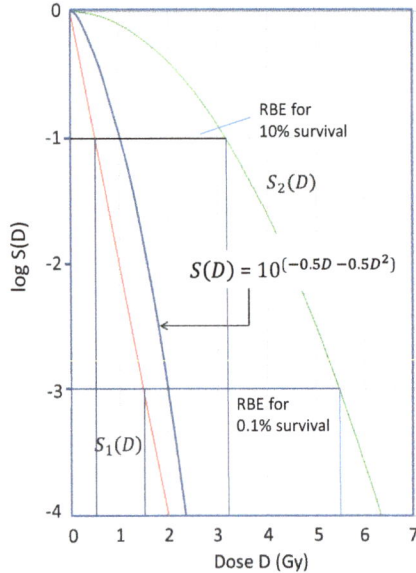

c. $S_1(D)$ is representative for high LET and high RBE radiation, which is the case for neutrons and ion beams (helium and carbon). $S_2(D)$ is representative for low LET and low RBE radiation, which is the case for x-ray radiation.

d. The combined curves $S(D) = 10^{(-0.5D - 0.5D^2)}$ with the coefficients $\alpha = \beta = 0.5$ are plotted in the above graph.

e. The RBE at 0.1% survival is $5.5/1.5 = 3.66$. At 10% survival, the RBE is $3.2/0.5 = 6.4$.

E1.3 Solution: DNA damage

In SSB, the DNA has a chance of self-repair. When both strands are broken at the same time, the DNA splits apart, repair is impossible, and replication has to stop immediately. The remaining DNA is not functional, which means that the necessary proteins cannot be synthesized.

E1.4 Solution: Tumor imaging

a. Confirmation of cancer cells always requires histological inspection, which uses a high resolution- optical microscope. Several methods exist for in situ recognition of tumor cells, depending on the type of tumor. For detecting skin cancer, usually, an optical lens with a magnification of 30 is sufficient. For intestinal, vaginal, and ovary cancer, endoscopy and colonoscopy including all modern versions (spectroscopic, capsule, etc.; see Chapter 2 in Volume 2) are the methods of choice. SPE or SPECT can recognize bone metastasis and thyroid cancer. FDG-PET is preferred for the recognition of a variety of cancers, such as brain, urinary, and prostate cancers.

b. The smallest tumor volume is seen in skin cancer, which can be detected in the submillimeter range with a good optical lens.

Chapter 2

E2.1 **Solution: Dose to the isocenter**

a. According to Figs. 2.4(a) and 2.5, the PDD at a distance of 10 cm from the skin is 73% for the anterior field.

b. After exposure of 100 MU from anterior, the dose at the isocenter is 100 MU × 0.73 = 73 cGy.

c. The dose delivered to the isocenter from the posterior is then 2 × 0.47 × 0.7 × 100 cGy = 2 × 33 cGy = 66 cGy.

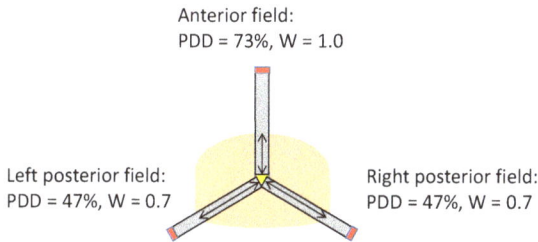

Anterior field:
PDD = 73%, W = 1.0

Left posterior field:
PDD = 47%, W = 0.7

Right posterior field:
PDD = 47%, W = 0.7

Tab. 12.2: Conversion of dose to monitor units for reaching defined total percentage photon dose (PDD) at the isocenter (IC).

Field	Starting MU (cGy)	PDD at IC (cGy)	Weight factor	Weighted PDD at IC (cGy)	Boosted up PDD at IC (cGy)	Rescaled MU (cGy)	Rescaled MU for 200 cGy at IC (cGy)
Anterior	100	73	1.0	73	73	100	114
Left post	100	47	0.7	33	51.1	155	177
Right post	100	47	0.7	33	51.1	155	177

E2.2 **Solution: Boosting up the dose to the isocenter**

a. Seventy percent of the anterior side to be delivered to the posterior side implies that the dose from each posterior side should be 0.7 × 73 cGy = 51.1 cGy or 102.2 cGy from both sides.

b. The boosting factor is 51.1/33 = 1.55. Therefore, 155 cGy must be delivered by the linac at both posterior sides.

c. The total dose delivered to the isocenter is 73 cGy + 102.2 cGy = 175.2 cGy.

d. By a factor 200/175 = 1.143.

e. This requires 73 × 1.143/0.73 = 114 MU from the anterior side and 51 × 1.143/(0.47 × 0.7) = 177 MU from each posterior side.

f. All numbers are presented in Tab. 2.2.

E2.3 **Solution: Exposure time**

The PDD at a distance of 10 cm from the skin is 73% according to Fig. 2.4(a). Therefore, the patient must receive a dose from the linac:

$$D_{\text{linac}} = \frac{D_{\text{patient}}}{\text{PDD}} = \frac{5 \text{ Gy}}{0.73} = 6.85 \text{ Gy}$$

corresponding to a dose density to the tumor of

$$\frac{D_{\text{linac}}}{A} = \frac{6.85 \text{ Gy}}{25 \text{ cm}^2} = 0.274 \text{ Gy/cm}^2.$$

As the beam intensity is 0.1 Gy/cm² min, the time of irradiation is then

$$t = \frac{D_{\text{linac}}}{\dot{D}_{\text{linac}}} = \frac{0.274 \text{ Gy/cm}^2}{0.1 \text{ Gy/cm}^2 \text{ min}} = 2.74 \text{ min}.$$

Chapter 3

E3.1 **Solution: Cyclotron magnetic field**

a. The proton energy required for a 27.5 cm deep cancer volume can be determined from Fig. 13.5 or using eq. (13.4): $E_p(\text{MeV}) = (R_{\text{water}}(\text{cm})/2 \times 10^{-3})^{1/1.8} = 200 \text{ MeV}$.

b. The maximum extraction energy of the cyclotron is (see eq. (5.30) in Volume 2)

$$E_{\text{kin, max}} = \frac{1}{2} \frac{(qBR)^2}{m_p}.$$

Therefore, the required field for the 200 MeV proton beam is

$$B(T) = \frac{\sqrt{2Em_p}}{qR}.$$

The circumference of the cyclotron is 12.56 m. Therefore, the radius is 2 m. The other required values are as follows:

Proton mass: 1.67×10^{-27} kg.

Proton energy: $200 \text{ MeV} \times 1.6 \times 10^{-19} \text{ J} = 3.2 \times 10^{-11} \text{ kgm}^2/\text{s}^2$.

Proton charge: 1.6×10^{-19} As.

Inserting these values yields:

$$B(T) = \frac{\sqrt{2 \times 3.210^{-11}\, \text{kg m}^2/\text{s}^2 \times 1.67 \times 10^{-27}\, \text{kg}}}{1.610^{-19}\, \text{As} \times 2\,\text{m}} \cong 1\frac{\text{kg}}{\text{As}^2} = 1\ \text{T}.$$

A magnetic field of 1T in the cyclotron is required to produce protons with an energy of 200 MeV.

E3.2 **Solution: Tuning the beam energy**

For a tumor volume seated at 10–15 cm below the surface, a proton beam energy is required stretching from 113.5 to 142.17 MeV. The delivered proton beam energy is 150 MeV from a cyclotron. In the case of a cyclotron, the beam energy can only be lowered by applying an attenuation wedge. Assuming that the wedge has the same density as water and tissue, then the thickness must range from 1.5 to 16.5 cm. With a wedge thickness of 1.5 cm, the proton beam is lowered from 150 to 142.17 MeV with a Bragg peak at 15 cm. Increasing the wedge thickness further from 1.5 to 6.5 cm will lower the beam energy from 142.17 to 113.5 MeV.

E3.3 **Solution: Beam accuracy**

The range is given by $\langle R_{\text{water}}(\text{cm})\rangle = 2 \times 10^{-3}\left(E_p(\text{MeV})\right)^{1.8} = \alpha E^\beta$. Therefore, we derive

$$\frac{dR}{R} = \beta \frac{dE}{E}.$$

At a depth of 10 mm, the error should not exceed 2 mm or

$$\frac{dR}{R} = 0.02 = 1.8 \frac{dE}{E}.$$

Then

$$\frac{dE}{E} = \frac{0.02}{1.8} \cong 0.01.$$

Or 1%. For a 100 MeV proton beam, a 1% error implies that the maximum energy fluctuation should not exceed 1 MeV.

Chapter 4

E4.1 **Solution: Dose**

Ten grams of boron contains 1 mol of boron atoms. Thirty micrograms of ^{10}B contain

$$N_{10_B} = \frac{30 \times 10^{-6}}{10} \times 6 \times 10^{23} = 1.8 \times 10^{18}$$

boron atoms. Each ^{10}B disintegration releases 2.3 MeV (E_{kin} = 1.47 MeV) and ^{7}Li-ion (E_{kin} = 0.84 MeV).

If all ^{10}B isotopes are activated, they would release a total of

$$2.3 \times 10^6 \text{ eV} \times 1.8 \times 10^{18} \times 1.6 \times 10^{-19} \text{ J/eV} = 6.6 \times 10^5 \text{J}.$$

This would correspond to 6.6×10^5 J/g or 6.6×10^8 J/kg = 660 MGy.

E4.2 **Solution: Neutron exposure**

The production rate is

$$R_{10_B} = N_{10_B} \sigma_{10_B} J_n.$$

This yields

$$R_{10_B} = 1.8 \times 10^{18} \times 3835 \times 10^{-28} \text{ m}^2 \times 10^{12} \text{ n/m}^2\text{s} = 6.903 \times 10^5 \text{ s}^{-1}.$$

Thus, about 7×10^5 boron atoms per second split apart. In 1 h irradiation, this yields a total of 2.5×10^9 disintegration with energy release of

$$2.3 \times 10^6 \text{ eV} \times 2.5 \times 10^9 \times 1.6 \times 10^{-19} \text{ J/eV} = 9.2 \times 10^{-4} \text{ J} \cong 1 \text{ mJ}.$$

This is the energy release per gram tumor volume. With respect to a mass of 1 kg, this radiation corresponds to 1 J/kg = 1 Gy.

In fact, this is too low. As this is high-LET radiation, one should achieve at least 2 Gy in a fraction of 30 min. A higher dose can only be achieved by increasing either the boron concentration or the neutron flux.

Chapter 5

E5.1 **Solution: Co activation**

According to eq. (5.23 in Volume 2), the number of activated ions B ($=^{60}$Co) after time t in the reaction ^{59}Co + n → ^{60}Co is given by

$$N_B(t) = \frac{R_B}{\lambda_B}\left(1 - e^{-\lambda_B t}\right),$$

where R_B is the production rate of isotope B by neutron capture of isotope A ($=^{59}$Co). Saturation is reached for $N_B(t=\infty) = R_B/\lambda_B$. For 75% saturation, we obtain

$$\frac{3R_B}{4\lambda_B} = \frac{R_B}{\lambda_B}\left(1 - e^{-\lambda_B t_{3/4}}\right)$$

or

$$\frac{1}{4} = e^{-\lambda_B t_{3/4}}.$$

Therefore,

$$t_{3/4} = \frac{\ln 4}{\lambda_B} = T_{1/2}\frac{2\ln 2}{\ln 2} = 2T_{1/2} = 10.52 \text{ years}.$$

E5.2 **Solution: Co activity**

a. $1\,\mu g$ Co contains $\frac{1}{60} \times 10^{-6} \times 6 \times 10^{23} = 10^{16}\ ^{60}$Co atoms.
   In saturation, the activity of $1\,\mu g\ ^{60}$Co is

$$A_{Co} = \lambda_{Co}N_{Co} = \frac{\ln 2N_{Co}}{T_{1/2}} = \frac{10^{16} \times \ln 2}{5.26 \text{ years}} = \frac{10^{16} \times \ln 2}{1.6588 \times 10^{8} \text{ s}} = 4.18 \times 10^{7}\ \text{s}^{-1}.$$

   The activity of $1\,\mu g\ ^{60}$Co is therefore 41.8 MBq.
b. The activity of 1 g of ^{60}Co is 42 TBq.
   The specific activity of ^{60}Co is 42 TBq/g.

E5.3 **Solution: Dose estimation**

Using eq. (15.8):

$$D(E_{ph}, r) = \frac{AE_{ph}t}{4\pi r^2}\left(\frac{\mu_{tr}}{\varrho}\right) = \frac{10^8 \text{s}^{-1} \times 0.66 \times 10^6 \text{eV} \times 1.6 \times 10^{-19}\text{J/eV} \times 3600\text{s} \times 25\,\text{cm}^2/\text{kg}}{4\pi \times 10^4 \text{cm}^2}$$

$$= \frac{9.5\,\text{J cm}^2}{4\pi 10^4\ \text{cm}^2\text{kg}} = 9.5 \times 10^{-6}\frac{\text{J}}{\text{kg}} \cong 10\,\mu\text{Sv}.$$

A 100 MBq source of ^{137}Cs generates a dose of 10 μSv at a distance of 100 cm from the source.

Chapter 6

E6.1 **Solution: Laser peak energy and peak power**

The peak fluence is

$$\frac{E_p}{A} = \frac{100\,\mu\text{J}}{1\ \text{cm}^2} = 0.1\ \frac{\text{mJ}}{\text{cm}^2}.$$

The peak intensity in P_p is determined by

$$P_p = \frac{E_p}{A\Delta t} = \frac{100\,\mu\text{J}}{1\ \text{cm}^2\,10\ \text{fs}} = \frac{10^{-4}\,\text{J}}{1\ \text{cm}^2\,10^{-14}\ \text{s}} = \frac{10\ \text{GW}}{\text{cm}^2}.$$

Note that this huge intensity in a single pulse is exclusively the result of the short pulse width.

E6.2 **Solution: LASIK eye surgery**
Power:

$$P_p = \frac{E_p}{\Delta t} = \frac{1 \text{ mJ}}{15 \text{ ns}} = \frac{10^{-3} \text{ J}}{15 \times 10^{-9} \text{ s}} = 66 \text{ kW}.$$

Power density (intensity):

$$P_p = \frac{E_p}{A\Delta t} = \frac{1 \text{ mJ}}{1 \text{ mm}^2 \times 15 \text{ ns}} = \frac{6.6 \text{ MW}}{\text{cm}^2}.$$

Again, note that the power and the power density are very high due to the short pulse length.

E6.3 **Solution: Average power of a pulsed laser**
The peak pulse energy is

$$P_p \Delta t = E_p = 5 \times 10^6 \text{W} \times 10^{-9} \text{ s} = 5 \text{ mJ}.$$

a. The peak fluence is, therefore,

$$\frac{E_p}{A} = \frac{5 \text{ mJ}}{1 \text{ mm}^2} = 0.5 \frac{\text{J}}{\text{cm}^2}.$$

b. The time average fluence is

$$\frac{E_p \, t_p}{A \, T} = \frac{5 \text{ mJ}}{1 \text{ mm}^2} \frac{1 \text{ ns}}{20 \text{ ms}} = \frac{5 \text{ mJ}}{1 \text{ mm}^2} 5 \times 10^{-8} = 2.5 \frac{\text{nJ}}{\text{cm}^2}.$$

E6.4 **Solution: Electric field**
The pointing vector is according to eq. (16.16):

$$I_p = S_p = \varepsilon_0 c \vec{E}_0^2.$$

Therefore, the electric field amplitude can be determined by

$$E_0 = \sqrt{\frac{I_p}{\varepsilon_0 c}}.$$

Substituting numbers:

$$E_0 = \sqrt{\frac{I_p}{\varepsilon_0 c}} = \sqrt{\frac{5 \times 10^8 \frac{\text{W}}{\text{m}^2}}{2.64 \times 10^{-3} \frac{\text{C}^2}{\text{N m s}}}} = \sqrt{1.9 \times 10^{11} \frac{\text{N}^2}{\text{C}^2}} = 4.35 \times 10^5 \frac{\text{N}}{\text{C}} = 4.35 \times 10^5 \frac{\text{V}}{\text{m}}.$$

This field amplitude is equivalent to the discharge voltage in air. Therefore, it is reasonable that pulsed laser with high peak amplitudes can ionize tissues and create plasma plumes.

E6.5 Solution: Flux

First, we convert the flux of $I_p = 10^{12}$ W/cm^2 to 10^{16} W/m^2. Inserting in $E_0 = \sqrt{I_p/\varepsilon_0 C}$ yields

$$E_0 = \sqrt{4 \times 10^{18} \frac{N^2}{C^2}} = 2 \times 10^9 \frac{V}{m} = 2 \times 10^7 \frac{V}{cm}.$$

E6.6 Solution: Spectroscopic terms

The angular momentum L has the value: $F = 3$, $I = 6$.

The overscript refers to the total spin quantum number: $2S + 1$.

The subscript refers to the total angular momentum $J = L - S$ for less than half-filled shells and $J = L + S$ for more than half-filled shells.

The spectroscopic nomenclature is: $^{(2S+1)}L_J$.

Using this nomenclature, we find

$$^4I_{9/2}: L = 6, \ S = \frac{3}{2}, \ J = L - S = \frac{12}{2} - \frac{3}{2} = \frac{9}{2};$$

$$^4F_{3/2}: L = 3, \ S = \frac{3}{2}, \ J = L - S = \frac{6}{2} - \frac{3}{2} = \frac{3}{2};$$

$$^4F_{5/2}: L = 3, \ S = \frac{3}{2}, \ J = L - S + 1 = \frac{6}{2} - \frac{3}{2} + \frac{2}{2} = \frac{5}{2};$$

$$^4I_{11/2}: L = 6, \ S = \frac{3}{2}, \ J = L - S + 1 = \frac{12}{2} - \frac{3}{2} + \frac{2}{2} = \frac{11}{2}.$$

E6.7 Solution: Degenerate states

The three degenerate states have the configurations:

In these three states, the m_L value $m_L = +3$ is identical. There are no further configurations that yield $m_L = +3$.

E6.8   **Solution: Temperature spread**

We start with the specific heat:

$$C_m = \frac{1}{m}\frac{dQ}{dT}.$$

Transforming:

$$dT = \frac{1}{mC_m}dQ.$$

Time derivative:

$$\frac{dT}{dt} = \frac{1}{mC_m}\frac{dQ}{dt}.$$

With heat per volume: $q = Q/V$:

$$\dot{T} = \frac{1}{\rho C_m}\frac{\dot{Q}}{V} = \frac{1}{\rho C_m}\dot{q}.$$

Using the continuity equation for heat flow: $div j_Q = -\frac{dq}{dt}$, we obtain

$$\dot{T} = -\frac{1}{\rho C_m}\operatorname{div} j_Q.$$

Now we combine with the equation for heat conduction: $j_Q = -\lambda \operatorname{grad}(T)$

$$\dot{T} = -\frac{1}{\rho C_m}\operatorname{div}(-\lambda \operatorname{grad}(T)),$$

yielding

$$\dot{T} = \frac{\lambda}{\rho C_m}\operatorname{div}(\operatorname{grad}(T)) = \frac{\lambda}{\rho C_m}\Delta T = \varepsilon \nabla^2 T,$$

where Δ is the Laplace operator and $\varepsilon = \lambda/\rho C_m$.

Chapter 7

E7.1   **Solution: Gd magnetic moment**

According to eqs. (17.1) and (17.2), the magnetic moment is in general:

$$m = g_J \sqrt{J(J+1)}\,\mu_B.$$

Gd^{3+} has a half-filled 4f shell. Therefore, $L = 0$ and $J = S$. Then $g_J = 2$. Therefore, the magnetic moment is

$$m = 2\sqrt{\frac{7}{2}\left(\frac{7}{2}+1\right)}\,\mu_B = 2\sqrt{\frac{7}{2}\times\frac{9}{2}}\,\mu_B = 7.94\,\mu_B.$$

E7.2 **Solution: NP macrospin**

Magnetite has a cubic crystal structure with a lattice constant of 0.84 nm and cubic volume of $V_{uc} = (0.84)^3 = 0.592$ nm^3.

The nanoparticle with a diameter of $d = 10$ nm has a volume of $V_{np} = 0.523$ $d^3 = 523$ nm^3.

There are $N = V_{np}/V_{uc} = 884$ magnetite unit cells in the nanoparticle.

Each cell contains eight formula units of Fe$_3$O$_4$. Each formula unit has a magnetic moment of 4.1 μ_B. Therefore, each unit cell has a magnetic moment of 8×4.1 $\mu_B = 32.8$ μ_B. This yields a macrospin moment of $m_{macro} = 884 \times 32.8$ $\mu_B = 2.9 \times 10^4 \mu_B$.

E7.3 **Solution: Magnetic induction**

The magnetic induction is $B = \mu_0(H + M)$. As $H=0$, we have $B = \mu_0 M$.

M is the magnetization of the NP, $\mu_0 = 4\pi 10^{-7}$ Vs/Am is the magnetic permeability of the vacuum, and $\mu_B = 9.27 \times 10^{-24}$ Am^2 is the Bohr magneton. The magnetization is therefore

$$M = m_{macro}/V_{np} = 2.9 \times 10^4 \mu_B/523 \, \text{nm}^3 = 55.4 \, \mu_B/\text{nm}^3.$$

The magnetic induction due to the magnetization is

$$B = \mu_0 M = 4\pi \times 55.4 \times 9.27 \times 10^{-4} \, \frac{\text{Vs}}{\text{m}^2} = 0.645 \text{ T}.$$

E7.4 **Solution: Blocking temperature**

The blocking temperature is related to the particle size via

$$T_B = \frac{K V_{np}}{k_B \ln(\tau_{ob}/\tau_0)} = A V_m.$$

We first determine the constant $A = T_B/V_{np}$ for the different particle sizes:

$$100\text{K}/0.523 \times 3.8^3 \text{nm}^3 = 3.48 \text{K}/\text{nm}^3;$$

$$150 \text{ K}/0.523 \times 4.5^3 \text{ nm}^3 = 3.14 \text{ K}/\text{nm}^3;$$

$$150 \text{ K}/0.523 \times 4.5^3 \text{nm}^3 = 3.14 \text{ K}/\text{nm}^3;$$

$$250 \text{ K}/0.523 \times 5.5^3 \text{nm}^3 = 2.87 \text{ K}/\text{nm}^3.$$

From this, we conclude that the constant A depends on the nanoparticle volume and can be described by $A = (a - bT_B)$ with $a = 3.8$ and $b = 0.004$ in appropriate units.

Now solving for $T_B = A V_{np} = (a - bT_B)V_{np}$, we find

$$T_B = \frac{a V_{np}}{1 + b V_{np}}.$$

Inserting numbers, we obtain for the 10 nm NP a blocking temperature of

$$T_B = \frac{3.8 \times 523}{1 + 4 \times 10^{-3} \times 523} = 642\,\text{K}.$$

The blocking temperature of the 10 nm nanoparticle is estimated to be about 640 K.

E7.5 **Solution: Positive and negative contrast agents**

CAs that shorten $T1$ are called positive, and CAs that shorten $T2$ are negative. In chelates, there is a single central Gd^{3+} ion with a magnetic moment of $8\mu_B$. The surrounding protons in the aqueous environment come close to the Gd^{3+} moment and the spin precession is strongly disturbed. Therefore, the $T1$ relaxation time is shortened in the presence of Gd^{3+}.

In the case of magnetite NPs, the magnetic moment of the macrospin can be much larger, on the order of $10^3 \mu_B$. However, the macrospin is inside the NP at the core. The surrounding protons cannot get close to the macrospin. As the magnetization drops rapidly with distance, the stray fields emanating from the magnetite NP disturb the long-range magnetic field distribution and have a dephasing effect on the proton spins. Therefore, $T2$ is reduced.

Chapter 8

E8.1 **Solution: Systematic error**

The probability of discovering a systematic error depends on the size of the shift. Larger deviations are easier to spot than small ones that still look "reasonable."

Therefore, the question focuses on small systematic shifts that are more difficult to detect. First, recorded data should never be simply "accepted," but critically reviewed and evaluated to see if they "make sense." Secondly, it helps to avoid errors if the measuring devices used are calibrated against a standard at regular intervals. Third, a measuring result should be confirmed by using another measuring instrument.

E8.2 **Solution: Cube volume**

The volume of a cube is $V = x \cdot y \cdot z$, with $x = y = z$. Therefore, the absolute error is

$$\Delta V = \sqrt{3(x^2 \Delta \bar{x})^2}.$$

Inserting numbers, we obtain for the absolute error

$$\Delta V = \sqrt{3\left(0.75^2 \times 10^{-3}\right)^2} = 9.74 \times 10^{-4} \ m^3$$

and for the relative error

$$\Delta V/V = 2.3 \times 10^{-3} = 0.23\%.$$

E8.3 Solution: Error propagation
In general, the error follows from

$$df = \sqrt{\left(\frac{\partial f}{\partial x}\right)^2 (\partial x)^2}.$$

Therefore, the absolute error is

$$df = \sqrt{(anx^{n-1})^2 (\Delta x)^2}$$

and the relative error is

$$\frac{df}{f} = an\left|\frac{\Delta x}{x}\right|.$$

E8.4 Solution: Probability
We seek the probability $P(X > 8.2)$, which is equivalent to the probability:

$$P(Z \le 8.2) = 1 - P(X > 8.2).$$

Evaluating $P(Z \le 8.2)$ is simpler because integration over the normal distribution always starts at $-\infty$ and goes up to the point Z.
With

$$z = \frac{x - \mu}{\sigma} = \frac{8.2 - 8.0}{0.1} = \frac{0.2}{0.1} = 2.$$

According to Tab. 18.1, $P(Z \le 2) = 0.97725$. Therefore, $P(X > 8.2) = 1 - P(Z \le 2) = 0.02275$. Hence, the probability to find a bag that weights more than 8.2 kg is merely 2.3%.

E8.5 Solution: Sensitivity and specificity
With a sensitivity of 80%, only 8 people out of 10 with the disease are tested positive. Two are tested false negative. Applied to the total population that do not have the disease, this implies that 10% of 990 people are tested false positive. So, a total of 99 + 8 = 107 are tested positive, but only 8 of those have the disease.
The test results are listed in the following table:

			Sum
Test is positive	TP=8	FP=99	107
Test is negative	FN=2	TN=891	893

b. The sensitivity of the test is

$$SE = \frac{TP}{TP + FN} = \frac{8}{8+2} = 0.8.$$

c. Specificity:

$$SP = \frac{TN}{TN + FP} = \frac{891}{891 + 99} = 0.9.$$

d. False-positive rate:

$$FP = \frac{FN}{FN + TP} = \frac{2}{2+8} = 0.2 = 1 - SE.$$

The sensitivity and specificity of the diagnostic method are acceptable. Nevertheless, the number of false-positive test is surprisingly high.

E8.6 **Solution: PSA test**

a. In a population of 1000, 210 develop prostate cancer, but only $210 \times 0.195 = 40.95 \cong 41$ test positive.

b. The false-positive rate is defined by

$$FPR = 1 - SP = 1 - 0.94 = 0.06 = 6\%.$$

This means that 6% of 790 men who do not have prostate cancer falsely test positive, which are 47.4 males.

c. The false-negative rate is

$$FNR = 1 - SE = 1 - 0.195 = 0.805 = 80.5\%.$$

This means that 80.5% of prostate cancer patients are not diagnosed positively, which are $210 \times 0.805 = 169$ males.

d. The positive predictive value of the PSA test is

$$PPV = \frac{TP}{TP + FP} = \frac{41}{41 + 47.4} = 46.4\%.$$

Therefore, the predictive value of the PSA test is about 50%. (This is about as good as the weather forecast in earlier times.)

11 List of acronyms (used in all three volumes)

ACD	annihilation coincidence detection		DBT	digital breast tomosynthesis
			DCE	dynamic contrast enhancement
ADC	analog digital converter		DES	drug elusion stent
ADC	apparent diffusional constant		DNA	deoxyribonucleic acid
ADP	adenosine diphosphate		DNP	dynamic nuclear polarization
ALARA	as low as reasonably achievable		DOV	depth of view
amu	atomic mass units		DSA	digital subtraction angiography
ANF	auditory nerve fiber		DSB	double strand break
ATP	adenosine triphosphate		DTI	diffusion tensor imaging
AV	atrioventricular node		DTL	drift tube linac
AVV	atrioventricular valve		DWI	diffusion weighted imaging
BBB	blood-brain barrier		EBRT	external beam radiotherapy
BF	breath frequency		ECC	extra-corporal circulation
BMD	bone mineral density		ECG	electrocardiography
BMR	basal metabolic rate		EDP	end diastolic pressure
BMS	bare metal stents		EDV	end diastolic volume
BNCT	boron neutron capture therapy		EEG	electroencephalography
BOLD fMRI	blood oxygen level dependent fMRI		EF	ejection fraction
			EM	electromagnetic
BPS	biodegradable polymeric stents		EMG	electromyography
BSA	beam shaping assembly		EPI	echo planar imaging
BSA	body surface area		EPP	end plate potential
BW	body weight		EPR	enhanced permeation and retention
CA	contrast agent			
CAP	cardiac action potential		ERBT	external radiation beam therapy
CBF	cerebral blood flow		ERV	expirational rest volume
CCC	continuous curvilinear capsulorhexis		ESP	end systolic pressure
			ESV	end systolic volume
CCD	caput-collum-diaphyseal angle		ETL	echo train length
CCD	charge coupled device		FCRM	fiber optic confocal reflectance microscope
CD	coincidence detector			
CIRT	carbon ion radiation therapy		FDG	18F-fluoro-deoxy-glucose
CK	cyber knife		FE	fractional excretion
CLE	confocal laser endoscopy		FEG	frequency encoding gradient
CNR	contrast to noise ratio		FET	18F-fluoro-ethyl-L-tyrosine
CO	cardiac output		FF	flattening filter
COE	caloric oxygen equivalent		FF	filtration fraction
c.m.	center of mass		FFDM	full-field digital mammography
CPA	charge particle activation		FFF	flattening filter free
CPB	cardiopulmonary bypass		FFT	fast Fourier transform
CS	Compton scattering		FID	free induction decay
CSF	cerebrospinal fluid		FLACS	femtosecond laser-assisted cataract surgery
CT	computed tomography			
CTV	clinical Target Volume		FLAIR	flid attenuated inversion recovery
cw	continuous wave			
CZT	CdZnTe		fMRI	functional magnetic resonance imaging
dB	decibel			

https://doi.org/10.1515/9783111168739-011

FOV	field of view	MHR	metabolic heat production
FRC	fractional rest volume	MHT	magnetic hyperthermia
FSE	fast spin echo	MHz	megahertz
FT	Fourier transform	MI	mechanical index
GFR	glomerular filtration rate	MLC	multi-leaf collimator
GTV	gross tumor volume	MNP	magnetic nanoparticle
Hct	hematocrit value	mpMRI	multiparameter MRI
HD	hydrodynamic diameter	MRI	magnetic resonance imaging
HDR	high dose rate	MRSI	magnetic resonance
HEP	hemi endo prosthesis		spectroscopy imaging
hMRI	hyperpolarization magnetic	MRgRT	magnetic resonance-guided
	resonance imaging		radiation therapy
HSLS	high spin–low spin	MRT	magnetic resonance
IAEA	International Atomic Energy		tomography
	Agency	MSFP	mean systemic filling pressure
ICRP	International Commission for	MSO	medial superior olive
	Radiological Protection	MUAP	motor unit action potential
IGRT	image guided radiotherapy	MV	minute ventilation
IHC	inner hair cell	NBI	narrow band imaging
ILD	intensity level differences	NCT	neutron capture therapy
IMRT	intensity modulated radiation	NIR	near infrared
	therapy	NIRS	near infrared spectroscopy
IRT	internal radiation therapy	NMJ	neuromuscular junction
IRV	inspirational rest volume	NP	nanoparticle
ITD	interaural time difference	NRT	neutron radiation treatment
Kerma	kinetic energy release in matter	NTD	non-target dose
kHz	kilohertz	OAR	organ at risk
kV	kilovolt	OCT	optical coherence tomography
kVp	peak kilovoltage	OER	oxygen enhancement ratio
kW	kilowatt	OHC	outer hair cell
LASEK	laser epithelial keratomileusis	PAH	para-aminohippuric acid
LASER	light amplification by	PCI	percutaneous coronary
	stimulated emission of		intervention
	radiation	PCI	phase contrast imaging
LASIK	laser-assisted interstitial	PCV	packed cell volume
	keratomileusis	PD	proton density
LCI	low-coherence interferometry	PDD	percent depth dose
LDR	low dose rate	PDR	proliferative diabetic
LED	light-emitting device		retinopathy
LET	linear energy transfer	PDR	pulse dose rate
Linac	linear accelerator	PDT	photodynamic therapy
LOR	line of response	PE	photoelectric effect
LQ	linear-quadratic	PEG	phase encoding gradient
LSO	lateral superior olive	PEG	polyethylene glycol
MC	Monte Carlo	PES	photoelectron emission
MEG	magnetoencephalography		spectroscopy
MEI	middle ear implants	PET	positron emission tomography
MET	mechanoelectric transduction	PHIP	parahydrogen-induced
MeV	mega-electron volt		polarization

PI	pulsatility index	SPE	single photon emission
PMT	photomultiplier tube	SPECT	single-photon emission
PPD	percentage photon dose		computed tomography
PRF	pulse repeat frequency	SPIO	superparamagnetic iron oxide
PRK	photorefractive keratectomy	SPR	surface plasmon resonance
PRP	pan-retinal photocoagulation	SSB	single strand break
PRT	proton radiotherapy	SSD	source-to-surface distance
PRT	pulse repeat time	SSG	slice selection gradient
PSA	prostate-specific antigen	SUV	standard uptake value
PSMA	prostate-specific membrane	SV	stroke volume
	antigen	TCE	transient charged particle
PTV	planning target volume		equilibrium
PWV	pulse wave velocity	TE	time of (spin, acoustic) echo
PZT	$PbZrTiO_3$	TEP	total endo prosthesis
Q	quality factor	TERMA	total energy released per unit
QA	quality assurance		mass
RBC	red blood cell	TGC	time gain compensation
RBE	relative biological effectiveness	THz	terahertz
RBF	renal blood flow	TIPPB	transperineal interstitial
RC	respiratory coefficient		permanent prostate
Re	Reynolds number		brachytherapy
RES	reticuloendothelial system	TLC	total lung capacity
RF	radio frequency	TMR	targeted muscle reinnervation
RF	respiratory fraction	ToF	time of flight
RMR	resting metabolic rate	TPFR	total peripheral flow resistance
RPF	renal plasma flow	TPS	treatment planning system
RT	radiotherapy	TR	time of repetition
RTR	real time radiography	TRIGA	Training, Research, and
RV	residual volume		Isotopes, General Atomics
SA	sinoatrial node	TT	transfer time
SAD	source to axis distance	TV	tidal volume
SATP	standard ambient temperature	TV	target volume
	and pressure	UHMWPE	ultra-high molecular weight
SAXS	small angle x-ray scattering		polyethylene
SBRT	stereotactic body radiation	ULFMRI	ultra-low field magnetic
	therapy		resonance imaging
SCI	spinal cord injury	US	ultrasound
SE	spin echo	VC	vital capacity
SID	source to image distance	VEGF	vascular endothelial growth
SERS	surface-enhanced Raman		factor
	scattering	VOR	vestibulo ocular reflex
SGRT	surface-guided radiotherapy	XRR	x-ray radiography
SLAC	Stanford linear accelerator	XRT	x-ray radiotherapy
SNR	signal-to-noise ratio	YAG	yttrium-aluminum garnet
SOBP	spread out Bragg peak	ZFC	zero field cooling

12 Selection of fundamental physical constants, conversions, and relationships

Speed of light c	$299792458 \ \mathrm{m/s} \sim 3 \times 10^8 \ \mathrm{m/s}$
Gravitational acceleration of the earth g	$9.81 \ \mathrm{m/s^2}$
Planck constant h	$6.623 \times 10^{-34} \ \mathrm{Js}$
Compton wavelength λ_c	$2.426 \ \mathrm{pm}$
Elementary charge e	$1.602 \times 10^{-19} \ \mathrm{As}$
Atomic mass unit u	$1.66 \times 10^{-27} \ \mathrm{kg}$
Electron mass m_e	$9.109 \times 10^{-31} \ \mathrm{kg}$
Avogadro number N_A	$6.022 \times 10^{23} \ \mathrm{mol^{-1}}$
Boltzmann constant k_B	$1.38 \times 10^{-23} \ \mathrm{JK^{-1}}$
Dielectric constant of the vacuum ε_0	$8.854 \times 10^{-12} \ \mathrm{As/Vm}$
Magnetic permeability μ_0	$4\pi \times 10^{-7} \mathrm{Vs/Am} = 1.256 \times 10^{-6} \ \mathrm{Vs/Am}$
Faraday constant F	$9.65 \times 10^4 \ \mathrm{C/mol}$
General gas constant R	$8.314 \ \mathrm{J/Kmol}$
Stefan-Boltzmann constant σ_{SB}	$5.66 \times 10^{-8} \ \mathrm{AV/m^2 K^4}$
Bohr magneton μ_B	$9.274 \times 10^{-24} \ \mathrm{J/T}$
Proton gyromagnetic ratio γ_p	$2.675 \times 10^8 \ \mathrm{T^{-1} rad \ s^{-1}}$
Proton mass m_p	$1.673 \times 10^{-27} \ \mathrm{kg}$
Neutron mass m_n	$1.675 \times 10^{-27} \ \mathrm{kg}$
Nuclear magneton μ_N	$5.05783 \times 10^{-27} \ \mathrm{J/T.}$

Conversions

1 eV	$= 1.602 \times 10^{-19} \ \mathrm{J}$
1 Joule	$= 6.242 \times 10^{18} \ \mathrm{eV}$
1 Tesla	$= 1 \ \mathrm{N/Am} = 1 \ \mathrm{Vs/m^2} = 10^4 \ \mathrm{Gauss}$
1 Pascal	$= 1 \ \mathrm{N/m^2}$
1000 hPa	$= 1000 \ \mathrm{mbar}$
1 A/m	$= 10^{-3} \ \mathrm{emu/cm^3}$
$\mathrm{emu/cm^3 p}$	$= \mathrm{emu/g}$

Relationships

Avogadro number $N_A = 1 \ g/1u$	$= 1 \ \mathrm{g}/1.66 \times 10^{-24} \ \mathrm{g}$
Faraday constant $F = N_A \times e$	$= 6.022 \times 10^{23} \ \mathrm{mol^{-1}} \times 1.602 \times 10^{-19} \ \mathrm{As}$
General gas constant $R = N_A \times k_B$	$= 6.022 \times 10^{23} \ \mathrm{mol^{-1}} \times 1.38 \times 10^{-23} \ \mathrm{JK^{-1}}$
Speed of light in vacuum $c = 1/\sqrt{\mu_0 \varepsilon_0}$	$= \left(\sqrt{1.25 \times 10^{-6} \ \mathrm{Vs/Am} \times 8.85 \ 10^{-12} \ \mathrm{As/Vm}} \right)^{-1}$

https://doi.org/10.1515/9783111168739-012

13 List of scientists named in this volume

https://doi.org/10.1515/9783111168739-013

14 Glossary

Afterloading machine: Used for placement of high dose rate wires into predefined tubes.

Antigen-antibody: Antigens are proteins on cell surfaces of bacteria and viruses. They are recognized by antibodies, i.e., special proteins produced by our body acting like a lock to the antigen as a key.

Apoptosis: Programmed cell death.

Applicator: Hollow tube that is inserted into the body and filled with radioactive capsules, seeds, or wires for brachytherapy.

Beam quality correction factor: Ratio of the dose delivered by a user beam in a gantry to the dose delivered by a standard reference beam.

Benign neoplasm: Abnormal cell mass that has not invaded surrounding tissue.

Bethe–Bloch equation: Expression for the energy loss of charged particles per distance traveled, assuming that ionization is the dominant energy loss process.

Blood–brain barrier (BBB): Dense endothelium in cerebral vessels that prevents solutes in blood from entering unselectively the extracellular fluid of the central nervous system.

Boron neutron capture therapy (BNCT): Boron administered to tumor cells splits by neutron capture, releasing energetic but short-range α-particles.

Brachytherapy (BT): Application of radioisotopes through direct contact with tumors, characterized as internal beam therapy in contrast to external radiation beam therapy (ERBT) such as x-ray radiotherapy (XRT), proton radiotherapy (PRT), and neutron radiation therapy (NRT).

Carbon ion radiation therapy (CIRT): Similar to proton beam radiation therapy, but featuring a much higher RBE and requiring much higher particle energies for comparable range.

Class of laser: Categorization of visible lasers according to their power output and potential danger.

Clinical target volume (CTV): Tissue volume containing the GTV and/or subclinical malignancies with some chance of becoming therapeutically relevant.

CO_2 laser: Emitting far infrared light (wavelength about 10 μm) for various medical applications.

Coherence length: Length of a wavetrain with a constant wavelength.

Compton scattering: Inelastic x-ray scattering by electrons, where part of the photon energy is converted into electron kinetic energy.

Cyber knife technology: Combines precision robotics, image-guided beam delivery, and respiratory compensation.

Dose fractionation: Total dose divided into smaller portions and administered in a sequence of time intervals.

Dose painting: Method of delivering the required dose to the PTV by scanning a pencil-like narrow photon or proton beam across the tumor volume.

Enhanced permeation and retention effect (EPR): Caused by leaky and ruptured blood vessels in porous tumor volume.

Fluence: Number of particles per area.

Flux: Number of particles per time.

fs-laser: Pulsed laser with 10–100 fs pulse width of high intensity and low pulse-repetition rate.

Gadolinium: Rare earth metal with a high magnetic moment acting as positive contrast agent in MRI.

Gamma knife: Radiation therapy of brain tumors with ^{60}Co radioisotopes emitting 1.2 MeV γ-radiation, arranged in a stereotactic head frame.

https://doi.org/10.1515/9783111168739-014

Gross tumor volume (GTV): Macroscopic tumor volume defined by imaging methods.

Intensity: Number of particles per time and area.

Intensity-modulated radiation therapy (IMRT): Combines MLCs to define PTV and modulates beam intensity to deposit the required dose at any location in the tumor volume.

Interphase: Phase of normal cell activity, including growth and DNA replication.

Kerma: Energy fluence of a photon beam converted into kinetic energy of charged particles.

Laser: Light amplified by stimulated emission of (electromagnetic) radiation, which is monochromatic, unidirectional, and phase coherent.

LASIK/LASEK: Methods of correcting the refractive properties of the cornea by changing its curvature. Both methods differ in the preservation or elimination of a flap containing the epithelium.

Linac: Linear accelerator for charged particles, consisting of a linear array of pillbox-shaped cavities in which standing high-frequency waves accelerate the particles.

Linear energy transfer (LET) of radiation: Rate at which a particular type of radiation deposits energy by passing through matter.

Linear-quadratic equation: Empirical expression describing cell survival depending on dose.

Magnetite: Iron-oxide crystal with ferrimagnetic properties and a high magnetic moment acting as negative contrast agent in MRI.

Malignant tumor: Abnormal cell mass that has invaded surrounding tissue and is metastasizing.

Mitosis: Phase in which the cell divides into two identical copies.

Multileaf collimator (MLC): Pneumatically operated and computer-controlled slit system to match the PTV and spare healthy tissue from radiation.

Necrosis: Sudden, accidental cell death.

Neutron production: Two methods: thermal neutron-induced fission of uranium isotopes; or acceleration of charged particles that trigger a nuclear fusion reaction in a beryllium target.

Neutron radiation therapy (NRT): PDD of fast neutrons is similar to that of photons; the RBE is high and comparable to that of carbon ions.

Neutrons: Charge neutral hadrons with large penetration depth in matter and very high RBE in tissues without requiring oxygen enhancement.

Oxygen enhancement ratio (OER): Dose to achieve a survival fraction without oxygen supplementation divided by the dose to achieve the same effect with oxygen supplementation.

Percent depth dose (PDD): The dose measured at a given depth in the tissue, normalized by the maximum dose delivered in the same volume.

Photoablation: High flux and short-wavelength laser impact that causes vaporization of tissue material without thermal lesions at the margins.

Planning target volume (PTV): Geometric concept introduced for treatment planning and evaluation.

Q-switch: Switch for the quality factor of laser resonators, which periodically interrupts the beam and increases the laser power.

Range monitoring: Verification of the range of proton beams in vivo with β-radiation or prompt γ-radiation.

Relative biological effectiveness (RBE): Defined as the dose deposited by photons into a tumor mass with a given survival fraction divided by the dose deposited by particles into the same tumor mass and for the same survival fraction.

Reticuloendothelial system (RES): Part of the immune system that recognizes and eliminates alien particles or cells.

Single field exposure: Single field irradiation of a PTV, only feasible with charged particles that have a rapid dose fall-off on the distal side.

Skin-sparing effect: Near surface dose that is considerably lower than in deeper regions.

Source-axis distance (SAD): Irradiation at a constant distance between the source and the isocenter of the gantry.

Source-to-surface distance (SSD): Irradiation at a constant distance between the source and the patient's skin surface.

Spread-out Bragg peak: Sequence of overlapping and weighted Bragg peaks with different proton energy and beam intensity, resulting in a Bragg peak plateau covering the entire cancer volume.

Staging: Oncological assessment and characterization of cancer development.

Stimulated emission: Occurs when the upper energy level is more populated with electrons than the lower one. Resonant photon frequencies can stimulate a transition between both levels and are thus amplified.

Synchrotron: Storage of high-energy charged particles in a closed, evacuated circular tube, where the particles are held in orbit by magnets.

Theranostic nanoparticles: Nanoparticles that fulfill at least a dual purpose for diagnosis and therapy.

Transient charged particle equilibrium (TCPE): Region below the surface where dose and kerma decrease simultaneously while their ratio stays constant.

YAG laser: Yttrium–aluminum garnet (YAG) crystal doped with neodymium ions and emitting in the infrared at, for instance, 1,064 nm.

15 Index of terms

https://doi.org/10.1515/9783111168739-015